Landscape Planning
for
Energy Conservation

Landscape Planning
for
Energy Conservation

Edited by

Gary O. Robinette

Charles McClenon
Director of Research
A.S.L.A. Foundation

VNR VAN NOSTRAND REINHOLD COMPANY
NEW YORK   CINCINNATI   TORONTO   LONDON   MELBOURNE

Paperback edition originally published by Environment Design Press.
Paperback edition copyright © 1977 by Environment Design Press.

Copyright © 1983 by Van Nostrand Reinhold Company Inc.

Library of Congress Catalog Card Number: 83-1093
ISBN: 0-442-22339-0

Manufactured in the United States of America

Published by Van Nostrand Reinhold Company Inc.
135 West 50th Street, New York, N.Y. 10020

Van Nostrand Reinhold
480 Latrobe Street
Melbourne, Victoria 3000, Australia

Van Nostrand Reinhold Company Limited
Molly Millars Lane
Wokingham, Berkshire, England

15 14 13 12 11 10 9 8 7 6 5 4 3 2 1

Library of Congress Cataloging in Publication Data
Main entry under title:

Landscape planning for energy conservation.

   Bibliography: p.
   1. Landscape architecture and energy conservation.
I. Robinette, Gary O.
SB475.9.E53L36   1983   712   83-1093
ISBN 0-442-22339-0

Contents

The footnotes beside each reference are identified as to source on page 216

Landscape Planning
for
Energy Conservation

Preface

There are at least two sides to a discussion of the land and the energy crisis. In a time of increasing concern about high energy costs from various sources, it is clear that extraction and restoration guidelines or controls are required to protect the natural environment while providing the needed fossil fuel resources. This concern is drawing into increasingly sharper focus. When major court decisions deal with the ownership of offshore oil sources needed to relieve dependence on foreign energy sources, it seems of vital importance to understand all of the relationships between the natural environment and the need for energy. Thus on the one hand, the manipulated natural environment is destroyed or at least disrupted for a certain period of time to provide certain types of energy.

On the other hand, this same landscape or natural environment may be manipulated to utilize the existing energy resources more effectively and to conserve the limited amounts of energy which are available.

The higher costs of energy and the need for national independence for sources of energy have caused:

— a greater need for proper land planning from a new perspective

— a greater need for understanding of the abilities of natural elements to assist in conserving existing energy and in utilizing the unlimited solar energy striking the earth

— a greater need for sensitivity of the abilities of natural elements to control or enhance architectural and engineering activities to utilize more efficiently and effectively existing energy sources and emerging solar radiation technology.

Much basic information for proper planning and utilization of site elements is currently available. It needs to be structured in this new way for these new purposes. It is also essential to establish a basis and a framework for future research and data collection. The emphasis on the design process or the design methodology needs to be continued and strengthened so as to utilize new research and information which is gathered with this particular emphasis or orientation.

However, a re-structuring of the relative weight of factors or parameters affecting the design process is necessary due to the increased price of energy. The study was undertaken by the American Society of Landscape Architects Foundation under contract with the Solar Energy Demonstration Program of the National Bureau of Standards and the Department of Housing and Urban Development. It was accomplished under the administration of the A.S.L.A. Foundation, utilizing the services of six landscape architectural consulting firms. These firms were:

Land/Design Research—Cyril B. Paumier
Edward D. Stone, Jr., Associates, P.A.—L. Rick Larson
Johnson, Johnson & Roy—Carl D. Johnson
Rahenkamp, Sachs, Wells & Associates, Inc.—John Rahenkamp
Sasaki Associates, Inc.—William Firth
Sasaki-Walker Associates—Gary E. Karner

The study was done in an extremely short time frame of from six weeks to two months, with a very limited budget for each of the consulting firms. The purpose of the study was to draw together the available information on this subject in one publication. Much of the data has been available in a variety of sources and it has been developed in this form to guide practitioners, pending the development of more detailed and complete information. Hopefully, it also will be used as a basis to solicit further support and activity in this area. Because of the time and money constraints as well as because of the limited available information, this study was by no means definitive. It gathered the limited experience of a relatively few landscape architectural firms and some of the most readily available literature on the particular subject.

One of the major problems involved in a study of this type is the difficulty of providing uniform presentation. Therefore, hopefully, with the new interest in the conservation of energy and the utilization of plentiful natural resources, work will be undertaken in landscape architecture to increase the body of literature with this particular orientation. At the same time, hopefully, a larger number of prototypical projects will be undertaken and will be able to be illustrated in time to come.

Introduction

There is little collected and organized information directly related to landscape planning or design for energy conservation or for solar energy utilization. This is so for a number of reasons. Primary among these is the fact that there has been very little data gathering or research about the ways in which any of the natural elements affect the human users of the outdoor environment. Most vegetation, landforms and water features are used for esthetic purposes and in subjective ways with little or no thought as to the impact these might have on climate and thus on human comfort or the use of energy in buildings. This is changing to a certain degree but has not changed yet to such an extent that there is a backlog of relevant quantified data for use by the homeowner, the environmental designer or the builder to adjust or modify small scale land areas with any assurance or certainty of easily measurable results. A secondary reason for the lack of a body of information on the use of landscape elements for energy conservation is the relative newness of the concern for the control and conservation of energy usage. A dramatic increase in the price of imported oil has altered the cost of many items and services. In an overall review of energy waste and methods of restraining it, the initial concern and thus the preponderance of research funding has gone for the improvement of mechanical systems to heat or air condition designed, placed and constructed buildings where quantification of effectiveness is easier and where there is an industrial component to exert political pressure. Thirdly, the "science" of measuring the environmental impact of natural materials as they relate to human comfort is not as sophisticated to the degree of agricultural research relating to crop production since there has not been the demand for it that there has been for other research activities.

There is, however, a great deal of information available concerning the ability of natural elements to modify microclimates in a variety of ways. Much of this research data was initiated as a result of interest in agricultural production and had only tangential concern for human health and comfort. Much of the existing data on the climatic impact of natural elements is located in the scholarly studies and writings of such diverse fields as meteorology, agronomy, crop production, forest influences, and soil conservation in a number of languages. Due to the fact that the basic research was not done for this purpose a great deal of interpretation and interpolation is necessary in using this data. This leads to variations in interpretation and application and thus inexact control of outdoor climatic conditions and indoor energy usage. All natural materials, especially plants and to a lesser extent land form and water bodies are unpredictable as to their ultimate shape and form. Plants generally grow into a predictable shape and form. However, in certain instances, because of soil, light or water, specific plants may be larger or smaller than the average or the form may differ from the predicted form which may be reached only at maturity, many years after planting. Earth forms may erode, water bodies may increase or decrease and thus those natural elements placed or envisioned by the original landscape designer may never be in the place or form to perform the climate control function in a number of decades.

In essence climatic factors affect energy usage. By controlling microclimate it is possible to control energy usage in heating and cooling of buildings. Natural processes and element control climate and they can be modified to a certain extent in a variety of ways; they are not given for a specific site or situation. Natural elements such as earth forms, at all scales and in all forms, vegetation of all types and water in all forms and at all scales all can be used to make the climate around a building warmer or cooler and thus lessen the load and the energy consumption of the internal mechanical systems.

The entire decision making process is a complex continuum leading from gross site selection, precise site selection for a building or a complex, building siting and orientation, building and site integration, site planning and design to the actual selection and specification of materials for use throughout the site. During the course of the decisions made in this process a number of them have energy conservation implications. One of each of the options at each step may be better able to either help in conserving energy or maximizing existing natural energy sources. The steps or points of decision making must be clearly identified, the options or consequences of each decision evaluated in light of increased energy costs and the importance of proper large scale decision making must be emphasized in order for the natural environment to be used most functionally and completely in the future.

Each site designer, whether amateur or professional, with his or her unique training, vantage point, special interests and distinctive site will, hopefully, be able to take some of the basic information contained in this book and utilize it to develop innovative and appropriate solutions for greater energy conservation and solar and wind energy utilization. This book gives a cursory overview of the site planning process as it is modified to accommodate the need for energy conservation and greater utilization of alternative energy sources. It also brings together much of the available information and material on the climatic impact of natural materials. It is by no means definitive but it does represent a large part of the information which is currently available. It also gives an objective basis for many of the design solutions suggested to assist in conserving energy and in using fully all natural energy sources.

Climate and Man

Since the time of the ancients, wise men have written that man's entire pattern of life is dependent upon the climatic conditions in which he finds himself. Environment dictates man's means of livelihood, his diet, his clothing, and his shelter.

Scholars from Aristotle to Montesquieu have believed that climate has a pronounced impact on human physiology and temperment. Authorities as late as the 19th century, the founders of modern sociology, and leaders in the early study or evolution suggested that proper climatic conditions were the main requirements, and main stimulus, for the development of civilization. More recently, industrial engineers have studied productivity in various countries at various times of year, and shown that worker productivity declines in times of excessive heat or cold.

In recent years, man has struggled to become independent of climate. Western man has established permanent research stations in Antarctica and Greenland, and has built cities on the equator. Through the use of his technology, he has separated the climate within buildings from that outside. Even factories are now air conditioned.

Today we are approaching a crisis. We have proven what can be done, and have accepted as second nature that it should be done. Add a heating and cooling system, add light, and we have a habitable building anywhere. Man can conquer the elements, we believe, and so, with seemingly unlimited power at our disposal, we rush headlong into combat with nature. Only rarely in the last two decades has man seen that it is the elements which have the unlimited power, and that man cannot always handle the extremes. When that occurs, we call it an "act of God" or a "natural disaster," implying that it is the exception, and that man's success is the rule.

Can man control his own environment? The more limited the scale, the closer he can come to success. The second law of thermodynamics, the concept of entropy, tells us that we cannot forever isolate order from chaos, that as we establish order on a small scale, we add more to the disorder on the larger scale.

One has only to look at his refrigerator to grasp this truth. While it cools the air inside, its vent warms the air behind it. As soon as we open the door, some of the warm air from behind it will get in, so the machine has made its own job more difficult.

On a larger scale, we find an urban heat island, even in the summer. One might suggest that cool air, leaking out from air conditioned buildings clustered togehter, should make the city cooler than the countryside in the summer, just as it is warmer in the winter. But instead, no matter how much we may use to cool our buildings, as we separate cool air from warm, we only make the conditions outside worse. If we built a giant dome over the entire city, we could then cool the whole city environment, using great amounts of fuel, and with drastic effects on the surrounding countryside.

An anonymous physics student once summed up the laws of thermodynamics thus: "You can't win, you can't even tie, and you can't get out of the game." Since we can't beat nature, the only approach is to use whatever it provides for us.

It is significant also that our fossil fuels are the product of the same forces that we are dealing with today; they can all be traced back to the sun. Hydrocarbons are created by photosynthesis. Over millions of years, under the forces of pressure and geothermal heat, the woody substances created by photosynthesis are broken down into pure carbon (coal), and other substances are broken down into petroleum oils. Because millions of years ago all life was sustained and built directly by the sun, today man is able to draw upon his vast store of fuel.

What does he do with it? He attempts to combat the solar forces. Could we not live with what the sun gives us today, instead of battling it with what it gave us years ago?

Primative man lived according to the sun. His only fuel was wood, the product of photosynthesis in his own time. His food he gathered himself, during daylight. His food was the product of daylight. His shelter, in whatever region he resided, was built to use the desired natural elements, and shield out the excess. He lived in balance with nature and with the natural processes and elements.

Even in historic times, man lived according to nature. This is what the ancient philosophers recognized when they said that each community, and each population, was built according to the local climate. Today, man is out of balance with nature. He draws on the suns of past years, while cursing the sun today.

This is ironic, because man's technology should make him better able to make use of what is available; yet he neglects the most powerful sources of energy around him.

The consideration in this work is how man may best use the natural processes and elements in housing and in site design. To the extent that the natural elements have been considered in modern landscape architecture, emphasis has been on isolating them, protecting against them, and shielding them out. By contrast, this book will take a positive approach to show how natural processes, elements and factors are able to be utilized and emphasized to a greater extent by site planning, site design and manipulation of site elements.

The Climatic Impact of Natural Elements

Climatic factors affect energy usage. Earth forms, at all scales, affect specific microclimates; vegetation of all types has an affect on microclimate. The acceptance or modification of water, in its various forms, affects climate. The use or inclusion of architectural elements of the modification of the basic architecture is able to affect climate. The use or manipulation of natural elements or natural processes then, can affect the microclimate of an area and hence the energy usage of buildings in that area.

Each of the major climatic regions in the United States has different needs or requirements in order to modify the basic climate to make it more comfortable to human users. In cool areas it is necessary to produce or utilize as much heat as possible in order to cut down on energy consuming mechanical heating. In warmer areas it is desirable to modify the climate to provide cooling and to cut down on the energy consuming mechanical air conditioning. In arid areas existing moisture may have to be maximized or other moisture may have to be introduced by means of natural processes in order to make the climate more habitable. In humid regions, excessive moisture may have to be minimized or lessened by natural processes in order to make the climate acceptable and comfortable.

As solar radiation moves toward the earth, it moves through the atmosphere before striking the earth's surface. In this movement a series of impediments cause a diminution and dissipation of the full impact of the original radiation. Some of it is reflected back into space; some is dissipated within the atmosphere; yet other portions are diffused throughout the atmosphere. A small portion of the original solar radiation strikes the earth's surface, the vegetation on the earth and buildings as well as man and animals as they move about on the earth. Natural elements and water bodies modify in a variety of ways the impact of the incoming solar radiation. How much solar radiation strikes the earth and how and where it strikes it materially affects both the micro and the macroclimate and hence the energy needed to heat or cool buildings in each region.

As winds move over the surface of the earth they encounter a series of impediments which detract, deflect, obstruct and lessen the impact or speed of the flow of the unobstructed wind. These obstacles do this in a variety of ways and to a variety of degrees. The climatic effect of natural elements is two sided—on the one hand, cutting down the impact of the solar radiation or wind, and on the other hand, accelerating or enhancing the impact. The speed, direction and type of wind in a specific location determines, to a large extent, the climate of an area or of a specific site. The degree to which the climate must be modified in order to bring it within the human comfort range determines the amount of energy which must be utilized or expended in order to make the proximate microclimate of the individual.

Often by cutting down on one climatic force while leaving another unaffected, the net effect will be the enhancement of the apparent impact of one specific climate force. For instance, providing shade, while not controlling the wind, will considerably cool an area, or to control the wind while not blocking or limiting the sun's rays will give a much warmer area.

Natural elements of all types, therefore, affect climate. The natural elements can be moved, manipulated, altered and shaped in order to control the affective climate more efficiently, effectively and completely. It is through the movement and modulation of these natural and introduced man-made elements that the site planner manipulates the perceived impact of the local microclimate as it affects people in a single building or a group of buildings.

An understanding of the effects of natural materials upon climate, on both the micro and macro-levels, is basic to any discussion of energy conservation and solar energy utilization. Information in this area has been developed as an adjunct to studies of food production and other areas of agriculture. Historically, little research has been undertaken to determine directly the effect of outdoor environmental manipulation or control for human comfort or productivity.

On the other hand, the planning, design and construction process is a continuum during which a great number of site related decisions are made which can lead to energy conservation and greater utilization of natural energy sources. It is therefore, essential, in any discussion of the climatic impact of natural elements, to point out where energy conservation related decisions may be made in the total process and what decision options affect energy conservation and to what extent they do this. Obviously the earlier in the process and the larger the scale at which the decision is made the greater will be the savings in energy and the greater the possibility of utilization of natural energy sources, such as solar radiation for natural heating and utilization of natural wind flow patterns for less energy-expensive cooling and heating.

Land and Landforms

Topography, or the modulation of earth's surface either in a natural undisturbed state or as manipulated by man, has the ability to modify, ameliorate or accentuate climatic variations in a wide range of ways.

It is patently obvious that the earth's surface is not flat. It consists of rises and depressions, the causes and results of geological forces of upheaval and the meteorological forces of weathering. The layer of air surrounding the earth is relatively thin and is usually laminar. Its movement or flow can be impeded or speeded up, or directed by rises or depressions in the earth's surface.

The sun's rays strike various sections of the earth's surface with various degrees of intensity depending upon the geographical location, the angle of inclination and the elevation of the height or depth of the landform.

It must be remembered, in any discussion of the effect of landforms upon climatic factors, that all landforms flow together—none exist in isolation. All landforms are fully three-dimensional and cannot be fully expressed by sections or elevations alone.

a. Large Scale Landforms

At the very largest scale, continental mountain ranges are diverters of air masses, affect the flow of moisture-laden air, assist in trapping and condensing moisture, accentuate the effects of the sun, or create inhabitable mountain peaks or valleys. The coastal ranges of North and Central America, particularly in the western United States, provide a buffer against which the moisture-laden Pacific air masses continually move. This causes a very moist environment on the west side of the mountains, while creating aridity behind the mountains. These same coastal ranges of the western United States serve to channel moist fog-laden air and at the same time assist in the creation of sunny and shady slopes on the western, southern and eastern slopes of the mountains.

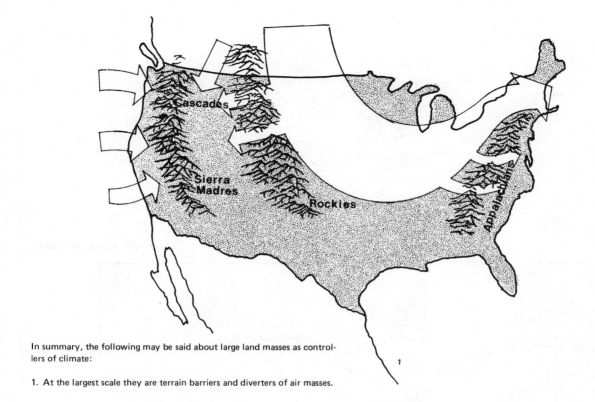

1

In summary, the following may be said about large land masses as controllers of climate:

1. At the largest scale they are terrain barriers and diverters of air masses.

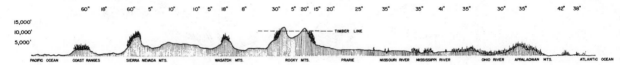

THE EFFECT OF TOPOGRAPHIC RELIEF ON PRECIPITATION AND FOREST COVER (+ AND−)
−NEGATIVE LANDFORMS, + = POSITIVE

2

2. Extremely large land masses cause rain shadows for the areas in their lee.

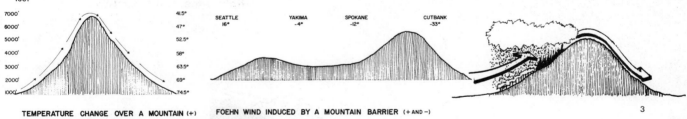

TEMPERATURE CHANGE OVER A MOUNTAIN (+)   FOEHN WIND INDUCED BY A MOUNTAIN BARRIER (+ AND −)

3

3. Temperature decreases with rise in height to a rather predictable degree at different seasons of the year.

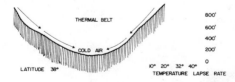

THERMAL-BELT CHARACTERISTIC IN A VALLEY (−)

4. Cold air flows downhill and settles in valleys.

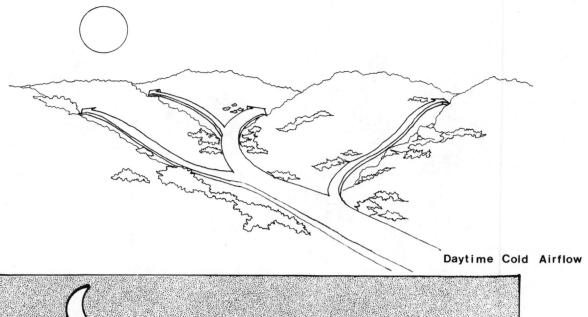

Daytime Cold Airflow

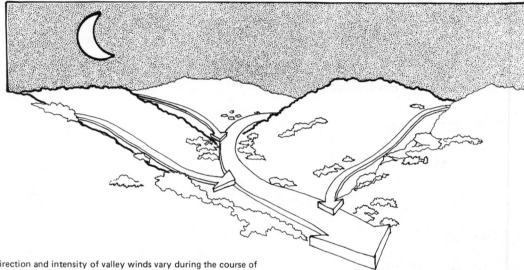

5. The direction and intensity of valley winds vary during the course of the 24-hour day.

Night time Cold Airflow

4

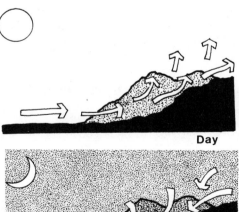

Day

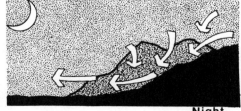

Night

Daily Cold Airflow Pattern

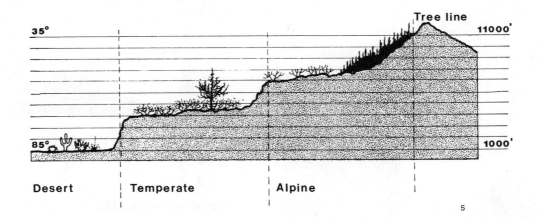

35°   Tree line   **11000'**

85°   **1000'**

Desert   Temperate   Alpine

5

6. Major landforms affect the types of vegetation which can be introduced and will survive at various elevations, both on the windward and leeward sides.

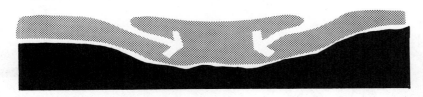

CONVECTION FROST

RADIATION FROST

7. Convection frosts may occur most commonly at the base of major landforms.

8. Radiation frosts may occur in depressions in otherwise level or relatively flat terrain.

SUNRISE   MIDMORNING   MID-DAY   LATE AFTERNOON

SLOPE AND VALLEY WINDS DURING THE COURSE OF A DAY
—NEGATIVE LANDFORMS + POSITIVE

9. Daytime winds will flow up slope, while night winds will flow down slopes of major landforms.

SUNSET   LATE EVENING   MIDNIGHT   EARLY MORNING   6

7

Basically, these smaller landforms may be involved in either solar radiation interception or in wind control through interception, deflection, or curtailment. Small scale landforms may be utilized, controlled or created for optimum utilization of solar radiation both in architectural structures and in the site areas surrounding buildings in the following ways:

1. The creation of either positive or negative small landforms within the larger land areas may be possible and advisable to provide either better orientation or site conditions for solar radiation utilization or interception.

2. Adjustments in major land areas or landforms may be advisable to pick up maximum solar radiation orientation and location, either for structures or functions.

3. Smaller scale landforms may be modified or created to provide assistance to vegetation used for climate control. This may be the use of landforms to increase the height or to improve the location of vegetation.

4. It may be possible through the manipulation or creation of small scale landforms to improve either the location, the total height or the orientation of architectural elements used in the landscape for solar radiation collection or utilization.

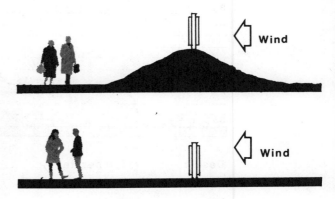

5. Landforms may be used to block unwanted wind or other cooling influences and to provide an area in which the potential solar radiation for specific area is maximized without shading or cooling devices, thus creating a usable sun pocket.

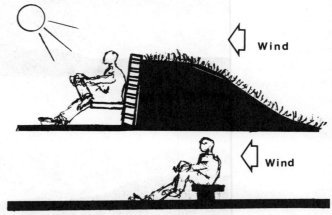

6. Occasionally, it may be necessary or advisable to locate solar collection devices apart from the building utilizing the solar energy gathered by the collectors. In such instances, small scale landforms may be created and utilized to integrate the solar collection device with the surrounding landscape, or to give it a better location without competition from surrounding landforms or vegetation.

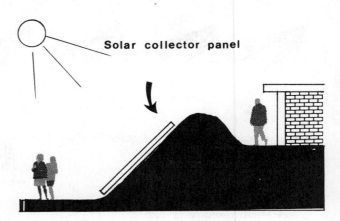

7. At other times, the necessary engineering or architectural devices used in solar collection either on the building or standing apart from it may be unsightly to the extent they need to be screened by small scale landforms.

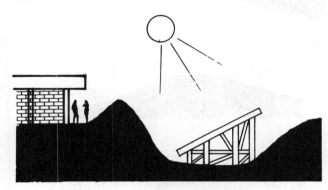

In addition to the ability of small scale landforms of various types to control solar radiation, these same elements may be used to control the wind, either through blockage, direction or creation of eddies on the leeward side of the specific landform. It is nearly impossible for small scale landforms to create extensive blockage; however it is possible for a "bluff" landform to create a certain degree of blockage immediately behind the landform itself. It is, of course, much easier to do this with the addition of either vegetation or architectural elements to the basic landform itself.

Landforms, especially small scale ones, if finely controlled or manipulated, may be able to direct wind patterns either toward or away from specific areas, either to provide cooling or to keep cooler winds from dissipating the desired collection of solar radiation for specific sites or projects.

By the careful creation and manipulation of small scale landforms it is possible to create beneficial eddies on the leeward side of the landforms which can be utilized by the environmental designer to supplement other environmental elements used in solar energy collection and utilization.

In summary, both positive and negative small scale landforms may be created, altered or utilized to control specific microclimates through the interception, deflection and curtailment of specific wind patterns. Landforms may be used either in conjunction with architectural elements, with vegetation or with the buildings or solar collection devices.

9.

WARM MOISTURE LADEN AIR COOLS AS IT
RISES AGAINST COASTAL MOUNTAINS. THE
COOLING AIR CANNOT SUPPORT THE
THE HEAVY WATER AND IS FORCED TO
RELEASE IT AS PRECIPITATION; MOST OF
WHICH FALLS ON THE WINDWARD SIDE OF
THE RANGE.

DRY AIR

MOIST AIR

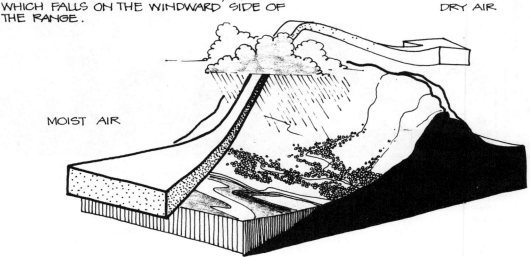

All of these factors affect the growth of vegetation which in turn further affects the microclimate of a specific area. These large-scale land masses affect the passage of wind through the openings, increasing it in velocity and intensity. The south slopes of large, land masses form sun pockets during much of the day. The east or west slopes receive large amounts of solar radiation, at least part of the day, while the north slopes of large land masses are in shade with little solar exposure during much of the day. The angle of inclination and direction of slope will affect, to a large degree, the microclimate on these major land masses.

Fog frequency is controlled by these large land masses through guidance and damming of the movement of fog inland. At the same time, major land forms control air flow by forming dams, channels, or solid or continuous impediments for the movement of either warm or cold air flow.

It must be remembered that cold air is heavier than warm air and thus cold air flows downhill, unless impeded. It must also be kept in mind that temperature decreases with altitude. This is usually one degree for each 330 foot rise in altitude in the summer and one degree decrease for each 440 foot rise in winter.

In effect, the climatic characteristics of large land masses must be accepted and can only be modified to a certain extent. The environmental designer must work with them where possible and modify the microclimates as much as possible and necessary in specific situations.

b. Small Scale Landforms

The distinction between large-scale and small-scale landforms, for our purposes, is an empirical one. A small-scale landform is any which may, to some degree, be altered by the site designer. They exist within the framework of large-scale landforms, and their basic effects differ only in scale, but for our purposes they are worthy of special attention because they are, to some degree, controllable.

Small scale as well as large scale landforms can be either concave or convex with respect to the horizontal surface of the earth. As found in natural landforms, these positive and negative landforms or changes of elevations are extremely complex. They flow together and they are continuous and unified. It may be possible to site buildings to take advantage of small scale landforms or it may be possible to create small scale landforms to control specific microclimates.

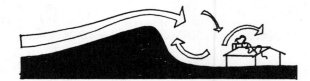

EFFECTS OF ROLLING HILL & BLUFF

7

Vegetation

Basically vegetation controls the sun's effect by filtration of the direct solar radiation, by control of the ground surface and hence the amount of heat radiated from these various surfaces, either daily or seasonally through the alteration of the ground temperature, through control of reflected radiation, and through total or major obstruction of the solar radiation itself. These are illustrated in the following drawings:

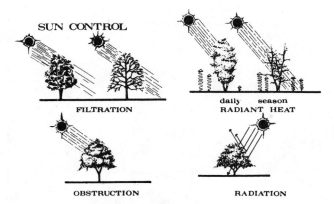

SUN CONTROL

FILTRATION

daily   season
RADIANT HEAT

OBSTRUCTION

RADIATION

Vegetation controls wind through obstruction, filtration, guidance and reflection. This is shown in the following illustrations:

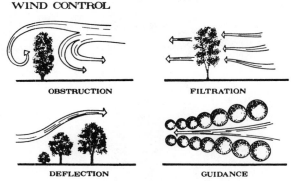

WIND CONTROL

OBSTRUCTION

FILTRATION

DEFLECTION

GUIDANCE

In much the same way, precipitation in different forms is controlled to various degrees by vegetation. Plant materials control the impact of rain, of sleet and hail, the position and amount of snow deposition, the intensity and location of dew and frost, and of the evaporation of moisture from the ground surface.

PRECIPITATION & HUMIDITY

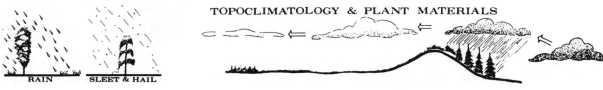

RAIN

SLEET & HAIL

TOPOCLIMATOLOGY & PLANT MATERIALS

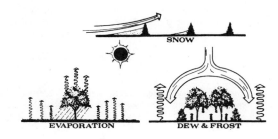

SNOW

EVAPORATION

DEW & FROST

Because plant materials control solar radiation, wind and precipitation and humidity, they control temperature in seasonal and in annual variations. As plant materials grow, they also control temperature variations, both during the day and at night.

TEMPERATURE

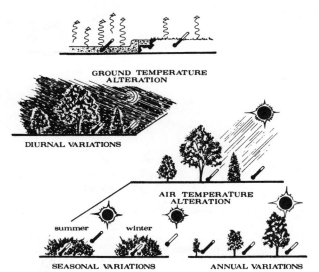

GROUND TEMPERATURE ALTERATION

DIURNAL VARIATIONS

AIR TEMPERATURE ALTERATION

summer   winter

SEASONAL VARIATIONS

ANNUAL VARIATIONS

The effectiveness of specific plant materials in climate control depends upon the form and character of the plant, the climate of the region, and the specific requirements of the site. This is obviously a complex matter, and it is outside the scope of this study to give specific recommendations. The experienced site designer will be familiar with the plant material of his own locale.

There exists an extremely scattered literature on the subject of the climatic influences of vegetation. Most of it is from research conducted in those subfields of forestry and meteorology concerned with the production of food and fiber. Little study has been done on the climatic effects of plants in urban or suburban situations, or in relation to human habitability and comfort.

Some of the raw data available on the climatic influences of plant materials has been digested, and is presented as an appendix to this study. The designer may find this information useful, in choosing materials, and in studying the exact ways in which he can manipulate the microclimate of his site.

The extent of the impact of vegetation upon the climate should not be underestimated. Vegetation may absorb over 90% of light falling upon it, reduce wind speeds in an area to less than 10% of that in the open, or increase them, reduce daytime temperatures by as much as 15 degrees Farenheit, and, in certain situations, raise nighttime temperatures.

Through a full understanding of these capabilities of plant materials, the site designer will find solutions to many of the climatic problems he faces, both in energy conservation and in the utilization of solar energy.

Summarized, the principal uses of vegetation of special importance to the energy-conscious designer are:

1) Large and small trees and shrubs may be used to screen out undesirable winds; conifers should be used to control winter winds.

2) Trees may be used to channel winds, to increase ventilation in specific areas.

3) Plantings will reduce the accumulation of snow on the ground, and so may be used to shield a solar collection unit.

4) Vegetation, especially needle-leaved trees may be used to capture fog, and thus increase sunlight reaching the ground or the collector unit.

5) Deciduous trees will screen out direct sunlight during the summer, to reduce required cooling loads, but allow it to pass through in the winter, reducing required heating loads.

6) Planted areas will be cooler during the day, and experience less heat loss at night.

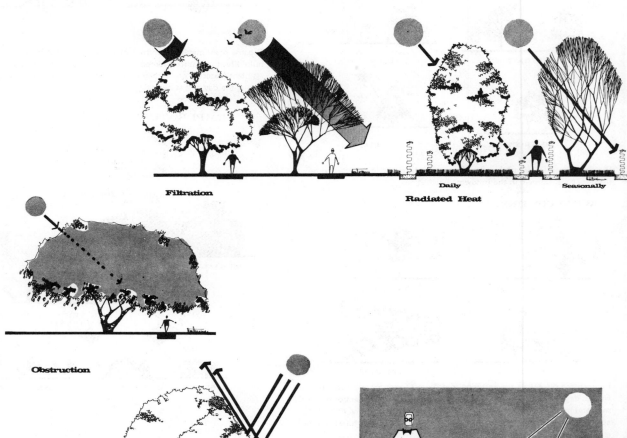

Filtration

Daily   Seasonally
Radiated Heat

Obstruction

Reflected Radiation

Solar Radiation Control

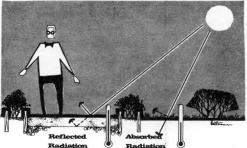

Reflected Radiation   Absorbed Radiation

Ground Temperature Alteration

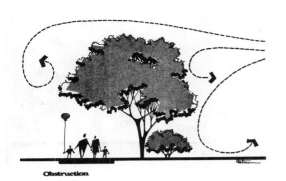

Obstruction

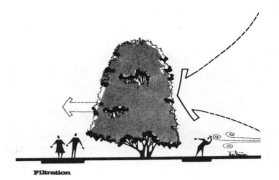

Filtration

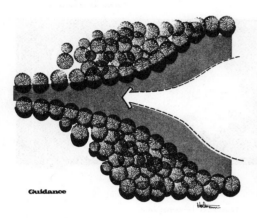

Guidance

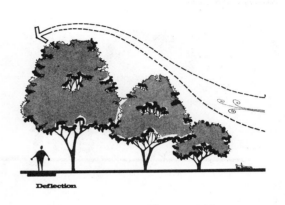

Deflection

Wind Control

Rain

Sleet & Hail

Snow

Dew & Frost

Evaporation

Precipitation & Humidity Control

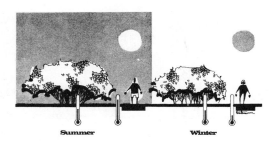

Seasonal variations

Annual variations

Day

Diurnal variations

Night

Diurnal variations

Temperature Control

a. Solar Radiation Control

The sun is the ultimate source of all of the earth's light and heat. As such it has immense effect on other climatic factors. The rays of the sun itself are welcome in northern temperate regions as well as in other cooler areas at certain times of the year. However, sometimes unmasked and unfiltered solar radiation, either apprehended directly or radiated or reflected from man-made surfaces, is too warm for human comfort. Trees and forests are among the best exterior solar radiation control devices. This has been one of the major functional uses of forests, both in tropical climates where solar radiation is oppressive requiring year-round control and in temperate climates where solar radiation is most oppressive in the summer, requiring seasonal control. Trees and forests play an important role in controlling excess or unwanted solar radiation. They provide this control in at least four ways. These are:

        1. absorbing
        2. reflecting
        3. radiating
        4. transmitting

C. A. Federer, in his paper on "The Effects of Trees in Modifying Urban Micro-Climate," has this to say about the microclimate effect of trees

    "The most obvious micro climatic effect of a tree is shade. Trees absorb and reflect solar radiation. The importance of solar radiation to the energy balance of an individual is demonstrated by the way we seek shade in hot weather and sun in cold weather. Evergreen trees provide shade year round; the deciduous trees have an appropriate seasonal variation; shade in summer and at least partial sun in winter. Buildings also provide shade but it is only where tall buildings are already crowded close together that increased shading by trees would not be a welcome benefit in summer."

                                           8

Numerous statements have been made concerning the sun and solar radiation and its relation to the forest environment. Following are two examples of such references:

    "The sun is the source of the radiant energy and light reaching the surface of the earth and, at times of clear sky, consists of 80 to 90 percent of direct sunlight and 10 to 20 percent of sky light."
                                                  9

Again, in one other source:

    "The forest is the principal organic recipient and storer of solar energy. It stores more energy annually than any other type of vegetation. The heating equivalent of wood and litter produced on 1 acre in 1 year by white pine, for instance, is equal to 3,325 pounds of coal, whereas grain or corn values range from about 1,200 to 1,700 pounds (Shirley 1936)."
                                                 10

1. Absorbing

A number of things happen to the rays of the sun as they strike the earth and more particularly trees and forests. The primary thing which happens to this solar radiation is its absoption by the forest canopy. One source has the following to say about the absorbing forest:

    "The forest stands out, dark in the landscape, a porous absorptive region that retains within itself, at ordinary stem densities, a half or two-thirds of the solar radiation incident upon it (Miller 1955).

    "The forest, its existence dependent on solar radiation, absorbs 60 to 90 percent of the total solar energy received. The amount absorbed by any one stand depends primarily on its density and the development of its foliage. Within a dense stand of conifers or hardwoods in leaf, 75 to 90 percent of the incident solar radiation

MEANS BY WHICH THE BODY EXCHANGES HEAT

A. ABSORPTION FROM SUN DIRECTLY
B. ABSORPTION FROM REFLECTIVE OBJECTS
C. ABSORPTION FROM GLOWING RADIATORS
D. HEAT CONDUCTION TOWARD THE BODY
E. HEAT CONDUCTED AWAY FROM THE BODY
F. OUTWARD RADIATION TO SKY
G. OUTWARD RADIATION TO COOLER OBJECTS
H. ABSORPTION OR LOSS FROM NON-GLOWING HOT RADIATORS

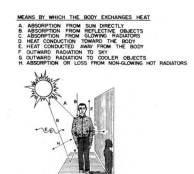

HEAT EXCHANGE - MAN

PER-CENT REFLECTION OF VARIOUS SURFACES

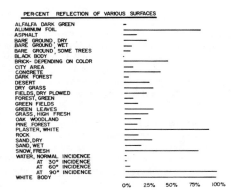

ALFALFA DARK GREEN
ALUMINUM FOIL
ASPHALT
BARE GROUND, DRY
BARE GROUND, WET
BARE GROUND, SOME TREES
BLACK BODY
BRICK- DEPENDING ON COLOR
CITY AREA
CONCRETE
DARK FOREST
DESERT
DRY GRASS
FIELDS, DRY PLOWED
FOREST, GREEN
GREEN FIELDS
GREEN LEAVES
GRASS, HIGH FRESH
OAK WOODLAND
PINE FOREST
PLASTER, WHITE
ROCK
SAND, DRY
SAND, WET
SNOW, FRESH
WATER, NORMAL INCIDENCE
    AT 30° INCIDENCE
    AT 60° INCIDENCE
    AT 90° INCIDENCE
WHITE BODY

0%   25%   50%   75%   100%

SURFACE REFLECTION

DARK SURFACE

LIGHT SURFACE

COURSE SURFACE

SMOOTH SURFACE

A. REFLECTED RADIATION
B. ABSORBED RADIATION
C. TRANSMITTED RADIATION

INCIDENT RADIATION AND VEGETATION

SOUTH SLOPE
20°
LEVEL
20°
NORTH SLOPE
944
122.6'
100'

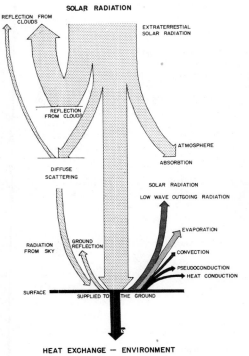

SOLAR RADIATION

REFLECTION FROM CLOUDS

EXTRATERRESTIAL SOLAR RADIATION

REFLECTION FROM CLOUDS

ATMOSPHERE

ABSORBTION

DIFFUSE SCATTERING

SOLAR RADIATION

LOW WAVE OUTGOING RADIATION

EVAPORATION

CONVECTION

PSEUDOCONDUCTION

HEAT CONDUCTION

RADIATION FROM SKY

GROUND REFLECTION

SURFACE

SUPPLIED TO THE GROUND

HEAT EXCHANGE — ENVIRONMENT

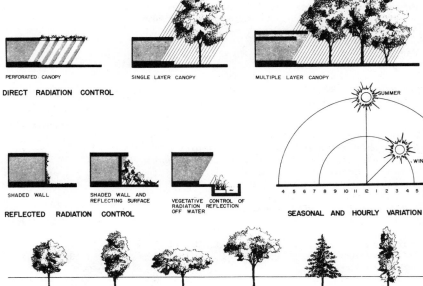

PERFORATED CANOPY

SINGLE LAYER CANOPY

MULTIPLE LAYER CANOPY

DIRECT RADIATION CONTROL

SHADED WALL

SHADED WALL AND REFLECTING SURFACE

VEGETATIVE CONTROL OF RADIATION REFLECTION OFF WATER

REFLECTED RADIATION CONTROL

SUMMER

WINTER

4 5 6 7 8 9 10 11 12 1 2 3 4 5 6 7 8

SEASONAL AND HOURLY VARIATION

ROUND

VERTICAL OVAL

HORIZONTAL OVAL

VASE

PYRAMIDAL

COLUMNAR

TREE FORM SHADOW PATTERNS

may be absorbed; for an open stand of small-crowned conifers, 60 percent may be absorbed.

"Thermal storage of heat in trees is not great. Well-stocked stands have a mass of about 1.5 grams per square centimeter of the forest area, and wood has a specific heat of about 0.3. Yet even with a change in temperature from day to night of 15°C., the amount of thermal storage is only about 7 langelys; this storage would have little effect on the nocturnal fall of temperature (Miller 1955).

"Baumgartner (1956) illustrated these partitions. He described the disposition of net radiation during a summer day for a young spruce plantation with an average height of 18 feet. Of the net radiation of 586 langleys, a little less than two-thirds was used in evaporation and transpiration, one-third was used in convection, and a small amount—5 langleys—was stored temporarily in the trees and ground. During the course of a day, there was a nearly complete turnover of the energy budget. As the author stated 'The climate within the plantation is newly formed by the weather every day.' "   11

The forest, the tree and the leaf all absorb solar radiation.

"Throughout the day the tree in the forest receives different amounts of radiation as continually changing areas of its foliage are presented to the sun."

"The leaf on the tree, continually moving and receiving radiation from a sun whose position is constantly changing above shifting cloud cover, absorbs fluctuating amounts of radiation.

"In the absence of wind and with restricted convection, leaf temperatures may be as much as 13°C. higher than air temperatures.

"Though a dense forest may absorb 75 to 90 percent of the solar radiation incident upon it, an individual leaf may absorb only about 25 percent.

"That the stand absorbs two to three times more radiation than the leaf is a measure of the effectiveness of interstices between needles, between branches, and between trees in capturing radiation (Miller 1955)."   12

2. Reflecting

In addition to absorbing some of the incoming solar radiation, the forest itself may reflect back into the atmosphere a certain percentage of the solar radiation. The same source has the following to say about the reflecting forest:

"If a dense forest cover absorbs 75 to 80 percent of solar radiation and transmits about 5 percent, it reflects 15 to 20 percent—a little less than reflection from meadows and grainfields. The only surface in the natural landscape that has a reflectivity as low as a forest cover is wet or dark-colored soils (Geiger 1957).

"Albedo of some natural surfaces, as summarized from Russian studies by Budyko (1956), are as follows:

```
Fresh, dry snow . . . . . . . . . . . . . . . . . . .0.80—0.95
Dry, light, sandy soils . . . . . . . . . . . . . . . .25— .45
Dry, clay, or grey soils . . . . . . . . . . . . . . . .20— .35
Moist, grey soils . . . . . . . . . . . . . . . . . . .10— .20
Dark soils . . . . . . . . . . . . . . . . . . . . .05— .15
Meadows. . . . . . . . . . . . . . . . . . . . . . .15— .25
Rye and wheat fields. . . . . . . . . . . . . . . . .10— .25
Deciduous forests . . . . . . . . . . . . . . . . . .15— .20
Coniferous forests . . . . . . . . . . . . . . . . . .10— .15
```

"The average albedo of the earth's solid surface is 0.14, and for the total global surface it is 0.40."   13

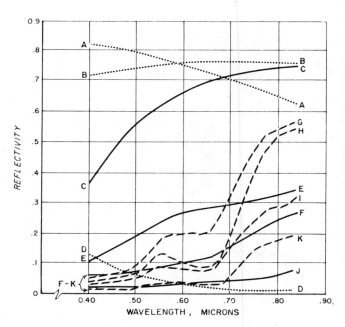

KEY

A FRESH SNOW
B SNOW COVERED WITH ICE
C LIMESTONE, CLAY
D WATER SURFACE, VIEWED OBLIQUELY
E DESERT
F PODSOLS, CLAY LOAM, PAVED ROADS

G DECIDUOUS FORESTS, AUTUMN
H DECIDUOUS FORESTS, SUMMER
I CONIFER FORESTS, SUMMER
J BLACK EARTH, SANDY LOAM, DIRT ROADS
K CONIFER FORESTS, AUTUMN

Reflectivity of forest cover, soils, and other formations (Krinov 1947).   14

Not only does the forest as a whole reflect a certain percentage of the suns ray's but the tree itself has a certain reflectivity which is best described in the following words:

"The tree reflects less solar radiation than its leaves. Light penetrating the crown receives repeated reflection; each time absorption depletes the beam by half or more (Miller 1955, 1959). It is this process that accounts for the high interception of light at low canopy closure value—for instance, 50 percent interception at a crown closure of 20 percent."   15

Not only do the forests and the trees but also the leaves reflect a certain percentage of the solar radiation. This phenomena is best described in the following words:

"The amount of solar radiation reflected from a leaf varies by wavelength, species, age, upper and lower surface, position in regard to incoming radiation, position within the crown, aridity of the site, fertility of the site, season and altitude."

"Miller (1955) computed an average albedo of 0.12 for conifers and 0.25 for hardwoods.

3. Radiating   16

Not only does the forest absorb and reflect, but it also radiates a certain percentage or a certain amount of the solar radiation. This phenomena for the radiating forest has been described as follows:

"A solid forest canopy approximates a black body in the longwave portion of the spectrum, absorbing and emitting almost all possible radiation. Long-wave emissivity (3 microns and up) from a closed cover, grass, or forest, and from dry, bare ground is at least 0.90 and probably more. For wet soil long-wave emissivity is about 0.95, and for snow it is 0.82 to 0.89 (Brooks 1959).

"The thermal capacity of the vegetation and the fact that the radiating surfaces are distributed throughout a considerable volume of air tend to maintain soil and near-surface air temperature at slightly higher levels at night than over bare soil. (Slayter and McIlroy 1961)."

The forest itself radiates but so does the tree.   17

"The radiating tree—Warmed by incoming short- and long-wave radiation, a tree canopy radiates heat energy in the long-wave spectrum in all directions. At night, outgoing radiation from leaves at the top can cool them 2.5°C. below the temperature of the surrounding air, when the air is calm and the sky half-clouded, while leaves below the crown cool off 0.4° or less. Leaves at the top radiate towards the slightly cooler leaves (Seltzer 1935, as given in Geiger 1957).

"The relative proportion of reflected, absorbed, radiated, and 'transmitted' radiation of a complete canopy on a clear midsummer day might be as follows:

	Short Wave	Long Wave
Reflected	10	———
Absorbed	80	(100)
Radiated	———	10
"Transmitted"	10	90

"Shade is the most obvious product of the forest under the sun. Ordinary continuous hardwood or conifer canopies in summer can screen out 90 percent or more of the visible light. The virgin forest shading early settlers in this country was described as rank and dank and 'darke' as in a cellar:

'While oak groves only loosely filtered the sunshine, maple forests cast a gloomy shade, and the interlocking branches in beech or hemlock woods shut out sun by day and the stars by night. Small streams ran their whole lengths without intersecting a sunbeam (Lillard 1947).'

"The reduction of short-wave radiation by 73 to 86 percent, as illustrated in the above table, is probably the greatest major effect of the forest on any climatic factor. In comparison, the forest can reduce monthly maximum air temperature in the summer by about 10°F. below that in the open, reduce annual rainfall (through interception) by 15 to 30 percent, and reduce wind velocities by 20 to 60 percent."   18

There is some slight differential between hardwood and conifer shade which has been described as follows:

"The amount of light that penetrates a forest canopy, expressed as percent of light in the open, is illustrated by the following studies and reviews of literature:

"Hardwoods in full leaf exclude as much light as the conifers. Leafless hardwoods let about half of the light through. In conifers there is a slight increase in shade as annual shoots are formed (Vezina 1961; Ovington and Madgwick 1955)."   19

Type	Percent of light in open	Reference
Conifers:		
Jack and red pine_____	7 –15	Shirley (1945b)
Conifers_____	.5– 6.7	Ovington and Madgwick (1955)
Eastern white pine_____	27	Gast (1930)
Western white pine_____	6 –15	Wellner (1948)
Scotch pine_____	11 –13	Czarnowski and Slomka (1959)
Fir-spruce-pine_____	2 –40	Geiger (1957)
Norway spruce_____	2 – 3	Vezina (1961)
Deciduous:		
Hardwoods (leafless)____	55	Shirley (1945a)
Do_____	30 –60	Ovington and Madgwick (1955)
Do_____	26 –80	Geiger (1957)
Hardwoods (in leaf)_____	1 – 5	Shirley (1945a)
Do_____	2 – 6	Ovington and Madgwick (1955)
Do_____	2 –60	Geiger (1957)
Tropical forest_____	.2– 1.0	Shirley (1945a)

20

What is the effect of the forest, the tree and the leaf on the amount of light and heat reaching the ground under a forest area? This has been dealt with by a number of experts in the following ways:

"Solar radiation and light intensity decrease from the top of the canopy to the forest floor; they decrease rapidly through the canopy, and then more slowly through the trunk space. Light reduction is indicated in figure 23 (Trapp 1938, as given by Geiger 1957). In the 120- to 150-year old red beech stand, 80 percent of the light on an overcast day and more than 90 percent on a sunny day was caught in the crown space. The amount that reached the forest floor—less than 5 percent—is then fairly independent of the prevailing weather."

"Skylight was progressively reduced as it passed through the crown and trunks, whereas sunlight was most effectively acted upon by the crown. The small difference at the forest floor in percent light intensity on clear and cloudy days is common to conifer stands, where, according to Roussel (1948), light intensities compared with those in the open have been found to be fairly constant, irrespective of weather or season of year.

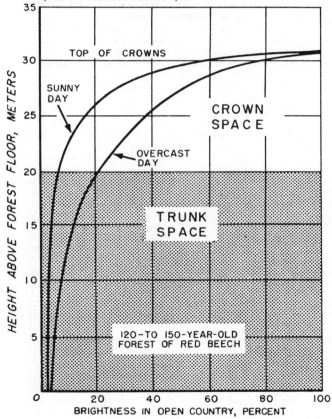

Light in the interior of a thick foliage of red beech (Trapp 1938).

"Miller (1959) has reported that light intensity in a lodgepole pine stand averaged 13 percent on days that were clear or nearly clear, 18 percent on days of moderate cloudiness, and 22 percent on overcast days. He also cites data from Brecheen (1951) to show that when insolation remains constant, transmission increases with cloudiness. A lodgepole pine stand, for instance, gave 14 percent transmission on an overcast day but 10 percent transmission on a day of almost equal insolation but with 0.2 cloudiness.

"The proportion of direct to diffused light and light intensity varies during the day with the angle of incidence of the light. In the above-mentioned 25-year-old Scotch pine stand, at 0 angle of incidence (the shortest distance covered by the light beam), 38 percent of   21 open-space intensity was measured; at 25°, 26.5 percent; at 45°, 10 percent; and at 55°, 3.5 percent. The lower the sun, the lower the intensity.

"Somewhat similar results were reported by Waggoner, Pack, and Reifsnyder (1959). On a clear day in August in red pine planta- tion, 4 percent of open-space intensity was measured at 6:30 a.m.; it ranged from 14 to 17 percent from 7:30 to 11:30 a.m., dropped to 10 percent by 1:30 p.m., and to 2 percent by 4:30 p.m. In the early morning and evening, most of the solar energy that reached the ground beneath the canopy was reflected and diffused light. Shortly before and after solar noon, a greater portion was from direct rays of the sun passing through the canopy." 22

One other source gives other data concerning the effectiveness of trees and forests in reducing the effects of solar radiation:

"Influences of Forest—for local comparisons between forest and open or between different types, densities, or sizes of vegetation, it is usually convenient to take the sun's radiation at noon on a clear day as a standard and express the radiation under different conditions of cover relative to that standard at 100 percent. The thermopile, an instrument that is equally sensitive to all wave lengths of light, is probably the most satisfactory instrument for the measurement of the effects in forest influences. It measures the radiant energy rather than the illumination, thus including the heating effects of the infrared rays. The duration of sunlight, the length of day, and the alternation of clear and cloudy periods with the resulting variable total amounts of radiation for any given period are all important factors that are modified by the forest canopy.

"The light intensity in percentage of full sunlight measured with a radiometer by Shirley (317) in a red pine forest in northern Minnesota was (a) for old-growth pine 17, (b) for dense pine re- production from 0.9 to 8, (c) for hazlenut and other brush from 0.7 to 16 percent. It is interesting that the dense deciduous under- growth, such as the hazlenut, may reduce the radiation reaching the ground to a greater extent than the forest or any other form of vegetation for which records are available.

"In white pine in New Hampshire, Li (220) and Smith (326), using the difference between black and white atometers, computed the light intensity reaching the forest floor on the basis of 100 percent in a denuded area. Certain relations to basal area and age are sug- gested by the following figures:

Basal area per acre, sq ft	0	110	127	155	316
Age, years.	...	30	50	55	Uneven
Radiation at ground, %	100	6	22	14	6

"The difference between forests of different types is illustrated by data from Nebraska by Holch (156) who reports summer intensities, using the Clements photometer, for the open prairie, 100 percent; for oak forest, 10.4 percent; and for basswood forest, 3.5 percent. The oak and basswood types represent preclimax and climax stages in the succession, respectively, and the indication is evident in pro- ducing shade increases as the forest succession progresses toward the climax.

"Perhaps the most interesting data for light intensities are those re- ported by Burns (71) from jack pine plantations 12 to 14 ft. high, differing only in the spacing between the trees, which was 2, 4, 6, and 8 ft. The crowns of the plantations spaced 6 and 8 ft. had not yet closed. The measurements were made with a Coblenz thermopile at a height of 35 in. above the ground, with the following results:"

Spacing	Trees per acre	Average solar radiation, %
2 × 2	10,890	15.9
4 × 4	2,722	36.0
6 × 6	1,210	46.6
8 × 8	680	55.4

23   18.

The larger scale effect of trees and forests on the city-country temperature differences and on the energy balance of cities are examined by the same author:

"City-Country Temperature Differences

Dozens of scientific studies have shown that the city is warmer than the surrounding countryside. This effect is known as the urban heat island. Cities are warmer in both winter and summer; mean annual temperatures are 0.5° to 1.5°F. higher (Kratzer, 1956). The winter difference, produced primarily by heat generated by burning fossil fuels; is considered an advantage by most inhabitants. But in summer the city often becomes too hot for comfort. This summer difference is due largely to the relative absence of trees and other vegetation in the city and the consequent lack of evaporational cooling.

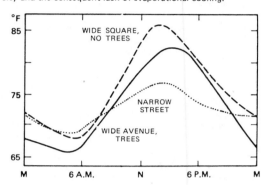

Diurnal temperature variation in Vienna on 4-5 August 1931 for a wide square with no trees, a wide avenue with trees, and a narrow street (from Kratzer, 1956). 24

"Mean monthly temperature differences between city and country in summer are about 2°F. (Dratzer, 1956). However, much greater dif- ferences occur in hot calm weather.

"Landsberg's (1956) data show the diurnal temperature variation for a city park and for the airport in Richmond, Virginia (Figure 1). (The difference shown might have been larger if the city location were not in a park and if the rural location were not at an airport.) This and other studies indicate that city-country differences are negligible dur- ing the day and the greatest—4° to 5°F.—in the early evening.

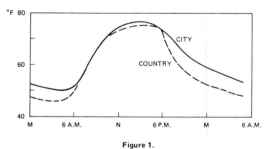

Figure 1.

Diurnal variation of city and country air temperatures for a clear day with little wind, Richmond, Virginia, 2 June 1953 (from Landsberg, 1956). 25

"The heat island has been studied in recent years by traversing the city with temperature sensors mounted on automobiles. Chandler's (1962) maps of London are classic examples of the results. Duckworth and Sandberg (1954) mapped San Francisco temperatures on 26 March 1952. They found night temperature differences of 19° to 20°F. between the built-up areas and the vegetated area of Golden Gate Park. . .These and other studies of the urban heat island have been reviewed by Kratzer (1956), Landsberg (1956), Munn (1966), Lowry (1967, 1969), Bornstein (1968), and Peterson (1969).

The pattern of city temperatures in San Francisco at 11 p.m. on 26 March 1952, as mapped by Duckworth and Sandberg (1954). The rectangle at left-center represents Golden Gate Park.

26

"The Energy Balance

The principles of the energy balance can be used to study the temperature of the city and the country, just as we have already used them to examine the temperature of human skin. The surface of the land and the surface of an individual have the following properties in common: absorption of most of the incident solar radiation, absorption and emission of longwave radiation, gain or loss of heat to surrounding air by convection, loss of heat by evaporation, and exchange of heat with the subsurface material. The generation of heat by human metabolism has its equivalent in the city in burning of fuels, but the vegetated environment has no such internal heat source.

"Solar radiation supplies the energy for warming during the day. Hand (1949) and others (Peterson, 1969) have shown that dust, soot, and other aerosols over cities reduce the incoming solar radiation to 80 to 85 percent of that received by the country. The city thus tends to warm more slowly than the country in the morning. Net longwave radiation loss is less in the city, because the atmospheric aerosols intercept part of the outgoing energy and re-radiate it back to the surface. This is one of the reasons for slower cooling of the city at night.

"The net amount of radiant energy absorbed by the country surface during the day is dissipated in three ways: by heating the air, by heating the soil and vegetation, and by evaporating water. When the soil is not too dry, transpiration from trees and other plants removes most of the available heat. In the built-up areas of the city, little water is available to evaporate, so almost all the energy must go into heating the air and the subsurface solid materials.

"The city is made largely of concrete, steel, brick, glass, and asphalt. All these are poor heat insulators. They conduct and store heat more readily than the soil of the countryside. Vegetation above the soil in rural areas provides insulation that further reduces the storage of heat in the soil. Thus in the city larger amounts of energy enter the subsurface materials during the day than in the country. This stored heat is then released at night. Consequently in the city, evening temperature remain high both indoors and out.

"The city also acts as a heat generator because of the burning of fuels in automobiles, factories, shops, and homes. The amount of heat involved is significant in the winter; air temperatures can sometimes be shown to be lower on weekends. However, in summer this heat source is insignificant except as a local modifier of the microclimate of a busy street or an industrial complex.

"Tag and Myrup (1969) have recently made the first attempts to quantitatively analyze energy-balance differences between a city and its vegetated surroundings. Myrup's results for a freely transpiring rural area and for a dry city are summarized in Table 1. The temperature data shows a larger heat-island effect at midday than in the evening. This is contrary to other observations, and Myrup cannot adequately explain the discrepancy. But in other respects the model behaves well. The major conclusions are clear. Reduced evaporation and higher

heat capacity and conductivity of the subsurface materials in the city allow much greater storage of subsurface heat during the day. This stored heat, released at night, keeps the city air temperature higher than that in surrounding areas.

"Complexity of City Microclimate

So far we have considered the city as a single entity, Actually the city is a complicated agglomeration of many types of natural and man-made structures: tall buildings, low buildings, factories, wide streets, narrow streets, parking lots, courtyards, parks, hills, lakes, rivers, and harbors. Each location has its own microclimate, determined by its local surroundings, by the weather, and by the character of upwind areas.

"Street-level microclimates can be separated into three broad classes. The first includes areas with extensive evaporating or transpiring surfaces—parks, wide streets with trees, and the vicinities of rivers or lakes. The second includes wide treeless streets, squares, and parking lots, which are open to the sky, but are dry. The third includes narrow streets and courtyards surrounded by relatively tall buildings; Kratzer (1956) showed diurnal temperature variations for three locations in Vienna: an avenue with trees, a large square without trees, and a narrow street (Figure 3). He said:

'The weather was calm, clear and hot. . .The broad streets and squares which do not have trees are very hot at noon, but cool off more noticeably in the evening. The avenues and squares containing trees are cooler and have a narrower daily range in variation. The narrow streets, however, are distinguished by much greater coolness during the noon and afternoon hours, amounting to from 5°C to 6°C. lower than the temperature of the surrounding area. At night the difference disappears or the sign is reversed. It is the high degree of blocking of outgoing and incoming radiation that causes the great reduction of the daily radiation curve.'

"For narrow streets, orientation also plays an important role. North-south streets will not be shaded from intense noontime solar radiations. They will have higher temperatures than narrow east-west streets.

"The importance of evapotranspiration is obvious in Figures 2 and 3. Golden Gate Park stands out as 15°F. cooler than the surrounding city. All cities have summer evening isotherms that parallel the boundaries between urban and rural or suburban areas.
"Wind plays an important modifying role in city microclimate. Two types of wind must be considered.

"Strong winds caused by large-scale pressure gradients tend to wipe out microclimatic differences. More of the available heat is carried away into the atmosphere, so air temperature near the surface remains cooler both day and night. Except for hot, dry, foehn or chinook winds, cities are cooler when the wind is stronger.

"When the pressure gradient wind is absent or light, local winds caused by temperature differences become important. Warm air tends to rise, causing a breeze by drawing in cooler surrounding air. For cities near oceans or large lakes, the sea breeze caused by warmer air rising over land helps keep temperatures down. Whiten (1956) has described local city street winds produced in the evening by warm air rising from hot dry squares and cool air sinking in park areas. Finally, topographically induced nighttime winds—cool air draining downhill—can produce temperature effects even in a relatively flat city like London (Chandler, 1962).

27

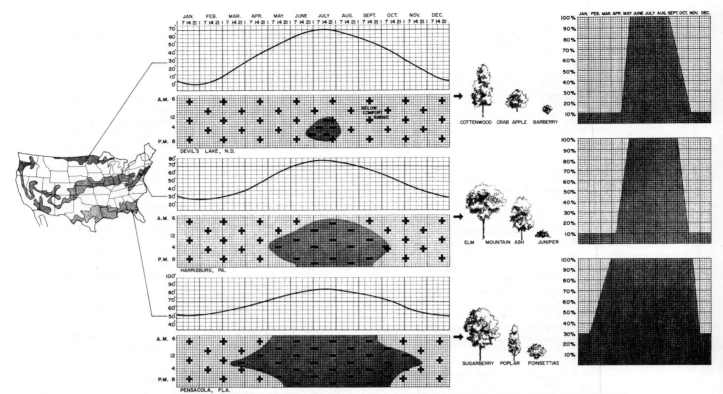

YEARLY TEMPERATURE VARIATIONS IN THREE LOCATIONS IN THE UNITED STATES WITH SHADING PERIODS OF NATIVE TREES

SOLAR RADIATION CONTROL

FROM DESIGN WITH CLIMATE BY V. OLGYAY

b. Wind Control

One of the ways in which vegetation affects the micro climate and hence the comfort of man is through its control of the flow of wind. Once again, much of the information currently available on the abilities of vegetation to control wind have come about as a result of agricultural studies. Consequently much more is known about the ability of wind to affect the growth of alfalfa, rye, strawberries, or winter wheat than the effect on human comfort. Much of the information has been done in wind control studies in conjunction with shelter-belt programs in the midwestern United States, in Russia, in Japan, in the Scandinavian countries, and in the Netherlands. There are however certain basic studies which, even though they were done with an agricultural or food production bias, still have certain pertinence to the use in improving the quality of the environment for man.

In rather predictable directions and cycles the wind moves vast bodies of air across the surface of the earth. Air movement or the lack of it controls human comfort to a large degree. Wind can control real or perceived temperature. The air movement or wind, of a low velocity and with a high enough temperature, may be desirable. However, when the velocity increases or the temperature decreases, wind is capable of causing great discomfort and even destruction to life and property.

The flow of air is caused by and related to a number of climatic factors far too complex to discuss in this limited space. Many times various obstacles or materials may be used not to just cut down wind flow but to control it in other ways, to make the environment more compatible to the human users. C. A. Federer, in his paper "Effect of Trees in Modifying Urban Microclimate," has the following to say concerning the relationship of wind alteration and vegetation to human comfort:

> "The most significant effect of wind is due to the increasing of convective cooling and evaporative cooling. We know wind plays an important role because we try to get out of it in cold weather and

into it in hot weather. Trees reduce wind speed by increasing the resistance to wind flow. Within the crown of a single tree or under a forest canopy wind is light and almost unrelated to the external wind. An isolated tree or stand of trees acts almost like a solid barrier to the wind forcing the wind over or around. Thus wind speed can be increased at the edges of a tree or stand (Reifsnyder 1955). Wind can also cause discomfort by making walking difficult or by driving rain and snow. The protection of trees is welcomed in these conditions. Deciduous trees have an adverse seasonal effect on wind, partially compensating for their beneficial shading characteristics. They decrease wind most in summer and least in winter. Generally, then, the wind effect of trees is beneficial but a disadvantage in summer.

28

1. Types of Wind Flow

There are a number of different types of wind flow which may be categorized depending on whether the wind is viewed from an elevational or sectional or a plan view of the wind flow itself and the potential wind screen. In the elevational view there are three basic types of wind or air flow when the majority of the movement is in a single direction. These are:

a. Laminar air flow—layers or streams of air flowing on top of another in parallel patterns (this occurs fairly regularly and is predictable).

b. Turbulent air flow—air masses moving in the same direction but in a random pattern (velocity of this type of air flow is unpredictable and is therefore difficult to understand or to control).

c. Separated air flow—layers of air varying in momentum (where a separation between layers occurs turbulence may be found).

Changes from laminar to turbulent flow are governed by the turbulence within the air stream and roughness of the surface over which it is flowing. Surfaces of buildings will always produce turbulent air flow and separation occurs when the air flows around sharp corners. When air is moving it exerts a pressure against any surface which tends to inhibit its flow. It is generally assumed that air has a viscosity co-efficient of zero. This means that air flows freely. Force and turbulence are produced between layers of flowing air whenever a blocking element is introduced. If the element is streamlined the air usually flows around it with a speed-up in the boundary layer of air next to it. A boundary layer of air generally follows the contour of the streamlined shape and a minimum of turbulence occurs between layers. If the element is "bluff" (or not streamlined) then the boundary layer cannot follow the contours of the element. Then a separation of air flow occurs and the force between the boundary layer and the other air layers becomes greater. Turbulence is likely to occur around a bluff body.

In plan view the air flows in a variety of ways, depending on the obstructions introduced in its path. A Japanese paper "On the Wind Tunnel Test of the Model Shelter-hedge," in the *Bulletin of the National Institute of Agricultural Sciences, Series A, No. 6* (March 1958), describes some studies undertaken to show the various configurations of plan view windflow around an obstacle as the velocity of the wind was increased. The following illustrations show the changes in the type of wind flow dependent on the speed of the wind movement past a barrier.

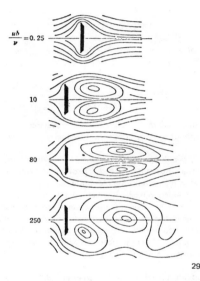

$$\frac{ub}{\nu} = 0.25$$

10

80

250

29

It must be kept in mind that windflow past any barrier is a three-dimensional phenomenon. So often various studies have shown wind flow as two dimensional elevational studies or occasionally even as two-dimensional plan view studies; however, the complexity of movement of air or wind is best understood when viewing it as it exists in reality in three dimensions.

2. The Basic Principles of Wind Flow

Wind flow or movement in the envelope of air immediately surrounding the earth is extremely complex. Generally the wind in the first 6 to 10 feet above the earth's surface is that of most interest to the human user of the environment. Occasionally winds up to 50 to 60 feet high above the earth's surface may have some effect on the human being walking or sitting at or near ground level.

Unhindered moving wind generally flows in parallel layers. Whenever the wind flows over a streamlined surface or over a bluff body the boundary layer of air generally speeds up creating a low pressure area between the boundary layer and the surface of the element. The low pressure area will either cause the element to move or attempt to pull the boundary layers of air back to its original position. When a barrier is introduced into an air stream the wind responds by flowing around or through the barrier, eventually returning to its original flow pattern. The greater the wind velocity, the greater the pressure differential on the leeward side of the barrier and the quicker the wind returns to its original flow pattern, lessening the area

of the wake or protected areas to the lee side of the barrier. The weather on the lee slope of a hill is usually quieter than on the weather slope. However, this may be reversed if the weather slope is steeper than the lee slope. Where the boundary layer of air is compressed as it passes over a ridge wind speed is usually 20 percent greater on the top of a ridge than on the slopes. A pierced screen allowing some wind to penetrate through it creates less pressure differential providing a larger wake on the lee side of the screen. A tree without foliage acts as an incomplete barrier with some modifying effect on wind. Very few, if any, studies have ever been conducted showing the complexity of wind flow around buildings and vegetation in combination or the wind flow around various combinations of vegetational obstructions. Very seldom is a wind screen of any type found on level topography with no other obstructing elements nearby or adjacent to it so as to have a multiple effect on wind flow. Much of the data currently available deals with wind flow obstruction in these artificial and limited ways.

3. Methods of Control

There are various methods of control of the wind's flow, moving along a continuum from complete blockage or obstruction to filtering, redirecting, channeling guidance, deflection or interception. Wind flow immediately behind any surface can be lessened from 0 percent or the maximum all the way to the very minimal effect. The amount of blockage will of course depend upon the character of the structure used to block the wind itself. In addition to the negative aspect of actually blocking or stopping wind, various materials may be used to channel or direct the wind into specific areas or locations.

4. Means of Control

The means of controlling wind flow or wind movement may vary from land forms or topography to various types of architecture, to walls, fences, vegetation and other natural or man-made obstructions which occur on or near the earth's surface. Land forms themselves are usually soft and rolling and have a tendency to pitch wind up or down. Many times cool air will flow down valley floors in late afternoon or early evening.

Architecture quite often is extremely abrupt and forms a very "bluff" object around which the air flows.

Many times wind studies have been done showing the flow of wind around individual pieces of architecture. Very little, however, has been done to indicate or show wind movement through multiple buildings of various sizes scattered through a particular area. Walls, whether of stone, brick, or other materials, whether they be pierced or solid, serve to curtail wind flow.

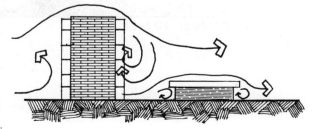

Fences of various heights, degrees of density, lengths and types of materials serve to control in various ways air flow or movement. Various studies have been conducted over the years, showing the relative effectiveness of fences in controlling or stopping wind.

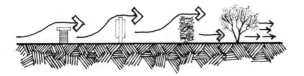

30

Vegetation, whether trees, shrubs, forests, or windbreaks have been rather extensively studied to ascertain their effectiveness in controlling the velocity, movement and direction of winds near the earth's surface. However, as mentioned previously, most of these wind studies concerning vegetation, whether they be on-site or in wind tunnels or laboratories, usually have to do with the agricultural aspects or in some cases the air conditioning aspects of vegetation in relationship to human life and buildings.

5. Degrees of Control

All types of barriers may provide various degrees of control depending upon the height of density, spacing, combinations of screening, or of obstruction materials at different distances behind the wind screen or at the edges of the wind screen. Careful calculation of the degree of control desired in a particular situation will determine to a large extent the precise height, density, and spacing of the wind control devices. Various barriers may be used in combination or with a variety of degrees of fenestration to give the desired degree of control in a specific situation or location.

6. Effectiveness of Various Forms of Vegetation

Though some study has been made of the effectiveness of other forms of wind control, trees and forests have been most extensively studied in their use as shelter-belts. The effectiveness of various forms of vegetation has been well chronicles in extensive literature to delineate the effectiveness of trees and forests in improving the quality of the environment through their ability to control winds. It seems advisable to make reference to some of this available literature. In the document "Windbreaks for Conservation" by Arthur E. Ferber, regional forester for the Soil Conservation Service, the statement is made that:

"A dense windbreak reduces a 30-mile per hour wind to 10 to 15 miles per hour."

31

In that same document, the following illustration is shown to indicate the effectiveness of the windbreak protecting a residence.

In Chapter 4 of the document entitled *Forest Influences,* published by the Food and Agriculture Organization, R. J. Van der Linde writes concerning the effects of trees outside the forest on wind velocities, and indicates in an illustration the influence of windscreens on the velocity of the wind measured at a height of 1.4 meters over the surface and chose the following illustration to indicate the effectiveness of four different degrees of density windbreak.

The influence of windscreens on the velocity of the wind, measured at a height of 1.4 meters over the surface.

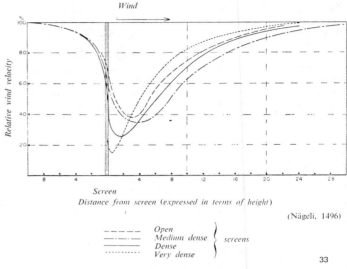

(Nägeli, 1496)

– – – – Open	
– · – · – Medium dense	} screens
———— Dense	
· · · · · · · · Very dense	

33

The same author also makes reference to the spacing and size of trees and the relationship of those factors to wind velocity in the following words:

"With many deciduous trees, though by no means all, the effects of the seasons are not quite marked. Flensborg and Nokkentved (1940) found that in the autumn the reduction percentages of the wind velocity had decreased in places just behind a dense screen. Further away from this screen, however, the reduction percentages had increased. In other words, the wind distribution curve, which is typical of a dense screen, had turned to one typical of a lower density class. The same phenomenon has been observed with medium dense screens, but with open screens the differences are small.

"While studying a barrier of spruce Woelfle (1939) found that this relative perviousness diminished as the velocity of the wind increased. Its flat horizontal branches, he assumed, acted like the slats of a Venetian blind; they have drawn closer to each other by the increasing wind.

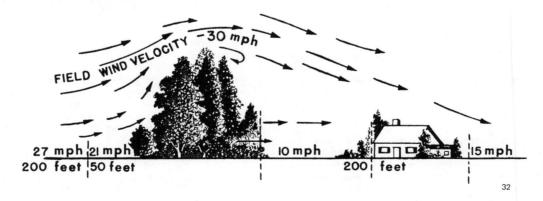

32

"There is no need to emphasize the influence which spacing has on density. Narrow spacing is naturally intended to produce a dense—and relatively high—screen when the conditions of the soil and the climate are favorable. Meanwhile, age has at least a twofold influence: it affects the height and the width of the screen and eventually it diminishes the branches and leaves in the lower-stem storey. Another positive influence on density is the number of rows. Even so, the density is not strictly proportionate to the number of rows which have been planted; for one thing, the inner rows soon lose their lower branches."

Van der Linde also shows two illustrations, one of which deals with the corresponding wind velocity curve showing the effect of different types of windscreens.   34

Corresponding wind velocity curves, showing the effect of different types of windscreens.

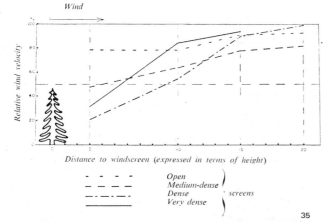

Distance to windscreen (expressed in terms of height)

- - - -   Open
— — —   Medium-dense   ⎫
— · — · —   Dense   ⎬ screens
————   Very dense   ⎭
                                                                                35

The second of these illustrations deals with the distribution of wind velocity over the fields adjacent to and between shelter-belts.

Distribution of wind velocity over the fields adjacent to and between the shelterbelts Epinette (E) and Champ-Bonnet (CH) in the Thoune plain in Switzerland.

(E) Height = 20 meters   (CH) Height = 23 meters

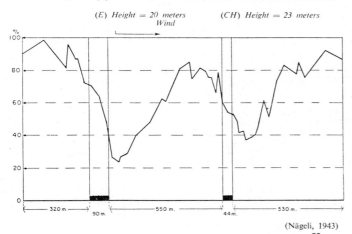

(Nägeli, 1943)
                                                                                36

Kittredge, in his book *Forest Influences,* also has a chapter on wind control or influences by trees and forests, makes reference in that document to an illustration dealing with the vertical gradients of wind velocities on the Shasta Experimental Forest.

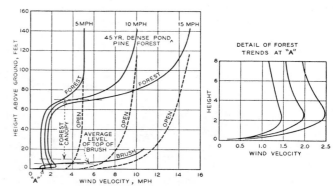

Vertical gradients of wind velocities on the Shasta Experimental Forest.
[*After Fons* (124).]
                                                                                37

One other illustration from that same source shows the actual and relative wind velocities at different elevations with respect to the canopy of Ponderosa Pine 70 foot high.

ACTUAL AND RELATIVE WIND VELOCITIES AT DIFFERENT ELEVATIONS WITH RESPECT TO THE CANOPY OF PONDEROSA PINE 70 FT HIGH

Elevation, ft	Position relative to canopy	Velocities, mph			Relative velocities, %		
142	Above	5	10	15	100	100	100
30	In	1	1.3	1.6	20	13	11
2.5	Below	1.4	1.9	2.5	28	19	17
                                                                                38

Once again the same author shows an illustration dealing with the relative wind velocities in percentages of those above the crowns for levels in and below the canopies of Pine and Pine with Spruce understory. This chart is accompanied by the following text:

RELATIVE WIND VELOCITIES IN PERCENTAGES OF THOSE ABOVE THE CROWNS FOR LEVELS IN AND BELOW THE CANOPIES OF PINE AND PINE WITH SPRUCE UNDERSTORY

Elevation, m	Wind class					
	Strong		Moderate		Light	
	Pine	P + Sp	Pine	P + Sp	Pine	P + Sp
13.7	75	46	83	48	77	47
1.1	50	17	58	17	62	27
                                                                                39

"When an air current reaches a forest stand, part is deflected upward with only a small change in velocity. Another part passes under the crowns with rapidly decreasing velocity and a third part passes through and among the crowns with very low velocity. When the wind strikes an extensive forest, there is a marked reduction in velocity close to the forest margin and only slight further reductions within the forest.

"The vertical distribution of wind velocity in and above the forest has been suggested previously for ponderosa pine. Confirmatory and more detailed studies on microclimate in Germany by Geiger (135) give the wind velocities at 1.1 m above the ground in the canopy at 13.7 m, and above the canopy at 16.9 m for pure Scotch pine stands and for Scotch pine with understory of Norway Spruce. The data were classified according to wind velocity above the canopy in three groups: (a) strong winds; (b) moderate winds; and (c) light winds. The velocities for the three groups and for the two kinds of stand relative to the velocities above the crowns as 100, are shown in Table II.

"The figures suggest several conclusions. The wind velocity in the crowns is much less than above them and somewhat greater than below them. An understory reduces the wind velocity in the canopy and even more below it. The reduction of actual velocities in the canopy and below it is greater in strong than in light winds. The relative reduction varies little with velocity."

23.

The same source shows a cross sectional drawing from Bates and Stoeckeler, indicating the relative effectiveness of a half-solid barrier in the reduction of percentages of full velocity of the wind.

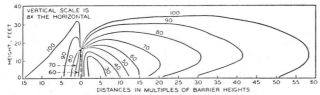

Fig. 9.—Distribution of wind velocities as percentages of velocities in the open at corresponding heights at different distances from a half-solid barrier. [*From Bates and Stoeckeler, United States Forest Service (38).*]

40

One deciduous and one coniferous shelterbelt system are compared in the following chart by the same author:

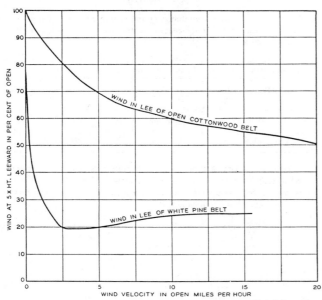

Reduction of wind velocity by shelterbelts. [*From Bates, United States Forest Service (36).*]

41

Kittredge also characterizes the effects of forests on wind reduction in the following words and with the following illustration:

"The wind velocities in larger forest areas as compared with those in the open are also illustrated from several sources in Table 12. In the Wagon Wheel Gap experiment at elevations of about 9,000 ft. in the Rocky Mountains, the wind movement after deforestation averaged 3.1 mph whereas before deforestation it averaged about 1 mph or three-tenths of the velocity in the open. A plantation of jack pine 40 ft. high, in the Nebraska sand hills, reduced the average wind velocity at 23 ft. above the ground to 28 percent of that in the open (275). As the trees increased in height from about the 23 ft. level of the tower in 1922 to 40 ft. in 1939, the velocities decreased at an essentially uniform rate from 6.5 to 1 mph or from 85 to 13 percent of the velocity in the open. Expressed in another way, as the trees grew 1 ft. in height, the wind velocity at 23 ft. among the crowns decreased by 1/3 mph or by 4.2 percent of that in the open. Apparently in this stand the point of minimum velocity was lower than 17 ft. below the tops, if it is assumed that the velocity increases again in the space below the living crowns.

"At 8,000 feet on the east slope of the Rocky Mountains the wind velocities in a mature stand of Douglas fir on a north slope measured at 1 ft. above the ground are reported by Bates (34) as follows:

Character of stand	Mean annual velocity, mph	Velocity, % of open	
		Mean annual	Growing season
Open....................	2.3	100	100
25 per cent left (shelterwood)....	1.4	63	54
35 per cent left (selection)......	2.0	89	75
Uncut....................	0.9	40	33

42

A summarization of the effects of forests and shelterbelts on wind velocity is shown in that same reference in the following way:

EFFECT OF FORESTS AND SHELTERBELTS ON WIND VELOCITIES

Forest	Miles per hour		Percentage relatives in forest, open =100	Reference
	Open	Forest		
Western white pine, Idaho........	1.4	0.2	12	(187)
Western white pine, half cut, Idaho.	1.4	0.8	58	(187)
Jack pine plantation, Nebr........	7.7	2.1	28	(275)
Aspen, Colo....................	3.1	1.0	32	(37)
Douglas fir, Colo................	2.3	0.9	33	(34)
Douglas fir, three-quarters cut, Colo.	2.3	1.4	63	(34)
Ponderosa pine, Ariz.............	5.3	2.7	51	(282)
Birch–maple–fir, cutover, N.Y.....	...	...	80	(332)

Shelterbelts	Distance to leeward multiples of height *h*	Open	Shelter-belt	Percentage relatives in shelterbelt, open =100	
Blue gum, S. Calif......	3h	21	7	34	(ms*)
Blue gum, S. Calif....	6h	21	10	49	(ms*)
Norway spruce, 4 rows, dense, Ind...........	2h	5	0.2	4	(111)
Norway spruce, 4 rows, dense, Ind...........	2h	10 / 15	2 / 3	20	(111)
Spruce and pine, 2 rows, medium dense, Ind. ..	2h	5 / 10	1 / 2	20	(111)
Spruce and pine, 2 rows, medium dense, Ind. ..	2h	15	4	27	(111)
Deciduous, in winter, density light, Ind. ...	2h	5	2	40	(111)
Deciduous, in winter, density light, Ind. ...	2h	10	5	50	(111)
Deciduous, in winter, density light, Ind. ...	2h	15	8	53	(111)
Willow, 1 row, medium dense, Ind.	15h	5	4	80	(111)
Willow, 1 row, medium dense, Ind.	15h	10	9	90	(111)
Willow, 1 row, medium dense, Ind.	15h	15	14	93	(111)

* From an unpublished manuscript, through the courtesy of Woodbridge Metcalf, Extension Forester, California Agricultural Experiment Station.

43

The chapter entitled "The Influence of the Forest on the Weather," in the book *Forest Influences,* by the Food and Agricultural Organization of the United Nations, makes the following statement and shows following illustration in regard to the effectiveness of wind control by trees and forests:

"Everyone has experienced the influence of a forest in greatly reducing the wind velocities when it is blowing hard in the open. The denser the forest, it is evident, the greater will be its influence on wind velocity. The density involves both the crowns of the individual trees and the stand.

WIND VELOCITIES IN FOREST AND OPEN IN METERS PER SECOND AND IN PERCENTAGES AT DIFFERENT HEIGHTS ABOVE THE GROUND IN NORTHERN CALIFORNIA, UNITED STATES, FOR VELOCITIES OF 2.2 AND 4.5 METERS PER SECOND ABOVE THE INFLUENCE OF THE FOREST AT 43 METERS [1]

Height	Velocity		Forest	Velocity		Forest
	open	forest	of open	open	forest	of open
Meters	m./sec.	m./sec.	°/₀	m./sec.	m./sec.	°/₀
43	2.2	2.2	100	4.5	4.5	100
27	2.2	2.1	95	4.4	3.8	87
21	2.1	1.3	62	4.3	1.7	39
12	2.1	0.5	24	4.0	0.6	15
6.1	2.0	0.5	25	3.6	0.6	17
3.0	1.9	0.5	26	3.3	0.7	21
1.5	1.8	0.6	33	3.0	0.8	27
0.8	1.7	0.6	35	2.6	0.9	35
0.5	1.6	0.6	38	2.3	0.8	35
0.2	1.5	0.4	27	2.1	0.6	29

[1] Fons, 1940.

"Wind velocities in a forest and above a ground surface are determined by the friction of the moving air masses with vertical and horizontal barriers or surfaces."
44

Evidence of this relation, as shown in the above table, . . .

"comes from the Shasta Experimental Forest at 1,220-meter altitude in northern California, United States. The forest was a pure stand of Pinus ponderosa 45 to 50 years old. The trees had an average height of 21 meters and the live crowns extended to 3 meters above the ground. There were 2,100 trees per hectare and no live ground cover. The open area—900 by 300 meters—had a sparse herbaceous cover which was seldom more than 15 centimeters in height. Tests indicated that wind velocities at the center of this open area were not influenced by the surrounding forest."
45

The above table . . .

"gives average velocities at different levels—two above the canopy, four in it and four below it—when the velocities at 43 meters, above the influence of the forest, were 2.2 and 4.5 meters per second. In the open, the vertical gradient becomes steeper as the distance to the ground decreases, and it is very steep below a height of 1 meter. The gradient becomes steeper as the wind velocity at 43 meters increases. Quite similar trends are evident as the height above the canopy—the friction surface—decreases.

"Continuing downward from the upper surface of the canopy, the trends reverse. The steepest gradients of decreasing velocity are close to the upper surface, where winds of 4.5 meters per second at 43 meters become 1.7 meters per second at 21 meters and 0.6 meters per second at 12 meters among the crowns. At the lower surface of the canopy, the velocity again increases to 0.7 meter per second at 3 meters. In the space beneath the crowns, there is a slight secondary maximum of 0.9 meter per second at 0.8 meter. Below that space the velocity decreases again more rapidly as the distance to the forest floor increases.

"The percentages in the table show clearly how much the forest reduces wind velocities as compared with those in the open. The greatest reduction to 15 percent is among the crowns in winds of 4.5 meters per second compared with 24 percent in winds of 2.2 meters per second."
46

The diurnal cycle of wind velocities is shown in the following chart from that same source:

Time of day	5.30	9	13	17	20.30
Velocity in open (m./sec.)	0.9	1.7	2.6	2.8	1.9
Velocity in selection forest (m./sec.)	0.2	0.3	0.4	0.4	0.3

The annual cycle is shown in the following chart which traces the influence 47 of an Oak forest on the mean monthly wind velocities in Tennessee, 1936-1939:

INFLUENCE OF *Quercus* FOREST ON MEAN MONTHLY WIND VELOCITIES [1] IN TENNESSEE, UNITED STATES, 1936-39 [2]

Month	Velocity		Forest	Departures Forest
	open	forest	of open	from open
	m./sec.	m./sec.	°/₀	m./sec.
January	3.3	0.4	12	− 2.9
February	3.4	0.4	12	− 3.0
March	3.5	0.5	14	− 3.0
April	3.4	0.4	12	− 3.0
May	2.4	0.2	8	− 2.2
June	2.1	0.1	5	− 2.0
July	1.8	0.04	2	− 1.76
August	1.7	0.04	2	− 1.66
September	1.8	0.04	2	− 1.76
October	2.4	0.2	8	− 2.2
November	2.6	0.3	11	− 2.3
December	2.5	0.3	12	− 2.2

[1] Measured at 0.6 meter above the ground. [2] Hursh, 1948.
48

The previously mentioned Japanese study conducted in wind tunnel tests had a variety of degrees of openings of both horizontal and vertical slate to simulate various possible windscreen patterns. This same study also made extensive study of various velocities of wind as it moves through these screens of varied openings.

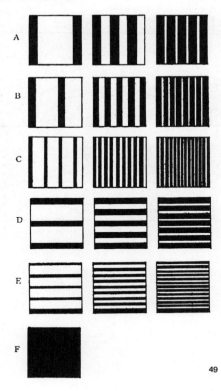

49

The possible results of these studies are shown in the following illustration:

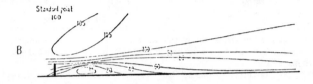

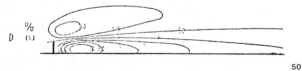

50

The relative effectiveness of these different configurations are shown in the following charts:

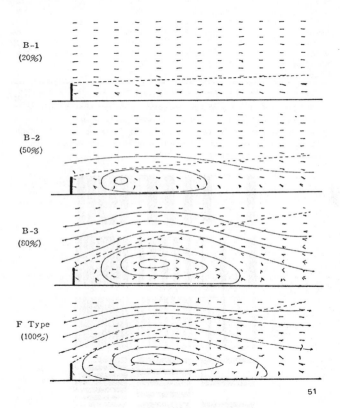

B-1
(20%)

B-2
(50%)

B-3
(80%)

F Type
(100%)

51

The height of the hedge and its effect on the ability to control the wind velocity leeward of the screen are shown in the following charts:

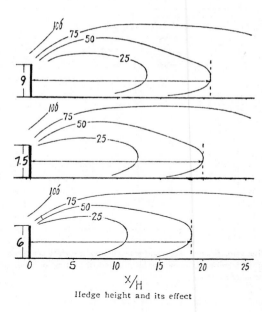

Hedge height and its effect

52

Guidance of winds is the subject of a number of studies on the placement of plants adjacent to buildings to enhance natural ventilation. The studies conducted by Robert F. White, landscape architect on the staff of the Texas Engineering Experiment Station, were primarily concerned with the wind control aspects of planting in relation to architecture to provide natural ventilation in warm areas of the world. White's research indicated that:

"(1) Planting can materially affect the movement of air through and about buildings; (2) dependent on the way it is used, planting may either augment or reduce the natural airflow through the building; (3) planting may cause actual change of direction of air-flow within the building; (4) planting on the lee side of buildings has little or no effect on the movement of air through the building, unless it is in such a position that it obstructs the outlet openings."[53]

The accompanying illustrations indicate air flow around and through a number of plants and building configurations:

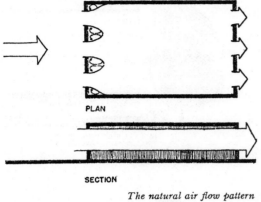

PLAN

SECTION

The natural air flow pattern of the test model with no planting.

AT BUILDING

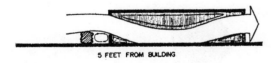

5 FEET FROM BUILDING

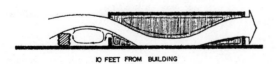

10 FEET FROM BUILDING

20 FEET FROM BUILDING

Low hedge (less than three feet high).

AT BUILDING

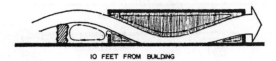

10 FEET FROM BUILDING

20 FEET FROM BUILDING

Medium hedge

5 FEET FROM BUILDING

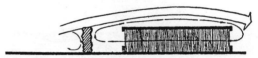

10 FEET FROM BUILDING

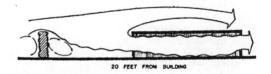

20 FEET FROM BUILDING

High hedge

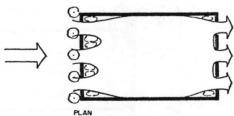

PLAN
SHRUBS 5 FEET ON CENTER

AT BUILDING

Shrubs

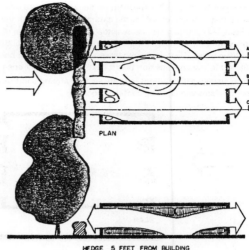

PLAN

HEDGE 5 FEET FROM BUILDING
TREE 10 FEET FROM BUILDING

SECTION A

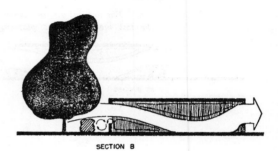

SECTION B

SECTION C

Hedge-tree combinations

54

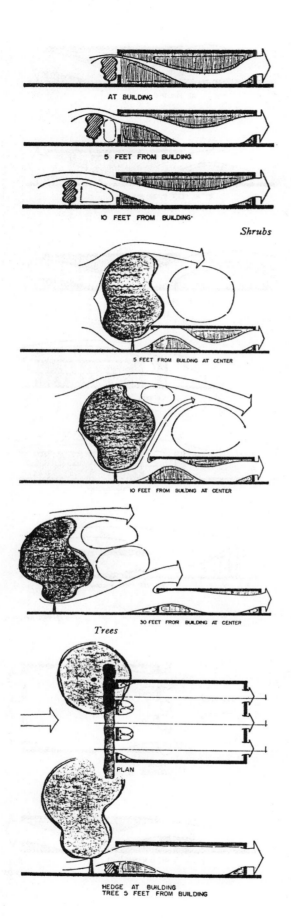

AT BUILDING

5 FEET FROM BUILDING

10 FEET FROM BUILDING·

Shrubs

5 FEET FROM BUILDING AT CENTER

10 FEET FROM BUILDING AT CENTER

30 FEET FROM BUILDING AT CENTER

Trees

PLAN

HEDGE AT BUILDING
TREE 5 FEET FROM BUILDING

28.

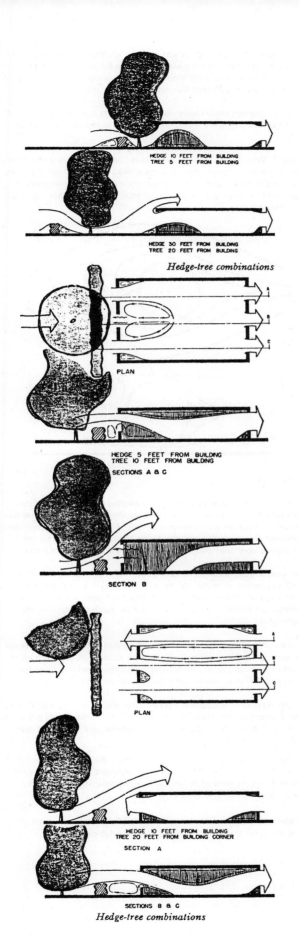

HEDGE 10 FEET FROM BUILDING
TREE 5 FEET FROM BUILDING

HEDGE 30 FEET FROM BUILDING
TREE 20 FEET FROM BUILDING

Hedge-tree combinations

PLAN

HEDGE 5 FEET FROM BUILDING
TREE 10 FEET FROM BUILDING

SECTIONS A & C

SECTION B

PLAN

HEDGE 10 FEET FROM BUILDING
TREE 20 FEET FROM BUILDING CORNER

SECTION A

SECTIONS B & C

Hedge-tree combinations

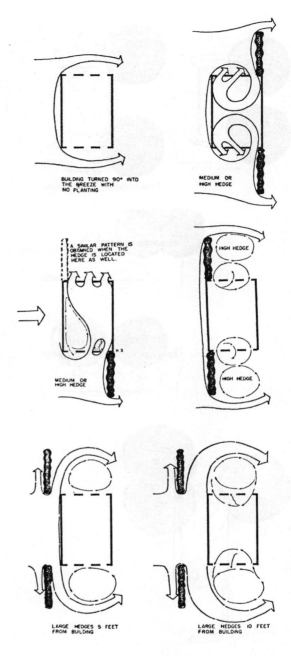

BUILDING TURNED 90° INTO
THE BREEZE WITH
NO PLANTING

MEDIUM OR
HIGH HEDGE

A SIMILAR PATTERN IS
OBTAINED WHEN THE
HEDGE IS LOCATED
HERE AS WELL.

MEDIUM OR
HIGH HEDGE

HIGH HEDGE

HIGH HEDGE

LARGE HEDGES 5 FEET
FROM BUILDING

LARGE HEDGES 10 FEET
FROM BUILDING

Hedge-building combinations

55

29.

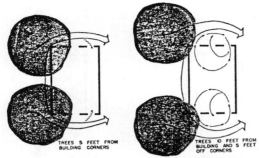

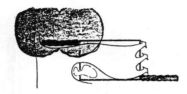

TREES 5 FEET FROM
BUILDING CORNERS

TREES 10 FEET FROM
BUILDING AND 5 FEET
OFF CORNERS

Tree-building combinations

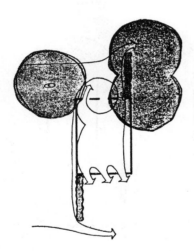

Tree-hedge-building combinations

When any bluff, non-streamlined barrier is introduced into an airflow, a pressure eddy is formed immediately in front of, and a suction eddy is created immediately leeward of the barrier. Beyond the barrier a turbulent wake is created. Wind is controlled a distance of from two to five times the height of a barrier in front of a wind obstruction, and from 10 to 15 times the height leeward of such a barrier.

Deflection of wind over trees or shrubs is another method of wind control. Plants of varying heights, widths, species, and composition, planted either individually or in rows, have varying degrees of effect on wind deflection.

For example, coniferous evergreen that branch to the ground are generally the most effective year-round plants for wind control; and deciduous shrubs and trees, when in leaf, are most effective in summer. Wind velocity is cut from 15 to 25 percent of the open field velocity directly leeward of

a dense screen planting, such as spruce or fir, while a loose barrier of lombardy poplars reduces leeward wind velocity to 60 percent of its open field velocity. Wind velocity is cut from 12 to 3 miles per hour for a distance of 40 feet leeward of a 20-foot Austrian pine.

Filtration of wind, passing under or through plants, is a method of control. There are instances where it may be desirable to speed up or slow down wind.

Robert F. White, in his studies at the Texas Engineering Experiment Station, concludes that the foliage mass of a tree serves as a direct block to passage of air; and that the speed of the air movement directly beneath a tree is measurably increased over speeds measured at the same height on its lee and windward side.

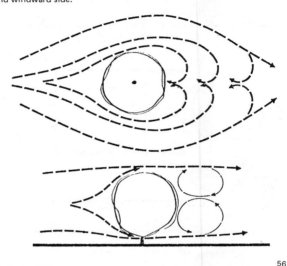

7. Shelter Belts

Studies have shown that shelterbelts and windbreaks are most effective when placed perpendicular to the prevailing winds. Wind velocity may be reduced by 50 percent for a distance of from 10 to 20 times the tree height downwind of a shelterbelt, and the degree of protection and wind reduction depend upon the height, width, and penetrability of the plants used.

Near a moderately dense shelterbelt, at the end of the windbreak, wind speed is increased more than 10 percent of the open field velocity prior to its interception. The leeward sheltered zone is not confined within lines drawn perpendicular to the ends of the barrier, but is broader than the lengths of the barrier.

Wind speed is also affected within or on the windward side of a windbreak. For example, the wind speed is reduced for a distance of 100 yards on the windward or front side of a 30-foot-high shelterbelt, and is reduced for a distance of 300 yards downwind or behind the shelterbelt.

Partially penetrable windbreaks have different effects on windflow than do dense windbreaks. Wind velocities immediately to the leeward of any windbreak are directly affected by the type of material used or kinds of plants used.

The more penetrable the windbreak is, the longer the distance behind the windbreak protection extends. Some wind, in passing through a penetrable windbreak, retains some of its laminar flow characteristics at a reduced velocity, thus inhibiting turbulence behind it.

The accompanying illustration shows the effectiveness of open, medium dense, and dense shelterbelts in reducing wind velocities.

Wind velocity	Clear cut	Half cut	Uncut
Miles per hour	1.4	0.8	0.2
Relatives (%)	100	58	12

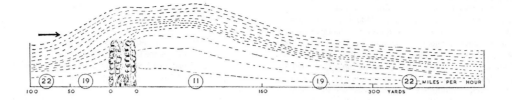

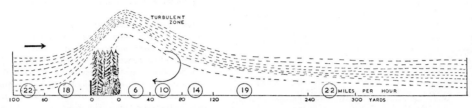

Effect of (a) moderately penetrable and (b) dense windbreaks on the flow of wind. Figures circled denote approximate wind speeds in the various sections. Horizontal distances (in yards) refer to windbreaks 30 ft tall (after Kuhlewind, et al.).

58

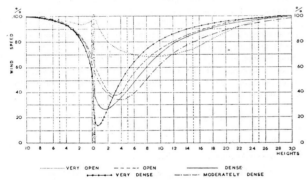

VERY OPEN ....   OPEN - - - -   DENSE ——
VERY DENSE +—+—+   MODERATELY DENSE —·—·—

Patterns of wind abatement in the vicinity of shelterbelts of different density (after Nägeli and Caborn).

59

Wind speed (percent of that in open) as shelterbelt is thinned.

Number removed from each six plants in a row	Distance from shelterbelt					
	-2.5h	0.5h	2.5h	5h	7.5h	10h
0	93	75	56	29	53	74
2	91	72	60	38	60	69
4	83	90	78	69	78	76
5	97	96	98	93	96	95
6	100	100	100	100	100	100

60

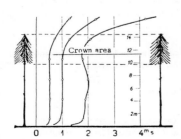

Wind profiles in a stand of pine for three ranges of wind speed.

61

WIND REDUCTION TO LEEWARD OF 30-FT HIGH SHELTERBELTS OF DIFFERENT DENSITY

DENSITY OF BELT	PERCENTAGE WIND SPEED REDUCTION			
	Averaged over first 50 yards	Averaged over first 100 yards	Averaged over first 150 yards	Averaged over first 300 yards
Very open	18	24	25	18
Open	54	46	37	20
Medium	60	56	48	28
Dense	66	55	44	25
Very dense	66	48	37	20

For extensive sheltering the moderately dense belt gives the best average (28 per cent). Some at the top of this class may

62

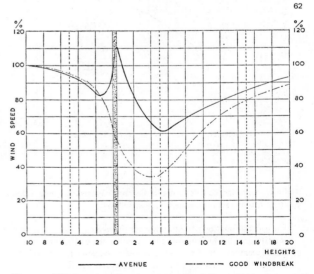

AVENUE ——   GOOD WINDBREAK —·—·—

Fig. 10 Wind accelerates under open avenue belts, reducing their value for shelter (after Nägeli).

63

Wind velocity	Clear cut	Half cut	Uncut
Miles per hour..........	1.4	0.8	0.2
Relatives (%)...............	100	58	12

64

The Agricultural Experiment Station at Kansas State University in Technical Bulletin 77 (December 1954) entitled "Shelterbelt and Surface Barrier Effects on Wind Velocities, Evaporation, House Heating, Snow-drifting" written by N.P. Woodruff deals with *Barrier Effects on House Heating*. Obviously the costs of fuel mentioned in this example is grossly out of date, however, the rationale and the approach is nevertheless valid in calculating potential energy savings through proper landscape planning. The effective velocity reductions of various man made and natural windbreaks and the evaporation reduction curves for leaved and defoliated 10 row shelterbelts and a solid wall are shown in the following illustrations from that Technical Bulletin.

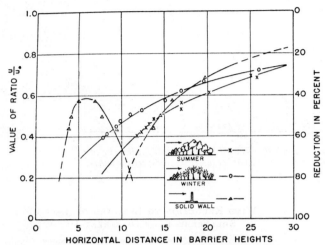

Dimensionless curves showing effective velocity reduction at ground level to the lee of a leaved and defoliated 10-row shelterbelt and a solid wall. Velocity reduction applies irrespective of wind velocity.

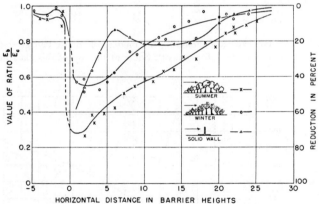

Evaporation reduction curves for a leaved and defoliated 10-row shelterbelt and a solid wall. Wind velocity was 27.3 mph measured at center of tunnel.

The text, formulae, calculations and diagrams concerning energy calculations from that Technical Bulletin are as follows:

"The general functional relationship for the heating load, the wind velocity, and the temperature difference for the house unprotected and at locations aft of the belt is expressed graphically in Figure 11. The general equation of the curve for no protection is:

$$\frac{Q}{T_\triangle} = 1.3 \, (10^{0.025\,u})$$

The curves for the various locations aft of the belt approximate a family of lines and, therefore, may be expressed in the following 3-variable equation:

$$\frac{Q}{T_\triangle} = 1.3 \, (10^{0.018\,L^{0.07}\,u})$$

Symbols in these two equations are defined as follows:

Q = heating load in BTU per hour.
$T_\triangle$ = difference between inside and outside temperature in °F.
L = distance from belt to house in barrier heights H.
u = wind velocity in mph.

The unprotected house function indicates that the heating load with a 20 mph wind is approximately 2.4 times as great as that for a 5 mph wind under the same temperature conditions. The heating load for the protected house exposed to a 20 mph wind was approximately twice as great as that for a 5 mph wind, thus indicating the belt to be slightly more effective at higher wind velocities.

Figure 12 shows reductions in heating load at various distances aft of the belt for wind velocities ranging from 5 to 35 mph in 5 mph increments. It is noted that the amount of reduction in heating load decreased with distance aft of the belt and that the largest percentage of reductions occur for the higher wind speeds. This analysis also shows that the percentate reduction in heating load is not a function of temperature, i.e., for a given temperature the reduction changes only with wind velocity. This is probably explained by data of Bates (1) where he shows only a 1°F. difference in air temperature behind a belt compared with the open. This does not mean, however that temperature is not important. Since the total heating load Q is a function of the temperature difference, it follows that as the temperature difference increases Q will increase. For example, assume a 32°F. outside temperature, a 30 mph wind and a house with a 76°F. inside temperature located 2 H aft of a shelterbelt. According to Figure 11, Q would equal 211 BTU per hour; but if the outside temperature dropped to 0°F. and the wind speed and location remained the same Q would equal 365 BTU per hour. For these conditions a drop in temperature of 32°F. increased the heating load 154 BTU per hour, or 73 percent.

Some idea of the savings in fuel consumption for house heating which may be obtained by using shelterbelts can be given by considering a hypothetical problem:

Problem: Assume the following:

Heating load on average home = 95,000 BTU per hr.
Inside temperature = 76°F.
Type of heating unit = gas furnace.
The house has shelterbelt protection from all winds.

Using design temperatures, average wind velocities, and degree days of heating for Topeka, Kan., as given in the Air conditioning manufacturers Interim Code (2) and the Degree Day Handbook by Strock and Hotchkiss (4), determine the savings in cu. ft. of gas for the heating season if a house were placed at 2 H aft of a shelterbelt.

Information Required:

Design temperature for Topeka, Kan. = −10°F.
Degree days of heating = 5,037.
Heating value of natural gas = 900 BTU per cu. ft.
Efficiency of gas burning furnace = 90%
Average winter wind velocity = 10 mph.
Percentage saving in heating load at 2 H aft of belt with 10 mph wind = 15.0

Solution:

Heating load for season for unprotected house

$$(\text{BTU per season}) = \frac{(95,000)\,(24)\,(5,037)}{76 - (-10)} = 133.5\,(10^6)$$

$$\text{Cu. ft. of gas consumed per season} = \frac{133.5\,(10^6)}{900\,(0.90)} = 164,815$$

Cu. ft. of gas saved per season for protected house at 2 H leeward of belt = 164,815 (0.15) = 24,722

32.

By assuming that the rate of consumption can be divided into 5 equal increments and the fuel bill is paid 5 times during the season, the dollar and cent saving for the season would be approximately $9.90 at present Topeka, Kan., gas rates.

These computations, of course, are only averages based on design recommendations for home heating; they serve only as guides for procedures in making calculations of this type.

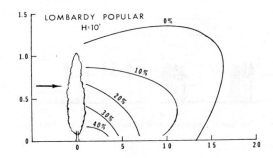

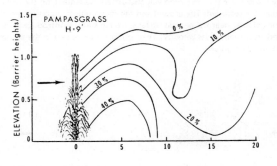

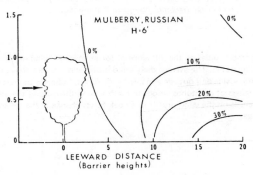

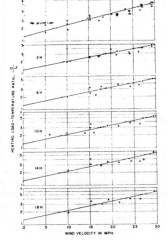

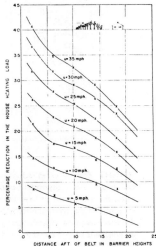

Relationship between house heating load Q, temperature difference T_λ, and wind velocity u for protected and unprotected house. All points are averages of 2 trials.

Percentage reduction in heating load for house at various distances aft of 10-row defoliated shelterbelt for velocities ranging from 5 to 35 mph.

Velocity reduction patterns for pampasgrass, foliated 3-year-old Lombardy poplar, and Russian mulberry barriers at Colby, Kansas.

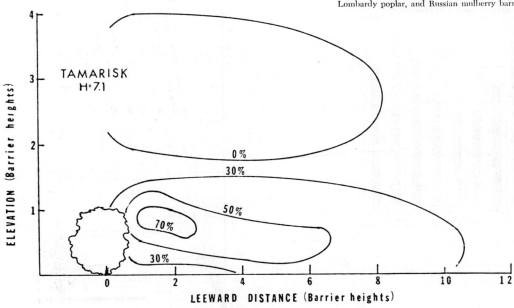

Velocity reduction patterns for a 4-year-old tamarisk barrier at Colby, Kansas.

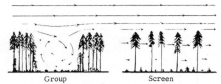

Group   Screen

Air flow in a regeneration area and under a screen of old trees.

65

The measurement of the wind velocity differs, depending on the height at which it is recorded. The ground and the rough objects upon it slow the wind velocity near the ground surface. Rough surfaces introduced into the windstream reduce its velocity. This concept is presented in the accompanying illustration.

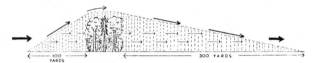

A 30-ft high shelterbelt affects wind speed for 100 yards in front of the trees and 300 yards down-wind.

66

Two other studies on the influence of forest cover on wind movement are also illustrated. The average daily wind movement, 15 feet above the ground, inside and outside a maple-beech forest, indicates the relative effectiveness of deciduous woods to control wind throughout the year. The influences on the wind profile in an oak forest change as the foliage emerges.

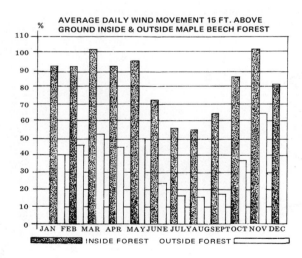

67

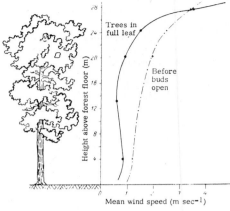

Influence of foliage on the wind profile in an oak forest.

68

HEIGHT. Zone of wind reduction on the leeward and windward side of a barrier are largely dependent upon the height of the barrier. The taller the trees, the more rows of trees are required for protection. With an increase in height of the trees, shelterbelts become more open. Instead of reducing the wind, avenues of trees open at the bottom increase wind speed, as the air stream is forced beneath the tree canopy and through the tree trunks.

According to N. M. Gorshenin (1934) the influence that trees have on reducing wind velocity extends from 30 to 40 times the height of the trees leeward of the windbreak, and the sharpest reduction in wind velocity extends from 10 to 15 times the height of the trees. Commenting on this same subject, B. I. Logginov (1941) stated that the maximum shelter from the wind is obtained at three to five times the height on the leeward side of trees. According to H. Iizuke in 1950, wind velocities recorded, behind an artificial windbreak in Japan at distances of 10, 20 and 30 times the height of the windbreak, were respectively 66.44, 69.33, and 77.44 percent of the open wind speed.

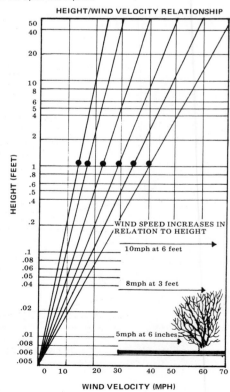

69

WIDTH: The field of effectiveness of a shelterbelt depends primarily on the height and penetration of the shelterbelt. Width of the planting is of secondary importance only insofar as it affects the degree of penetrability. The width of the windbreak has a negligible influence on reducing wind velocity at its leeward edge, but can cause notable variation on the microclimate within the sheltered area. With a wide shelterbelt or forest, the maximum reduction in the velocity of the wind occurs within the area of the shelterbelt or forest itself. Therefore, the wide shelterbelt or forest block actually consumes its own shelter to the extent that the wind velocity is reduced within the shelter itself.

Shrubs used to fill the space below heavily-crowned trees must be capable of closing the gap effectively.

70

Character of stand	Mean annual velocity, mph	Velocity, % of open	
		Mean annual	Growing season
Open....................	2.3	100	100
25 per cent left (shelterwood)....	1.4	63	54
35 per cent left (selection).......	2.0	89	75
Uncut....................	0.9	40	33

71

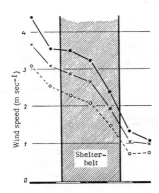

Braking of wind in a shelterbelt of trees 20 m high. (After W. Nägeli)

72

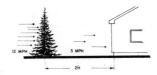

73

According to Smal'ko (1954) temperature stratification influences wind velocities in a tree strip of dense structure. The distance on the leeward to which it will affect the wind by reducing its velocity varies accordingly from 18 to 37 times the height of the windscreen, depending upon temperature stratification.

74

The accompanying diagrams, which show the flow of wind over a forest block and a narrow shelterbelt, illustrate the influence width of a windbreak has on reducing flow.

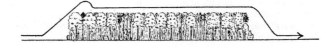

The flow of wind over a narrow shelterbelt and a wide forest block.

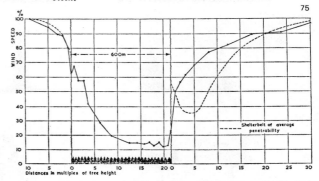

Wind abatement behind a forest block compared with that leeward of an efficient shelterbelt (after Nägeli).

76

An irregular windbreak, such as the top of a picket fence, is more effective in breaking up a portion of the airstream deflected over it. A mixture of species and sizes of plants within the windbreak, therefore, produces a rough upper surface and is more effective in controlling wind.

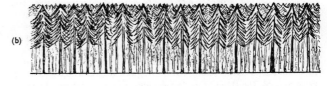

An irregular profile is more effective in reducing eddying (c), than more uniform structures (a) and (b).

77

Belts with a 'pitched-roof' cross section (a), are less effective in halting winds than belts with more or less vertical edges (b).

78

Configurations of windflow over windbreaks were studied in the hydraulics laboratory wind tunnel at the University of Wisconsin in 1954 and 1964. Findings revealed that when a windbreak is completely impenetrable, practically the entire wind force is deflected upward, over the barrier. Pressure behind the barrier is low because the wind does not pass through the barrier, causing a suction effect and forcing the air currents above the windbreaks downward. Shelterbelts with a cross section resembling a pitched roof are less effective in blocking winds than the vertical edges.

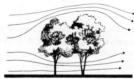

THE PATTERNS ABOVE WERE OBTAINED AS AN AID TO THE PLANNER TO HELP HIM DIRECT WIND THROUGH OR AROUND HIS PROJECT. THE PATTERNS WERE OBTAINED BY DRAWING CIGAR SMOKE THROUGH STEEL WOOL TREES WHICH WERE MORE DENSE THAN ACTUAL TREES. THE WIND WAS TRAVELING THROUGH THE MODEL AT APPROXIMATELY ONE MILE PER HOUR (31 MPH TO SCALE). THE EFFECTIVE MODEL SIZE WAS 4" X 6", WHICH WAS A LIMITING FACTOR. THE EXPERIMENTS WERE CONDUCTED IN THE HYDRAULICS LAB WIND TUNNEL AT THE UNIVERSITY OF WISCONSIN.

WINDBREAK PATTERNS

Panfilov's studies (1936) list the types of windbreaks as follows: windbreaks open throughout their height (partly penetrable to wind); windbreaks dense throughout their height (impenetrable to wind); windbreaks of medium density below and high density above (slightly penetrable); and windbreaks of medium density above and open below. He indicates that wind turbulence behind the windbreak increases with the density of the belt. Although it may provide a greater degree of shelter from the wind on the leeward, a dense shelterbelt provides a restricted zone of effective shelter downwind. A penetrable shelter has a lower percentage of actual wind reduction near the screen; but the overall effects extend a greater distance beyond. The effect of suction immediately behind the penetrable shelterbelt is less than that of a dense shelterbelt, and the acceleration of the wind back to its original force is more gradual behind a penetrable shelterbelt.

The optimum density for a shelterbelt is about 50 to 60 percent. This means that the leaves, branches, twigs, and trunks should cover 60 percent of the frontal area of the belt. With this density, narrow belts will afford as much shelter as wider belts of the same overall penetrability.

Wind-speed effects, leeward of a 30-foot-high shelterbelt of different densities, are shown in these illustrations.

Factors of penetrability and patterns of wind abatement in the vicinity of shelterbelts of different densities are also shown.

A diagram by Werner von Nageli, showing the effects of a 42-foot wide avenue of 72-foot poplar trees, spaced 21 feet apart and bare of branches for the first 25 feet, also indicates that wind accelerates under open-avenue belts, and reduces their value as wind barriers.

An accompanying chart illustrates the differences between open and closed wind barriers based on the percent of openfield wind velocity.

In selecting plants for a windbreak and in the design of the windbreak configuration, the designer should be aware of these effects.

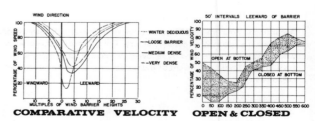

COMPARATIVE VELOCITY   OPEN & CLOSED

79

LENGTH: Research shows that the lines of equal wind velocity (isotacks) have a tendency to deviate toward the center of the barrier in the field of protection afforded by windbreaks, and to adapt courses parallel to it. An extension of the barrier changes nothing except that the zones of isotacks parallel to it are widened. The most efficient ratio of height to length for windbreaks is 1:11.5. According to Werner von Nageli, this ratio produces the greatest possible shelter effect, at least near the center of the shelter-belt. As a result of the air current sweeping around the shelterbelts, increased wind velocity, higher than that found in the open field, occurs at both ends of windbreaks. According to Kreutz (1950), smoke experiments have confirmed this.

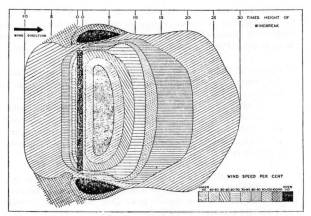

Plan of wind conditions in the vicinity of a moderately dense shelterbelt. Wind speed expressed as a percentage of the free, unobstructed wind.   80

Frequent gaps in a system of shelterbelts, particularly where belts intersect one another with a gap of 12-meter widths (40 feet), are advocated by the Russian scientists. According to one study, as a zone of wind-veolcity increase develops within the gap, there is practically no lateral extension of this "draught" zone, and at a short distance from the gap the wind abatement is normal. Findings also were made by J. van Eimern in 1951 who measured a wind velocity of 3.6 meters per second in the open and wind velocity of 4.5 meters per second within the gaps between shelterbelts.

Ways through cross-wind belts may be angled or curved to keep out the wind but are liable to fill with drifted snow.

Parallel shelterbelts, spaced widely apart, have no cumulative effect on reducing wind velocity. According to Carlos G. Bates, however, each shelterbelt has a larger coherent mass of stilled air within a zone of seven to 12 times the height of that belt, stretching laterally from the leeward edge of it, creating some small degree of protection. A series of shelterbelts, spaced closer than five times their height, have some cumulative effect. It would appear that when the distance between two shelterbelts is less than 30 times their height, the velocity of the wind between the two never reaches the velocity of the free wind; scientific evidence does not substantiate this.

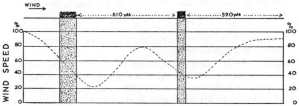

Parallel belts have no apparent cumulative effect on wind abatement (after Nägeli).   81

A windbreak has a protective effect for a distance equal to the height of the barrier, and leeward of it. If the barrier is a dense planting, it will remain relatively stable, and the percentage value of minimum wind velocity will remain relatively unchanged.

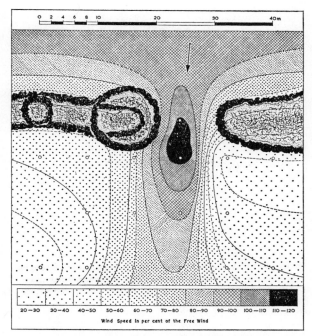

Plan of wind conditions near a shelterbelt, showing acceleration of the wind caused by a gap (after Nägeli).   82

In a penetrable windbreak, the protective effect is somewhat less next to the barrier and is felt at a distance equal to its height, and the zone of overall protection is more pronounced. The increased shelter effect with increased free wind velocity has been mentioned frequently. One study indicates that wind measurements made during a period of 30 years demonstrated that a spruce belt reduces the wind velocity by 30 percent and up to 47 percent during heavy gales (Geiger 1931). The relative shelter effect behind a barrier of rigid form remains more or less the same for varying speeds of wind. However, when the windbreak changes its form according to wind pressure, as illustrated by trees bending and swaying, the penetrability and vertical structure of it are affected, and the zone of reduced velocity is decreased. A row of trees, being somewhat elastic, will change its form according to the prevailing wind speed, thereby affecting the resistance to the wind and the degree of penetrability (Bates). The percentage of velocity reduction and the zone of quieter air increase as the wind becomes stronger and the center of the quieter air tends to move farther away from the windbreak on the leeward side.   86

A moderately penetrable shelterbelt becomes more impenetrable in high winds. Similarly, a belt too sparse to give protection in light winds affords a greater degree of shelter as the wind velocity increases.

According to M. Jensen, sheltering efficiency of a treebelt is reduced when the wind passes over a very rough surface before it strikes the belt. The character of wind is very important in determining the type, size, and characteristics of the windbreak (trees) required.   87

8. Intermediate Control

Prevailing winds may change direction seasonally: therefore, plants used should be of the type best suited for controlling winds from all directions. For instance, a dense coniferous windbreak of pyramidal aborvitae on the northwest side of a structure may protect it from harsh winter winds, and yet direct summer breezes around it. Not only may wind be slowed down or deflected for preserving a degree of warmth, it may also provide a degree of coolness by acceleration and channeling through an opening in a planting, or be guided in a desired direction by the angle of the planting. The advantageous base of plants to control the wind can reduce the cost of interior heating and cooling.

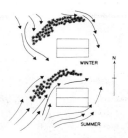

WIND & BUILDING

83

Cooler winds flow downhill at night. Dense evergreen plants placed on a slope, trap and hold cold wind flow upwind, thus creating cool spaces. Deciduous or "loose" plants create cool spaces on the downwind side by filtering the air.

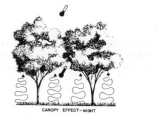 

CANOPY EFFECT – NIGHT   COLD AIR POCKET

84

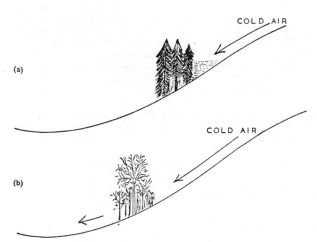

Dense windbreaks (a) create frost pockets. Penetrable windbreaks (b) allow cold air to drain away and are important at the foot of sloping gardens and orchards.

85

Wind gusts may be effectively controlled by planting along highways, to break the force of wind gusts blowing against the sides of automobiles. According to J. M. Caborn:

"There are pockets along many highways, particularly when emerging from road cuts, where furious gusts can catch a car broadside and have a disconcerting effect on the driver—even a dangerous one at high speeds. Not all these hazardous spots could be improved by shelterbelts, particularly in a wild broken country . . . but some could."

88

9. Ventilation

All functional uses of plant material are interconnected; therefore, it is difficult and artificial to attempt to separate one use from another. Dr. Wilfrid Bach and Dr. Edward Mathews, in a paper on "The Importance of Green Areas in Urban Planning," state in regard to ventilation in urban areas:

"Wind is one of the most important climatic elements in urban planning, since dispersion of air pollution and human comfort are largely dependent on it. Winds that are too strong cause the funnel effect in our canyon-like streets. This may locally lead to high air pollution potential through lifted street dust and strong wind fumigation from elevated sources. Winds which are too light may lead to a stagnation condition which may also produce high air pollution potential and in summer, muggy conditions.

"The ideal ventilation system would prevent the funnelling effect but favor the country breeze, i.e., wind blowing from relatively cleaner and cooler country, suburban, or green areas. This could be achieved by a properly spaced system of green areas cutting through the entire built-up area.

89

10. The Effect of Windflow on Other Factors

None of the abilities of trees and forests to affect the human environment exist in isolation. By the same token, none of the abilities of plants to control climate exist by itself. The windflow obviously affects the temperature, it affects the evaporation rate and speed of moisture loss from the vegetation itself as well as from the soil; it affects transpiration of moisture through the leaves of the trees; the wind obviously affects soil erosion; it affects snow drifting; it affects soil moisture and a number of references have indicated that the wind affects the carbon dioxide level in the air behind windbreaks. The necessity for scientific reductionism in isolating the functional abilities of trees and forests sometimes clouds the fact that each of these factors is mutually interactive. Space does not allow in this location the complex tracing of these interactivities. The designer, the forester, or the researcher must, of course, keep these in mind in manipulating or using the above isolated facts.

Summary

Vegetation of all types in a variety of configurations affects the flow of wind over the earth's surface. Kittredge, in his book, *Forest Influences,* has summarized to some degree the ability of trees and forests to influence or affect the movement of wind in the following words:

"The degree of influence of forest cover on wind movement varies with the height, length, width and density of individual crowns and with the density of the stand.

"The velocity may be reduced even by a shelterbelt to only 20 percent of the velocity in the open. In the forest, it is usually 20 to 60 percent of that in the open.

"Velocities in the forest are characteristically low, usually ranging from 1 to 2 mph on the average."

90

Because of the fact that trees and forests do affect materially the flow of wind on the earth's surface, extreme care must be taken in choosing and placing them in order that the desired effect may actually be realized in the location desired. By the same token, an equal amount of care should be exercised so that trees or forests are not placed in such a way or in such a location or with such a configuration that it would inhibit or misdirect the flow of wind where it is actually desired. At times it may be necessary to call for selective thinning or clearing of vegetation so that a greater amount of desired wind may move into a specific area in a manner carefully determined to improve the quality of the environment in a specific location. This entire matter requires much more research in order to precisely understand the use of wind and windscreens to improve the human environment.

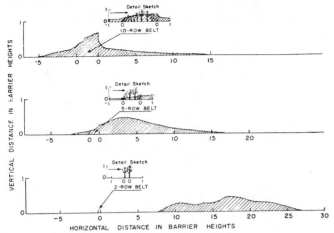

Fig. 13.—"Snow catch" profiles developed by 2-, 5-, and 10-row defoliated shelterbelts. Wind velocity constant at 15.9 mph measured 5 inches above tunnel floor.

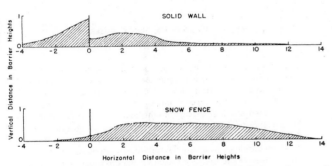

"Snow catch" profiles developed by a single snow fence and a solid wall. Wind velocity constant at 15.9 mph measured 5 inches above tunnel floor.

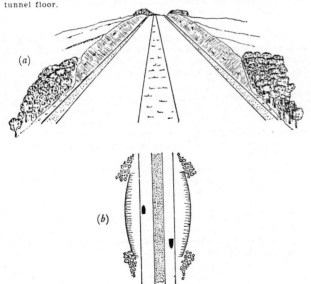

Suggested arrangement of planting to check cross-wind gusts near cuttings on highways. (a) general view; (b) plan of arrangement.

c. Precipitation and Humidity Control

As plants control solar radiation and wind, they also control precipitation and humidity in the atmosphere and as it moves to and through the earth's surface. There is a very close tie between the precipitation and humidity control of plant materials and the entire matter of water pollution control. Once again none of the abilities of trees and forests may be separated in actuality, but for the purposes of more intense study it is necessary to artificially isolate and treat as separate processes each of the various activities of vegetation as it relates to human comfort and environmental control. In understanding the effects of plant material on precipitation and humidity, the role of the plant in the water cycle must be clearly understood. The trees in the forest act not only as interceptors of falling water in all forms, but also as water pumps bringing moisture out of the soil and transpiring it back into the atmosphere. Therefore, the tree or forest is a way station for moisture as it moves from atmosphere to soil and from soil to atmosphere. Water in all forms is essential for the growth of the plant, but at the same time moisture in all forms moves through the leaves, branches, trunks, and roots of the plant in its passage between sky and earth.

1. Principles

Water falls in various forms, depending on the air temperature; at the same time, moisture is transpired or evaporated from the earth's surface and from the leaves of trees and forests. As the moisture moves through the trees and the surrounding air layer it affects humans moving through or working on the earth's surface. By the same token, plant materials affect and control in a variety of ways the precipitation and the humidity in this air layer. In this way they control by another means the comfort or discomfort level of those persons living or moving on the earth's surface.

The question may be asked, "In what ways does this ability of vegetation to control precipitation and humidity have any effect on humans?" The answer may be in various forms of more or less importance to various people in various geographical areas. The microclimatic control of rainfall, for instance, gives protection to people standing, sitting or working under the forest or tree canopy. The ability of a forest to control or direct fog banks may assist materially in the extended use of certain areas of the environment by careful planting to give either fog-free areas or areas substantially covered by fog. The use of trees or forests to control snow drifting may be utilized in planting along roads or when locating roads initially. It may also assist materially in the location of ski areas. The abilities of vegetation to control dew and frost may be manipulated so as to provide either frost-free areas or sections of the landscape in which the dew and frost are heavier or more substantial than normal. Certainly tennis courts, putting greens or driving ranges which are to be used either early in the morning or early or late in the season would be planted in such a way as to protect them as much as possible from excess dew and frost. Atmospheric humidity may be either increased or decreased through the proper use, preservation or removal of certain plants or plantings. Sleet and hail are not only unpleasant but dangerous to human beings out of doors. Trees or forests may provide for the human users some areas of protection from such natural phenomena.

Once again, much of the available research has been done to show the ability of plants to protect agricultural crops or livestock from the effects of sleet and hail or to show the abilities of trees and windbreaks to protect growing plants from snowdrifting; but very little information is available on how these abilities of trees and forests impact on human beings. Some extrapolation, however, in the following data may be able to be made to show these functional uses of plants. This is done in the hope that additional research will be undertaken in the years to come to cover these gaps in our available knowledge.

2. Rain

"Some precipitation that falls on a tree does not reach the ground. The amount of rainfall which reaches the ground varies not only according to the species of the trees, but also according to the particular zone under the tree canopy. Some studies have indicated that only 60 percent of the rain falling on a pine forest canopy reaches the ground, and 80 percent of the rain falling on a hardwood forest canopy reaches the ground. H. W. Beale (1934), H. F. Linskens (1951), and J. D. Ovington

(1954) demonstrated the difference in the quantity of rainfall reaching the ground under softwoods (conifers) and hardwoods. The annual average of water reaching the ground is greater under hardwoods than conifers. The reasons are because softwoods have leaves with a greater number of sharp angles and trap water droplets in their numerous cavities; they absorb moisture resulting from precipitation and transpiration.

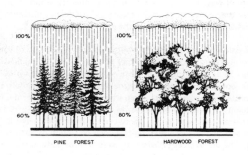

PRECIPITATION INTERCEPTION BY
DECIDUOUS LEAVES

PRECIPITATION INTERCEPTION BY
CONIFEROUS LEAVES   91

"Beale (1934) conducted an experiment which showed variations in the percentage of rainfall reaching the ground under a tree in different locations.

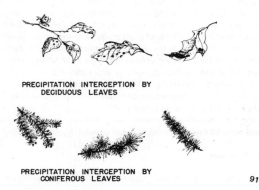

RAIN PENETRATION   92

"The percentage of precipitation actually reaching the ground under particular parts of various species of trees was measured by Linskens (1951).

"The dense shrubby vegetation may intercept more than the coniferous forest and that a stand of fewer tall trees on a good site may intercept more than a denser but shorter stand on a poorer site. A 28-year plantation of *Pinus canariensis* with 0.8 crown coverage, in California, intercepted between 17 and 28 percent of the seasonal rainfall in different years; and of this, 1 percent was returned to the soil as stemflow. A 10-year plantation of *Pinus taeda* with 58 percent crown coverage, in South Carolina, showed 29 percent interception and 20 percent stemflow, so that the interception loss was only 9 percent. A 70-year dense stand of *Pinus ponderosa* of 40 percent density at 1,020-meter elevation, in California, intercepted 16 percent; and of this, 3 percent was stemflow.

"The illustrations indicate the percentage of precipitation in the form of rain and snow reaching the ground under beech, apple, oak, maple and conifers.

"Intensity of Rainfall. The intensity of the rainfall is a key factor in the ability of plants to control precipitation. In light rain, softwoods have a greater water retention ability than do hardwoods.

"Ovington (1954) said that in light rainfall, conifers retain as much as five times the quantity of water as that intercepted and retained by broad-leaved trees. After a long rain shower, the tree canopies and the holes in the branches and trunks become saturated, and an appreciable quantity of water runs down into the soil.   93

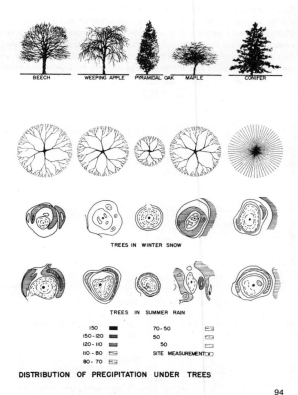

DISTRIBUTION OF PRECIPITATION UNDER TREES

94

"The amount of water reaching the ground is dependent upon the tree canopy structure, rather than upon the size of the tree. The rain reaches the ground under plants in two forms; raindrops which are not intercepted and pass freely through the leaves; and the dripping of previously intercepted raindrops from the leaves and branches of the tree. The presence or absence of foliage in the hardwoods does not, with the probable exception of very light rainfall, affect the percentage of water penetrating the canopy (Ovington). One study demonstrated that the leaves were the main reason for the variations in percentage of water reaching the ground (Linskens). Another showed that the intensity of rainfall is of greater importance than the time of the year (Ovington).

"The amount of rainfall which reaches the ground is influenced by the intensity and duration of the rainfall, the type of tree (coniferous or deciduous), and the structure of the tree canopy rather than the season (such as hardwoods with or without leaves).

"Transpiration and Evaporation. Various parts of plants intercept precipitation and thus are able to modify or control climate; trees and shrubs tend to cause an increase in precipitation above them. Leaf surfaces cause excess transpiration of water from the soil through the plant into the atmosphere immediately above the plant.

"Dr. John Carew of Michigan State University estimated that on a single day in summer, an acre of turf will lose about 2,400 gallons of water to transpiration and evaporation.

"The tree or shrub canopy serves to prevent evaporation of moisture from soil into the atmosphere. In this way, plants preserve and retain moisture in the soil. Because of these factors, a diurnal temperature-humidity relationship exists.

"Humidity and Temperature. Because plants block and filter solar radiation, inhibit windflow, transpire water into the atmosphere, and reduce evaporation from the soil, a microclimate of controlled humidity and temperature exists under plants, particularly in a forest-like cover of plants. The relatively high humidity and low evaporation rate acts to stabilize temperature keeping it lower than the surrounding air during the day and preventing it from dropping greatly at night. Three accompanying illustrations depict temperature-humidity relationships within a forest, on a cloudy day, a clear night, and a windy day."

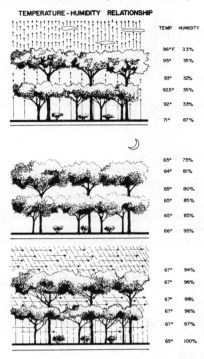

TEMPERATURE - HUMIDITY RELATIONSHIP

TEMP	HUMIDITY
96°F	33%
95°	35%
93°	32%
92.5°	35%
92°	33%
71°	87%
65°	75%
64°	81%
65°	80%
65°	85%
65°	85%
66°	93%
67°	94%
67°	96%
67°	96%
67°	96%
67°	97%
65°	100%

95

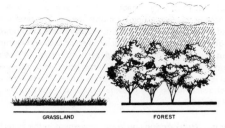

GRASSLAND   FOREST

PRECIPITATION INCREASES IN FOREST AREAS

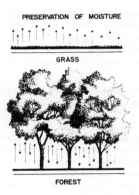

PRESERVATION OF MOISTURE

GRASS

FOREST   98

In the paper on the "Influence of Forests on the Weather" in the book, *Forest Influences,* a number of references are made to atmospheric moisture. These statements are as follows:

"The influence of the forest on the humidity of the air has most often been expressed as relative humidity. Relative humidity varies inversely with temperature. Temperatures in forest and open are usually different.

"For this reason, the relative humidities will be different, even if the amount of water vapor in the atmosphere is the same. If the amount of water vapor in the atmosphere is the same. If the amount of water vapor is different, the relative humidities give no evidence to distinguish the part of the difference attributable to water vapor and the part attributable to temperature."

96

VERTICAL GRADIENTS OF VAPOR PRESSURE AND RELATIVE HUMIDITY
IN THE OPEN AND UNDER FOREST [1]

	Vapor pressure			Relative humidity		
	30 November	3 October		30 November	3 October	
	Soil moist	Soil dry		Soil moist	Soil dry	
height	open	open	forest	open	open	forest
Centimeters	Millimeters	.... Millimeters		.... Percentage		
305	10.2	11.8	11.0	61	45	48
213	10.5	12.2	11.0	61	44	46
152	10.6	12.2	11.2	59	46	47
122	10.6	12.2	11.6	57	45	48
91	10.8	11.8	11.4	57	45	48
61	11.0	12.2	11.2	58	44	47
46	11.0	12.2	11.2	58	43	47
30	11.4	12.7	11.2	59	41	47
15	12.5	12.2	11.4	61	39	47
6	13.6	12.2	11.4	63	35	47

[1] Data hitherto unpublished, supplied through the courtesy of J. S. Rothacher. 97   41.

"Moisture Retention. Moisture reaching the earth through a shrub or tree canopy is retained longer than moisture falling upon exposed soil. The ability of plants to intercept precipitation and slow it down helps to control surface water runoff and resulting soil erosion. Through the addition of organic material, plants loosen and maintain the porosity of the soil, which helps it retain water. The protection from sun and wind provided by plants reduces evaporation of soil water. Therefore, large-scale plantings which entrap water and help the soil retain it are used on watersheds above reservoirs. These plants serve several purposes. They slow down water runoff, prevent soil erosion and reservoir siltation, and reduce evaporation of soil water. This also applies to critical areas over natural underground aquifers used for water supply sources.

"Plants and Fog and Dew. Fog condenses on the needles of conifers and on the upper and lower surfaces of leaves on deciduous trees. This condensation of fog then falls to the earth as drip water from the various parts of the plant.

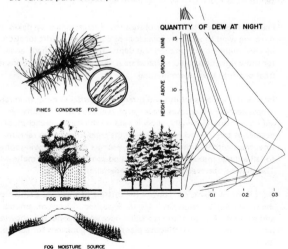

PINES CONDENSE FOG

FOG DRIP WATER

FOG MOISTURE SOURCE

QUANTITY OF DEW AT NIGHT

HEIGHT ABOVE GROUND (MM)

99

"When the crown of a tree is covered with dew the ground below it is generally free of it. Geiger made a study of dew measurements under an isolated beech tree during a 1 month period. The results are summarized in the accompanying chart. He also refers to a comparative study of the deposition of dew upon open ground and within a young fir plantation, and notes that the greatest deposition of dew within the fir stand occurs at the crown level of the trees, while above open ground it occurs just above the ground surface. The dew on a stand of trees, on some nights, may become quite heavy and drip through and release a brief shower." 100

The paper on "Water Action and Water Movement in the Forest," in the book *Forest Influences,* published by the F. A. O., has the following to say about the use of forests to control the movement of sea fog:

"Forests are relied upon to prevent invasions of sea fog in the south-eastern coastal area of Hokkaido in Japan. Investigations there have shown that six to ten times as much fog water is caught by forest as by an open field. The most effective type of forest for fog capture was a comparatively sparse forest of needle-leaved trees which had no lower branches." 108

The paper on "Influence of Forest on the Weather," in the same publication, indicates the following information concerning dew and frost in forested areas:

"Dew and frost are deposits of water on the exposed surfaces of plants or soil which result from the condensation of water vapor from the atmosphere. The usual cause is nocturnal cooling by outgoing radiation, and forest cover influences radiation as well as the deposition of the dew and the frost. According to earlier studies, dew and frost would not add more than 25 millimeters a year in northern latitudes; but recent records from the weighing monolith lysimeters in Ohio, United States, show a total of 199 millimeters as the annual average for three lysimeters during a period of six years. This constitutes 18 percent of the annual precipitation of 1,090 millimeters; and it represents a substantial addition. The mean monthly totals show the following annual cycle: January, 24 millimeters; February, 25; March, 23; April, 14; May, 9; June, 7; July, 8; August, 12; September, 12; October, 197; November, 19; and December, 28 millimeters. The highest in any one lysimeter and month was 53 millimeters in February; the lowest was 2 millimeters in July and August. The lysimeters were covered with grass or with annual crops during the growing season. 109

"In the forest, dew is deposited chiefly on the exposed upper surface of the crowns, and the amount decreased rapidly with the distance into the crowns. In Europe and North America, the deposition of dew on the forest sometimes amounts to 1 millimeter in a single night, while in the open—as indicated by the data from Ohio—the range is from 0.1 to 0.4 millimeter; but in the tropics, there may be ten times as much dew as in northern latitudes.

"Plants and Snow. Plants control snow by intercepting snowflakes, by directing wind to scour an area clean or control snowdrift location, by determining conformation and depth, by providing shade areas for snow retention and controlled melt, and by causing variations in frost depth to slow down melting.

"Interception. Plants inhibit and intercept snowflakes more than raindrops, because snowflakes are larger, travel at a lower velocity, and are not viscous. Generally, snow is held on the leaves, branches, or needles of trees. Snow is retained on the trees longer than rain; therefore, trees withhold the moisture for a longer period, allowing some to fall as large masses before melting and some to fall as it melts. This may be observed after a heavy snowfall in a forest.

"Drift Control. Plants can control the drifting of snow. As plants slow the wind velocity, snow particles are deposited in front of, among, and leeward of them. Various studies have been conducted, showing patterns, and the most effective planting configurations for controlling them.

"Caborn refers to windbreaks designed for collecting snow. He feels that windbreaks should include shrubs to form snowdrifts on the windward or leeward sides of the windbreak.

103

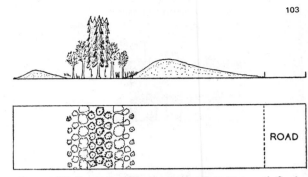

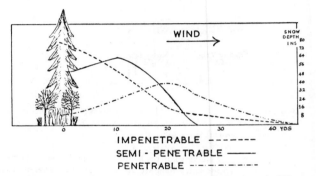

Fig. 23   Windbreak designed to trap snow adjacent to a road, Czecho-slovakia (after Luncz).

104

"He refers to studies conducted in Czechoslovakia by Luncz, showing the effect of snowdrift patterns, adjacent to windbreaks with different degrees of penetrability. This effect is illustrated in the accompanying chart.

Snow drift patterns adjacent to windbreaks of different penetrability, Czechoslovakia (after Luncz).

105

"Windborne snow is deposited wherever there is a local drop in wind velocity. Wind velocity is reduced wherever the airflow encounters a barrier which will project a "wind shadow" area. A great deal of effort has been directed toward understanding the pattern and distribution of snow for agricultural purposes and for the protection of transportation systems. Two ways of protecting an area against snowdrifts are snow fences and windbreaks. The fence is standardized, it gives immediate protection, and raises objections from landowners. The main disadvantage of a snow fence is cost and yearly maintenance.

"Windbreaks, with height as a prime advantage, are also important assets when there is a variation in topography. A good windbreak should filter the wind and trap the snow. Tall trees are not sufficient to initiate drifting. They must be augmented with shrubs and ground covers. The depth of the drift and the extent of the drifting depend upon the penetrability of the plant barrier or fence. Solid fences or plants produce drifts on both sides, while an open structure keeps the drift to the leeward side. The leeward drifts near a solid screen are generally deep, do not extend to a great length, and reach their maximum saturation point a short distance from the barrier. The leeward drifts near a penetrable screen are shallow, extend to a greater distance from the barrier, and absorb more snow. The greater the velocity of wind, the closer the drift is to the barrier itself. The optimum efficiency of a barrier is attained with a 50 percent density. A barrier with a density of approximately 50 percent and 4 feet high will initiate the formation of a drift extending up to 56 feet in length. According to Caborn in his studies in 1963, a 3-foot-high barrier results in a drift extending to 51 feet. 106

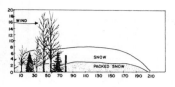

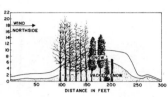

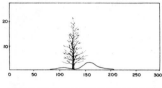

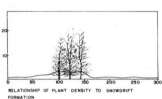

RELATIONSHIP OF PLANT DENSITY TO SNOWDRIFT FORMATION

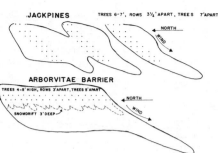

JACKPINES   TREES 6-7', ROWS 3½' APART, TREES 7' APART

NORTH

WIND

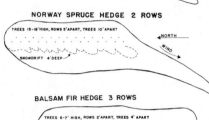

ARBORVITAE BARRIER

TREES 4-5' HIGH, ROWS 3' APART, TREES 5' APART

SNOWDRIFT 3' DEEP

NORTH

WIND

NORWAY SPRUCE HEDGE 2 ROWS

TREES 15-18' HIGH, ROWS 5' APART, TREES 10' APART

SNOWDRIFT 4' DEEP

NORTH

WIND

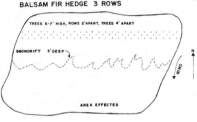

BALSAM FIR HEDGE 3 ROWS

TREES 6-7' HIGH, ROWS 2' APART, TREES 4' APART

SNOWDRIFT 3' DEEP

N

WIND

AREA EFFECTED

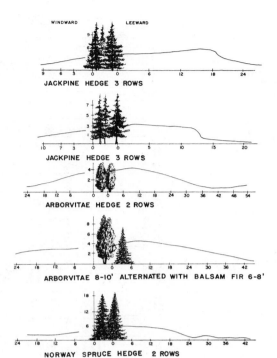

WINDWARD   LEEWARD

JACKPINE HEDGE 3 ROWS

JACKPINE HEDGE 3 ROWS

ARBORVITAE HEDGE 2 ROWS

ARBORVITAE 8-10' ALTERNATED WITH BALSAM FIR 6-8'

NORWAY SPRUCE HEDGE 2 ROWS

"M. Jensen showed that the length of drift is related to the height in feet of the screen in the following fashion:

$$L = \frac{36 + 5h}{K}$$

L = the length of the drift in feet
h = the height of the screen in feet
K = the function of the screen density 1 for a 50% density; 1.28 for a 70% density.

"This should not be rigorously applied and a safety margin (20 feet) should be allowed for between the screen and the end of the snowdrift.

"J. H. Stoeckler and E. J. Dortignac showed that narrow belts of tall trees, devoid of branches near the ground, allowed snow to sweep under the trees. The snow was deposited in thin layers on the leeside of the belt, in a band extending between 60 and 120 feet beyond the belt. The same writers reported that shelterbelts, with one or more dense-growing shrub rows at least 8 feet high, were very effective in trapping snow and drifts from 5 to 8 feet or more in depth, and that the snow was practically all deposited in a band from 30 to 40 feet wide on the leeside of the first row of shrubs.

107

Exposure	N	E	S	W
Rainfall, in.	26.3	27.8	22.5	22.5

EVAPORATION

	In open	Under forest	Difference
	.... Millimeters		
Total for mean year	702.2	297.5	– 404.7
Highest monthly total	126.4	67.1	– 59.3
Lowest monthly total	26.2	6.6	– 19.6
Mean total per 24 hours	1.9	0.8	– 1.1
Maximum per 24 hours in whole period	8.3	----	----

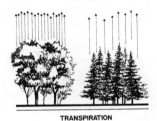

TRANSPIRATION

110

"The effectiveness of plants in conjunction with snow drifting, is well illustrated in a quote from an article by Harold E. Olson, in the March 1963 issue of *Park Maintenance:*

'Where the design or location of highways cause or permit the drifting of snow, it is sometimes considered desirable to plant trees and shrubs rather than to erect and dismantle the slat-type snow fencing. This item results in an expenditure in one State of over $300,000 per annum. In this same state the cost of the removal of snow ran to $701,400—a considerable expenditure of money with no resulting betterments.

'The natural barrier, or snow fence, consists of trees or shrubs planted in rows or groups in such a way as to slow down the normal velocity of the wind and cause the snow to be deposited before it reaches the traveled way. This type of snow control is best adapted to wind right-of-ways and localities where permanent snow fences may be left in place the year round. The natural barrier fits in very nicely with the modern trend toward roadside development.'

101

"The accompanying charts show the effectiveness of various windbreak configurations in snow-drifting control in North Dakota.

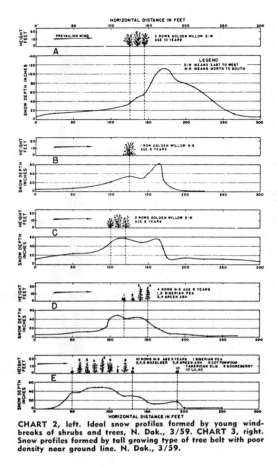

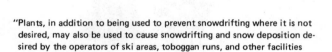

CHART 2, left. Ideal snow profiles formed by young wind-breaks of shrubs and trees, N. Dak., 3/59. CHART 3, right. Snow profiles formed by tall growing type of tree belt with poor density near ground line. N. Dak., 3/59.

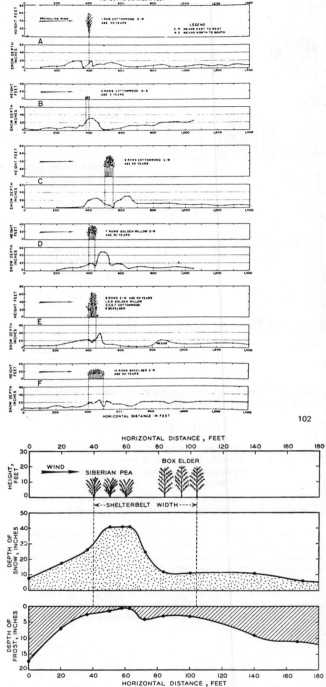

"Plants, in addition to being used to prevent snowdrifting where it is not desired, may also be used to cause snowdrifting and snow deposition desired by the operators of ski areas, toboggan runs, and other facilities which need snowfall.

"Prudent planting greatly reduces the cost of snow removal on parking lots and other areas. As noted previously, the speed of the wind is accelerated through openings in shelterbelts, as well as at the ends of windbreak planting. This information may be utilized to provide snowfree parking lots, roadways, or walkways."

The book, *Forest Influences*, by Kittredge gives the following statement and the following illustration concerning the ability of a Siberian Pea Shrub and a Box Elder shelterbelt to control snowdrifting as well as to control frost penetration into the soil beneath the snow drift.

"Striking differences in penetration of frost in relation to a shelterbelt are illustrated in the following chart. However, it is evidently the snow drift trapped by the shrub rows that has more influence on the depth of freezing than any direct effect on the shade or litter. Where the snow was 41 inches deep, only 1 or 2 in. of soil were frozen. Away from the shelterbelt where the snow was only 6 to 8 in. deep the freezing extended 12 to 17 in. deep."

Accumulation of snow and penetration of frost as affected by a multiple-row shelterbelt. [*From Stoeckeler and Dortignac (335).*]

"Shading. Snowfall which has drifted onto the north side of large conifers, such as spruce, fir, or pine, is shaded from the late winter sun, and lasts longer. This may be undesirable on a driveway or golf course, but is of great benefit on a ski or tobogganing hill. This also is desirable for controlling and slowing down snow melt, permitting it to enter the ground rather than run off.

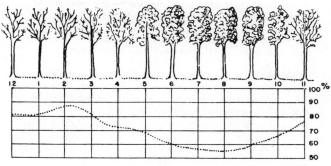

"Frost Line. Undisturbed litter on the forest floor or on the ground under groups of trees enables the frost line to penetrate deeper during the colder months when freezing occurs. This makes the ground cooler, ensuring that snow which falls on frozen ground will not melt rapidly. This will affect the retention of fallen snow, and it will remain in a frozen state upon the earth for a longer time. This may be desirable or undesirable, depending upon the circumstances. However, it should be recognized that plants do provide an insulating layer."

111

(after Schubert)

116

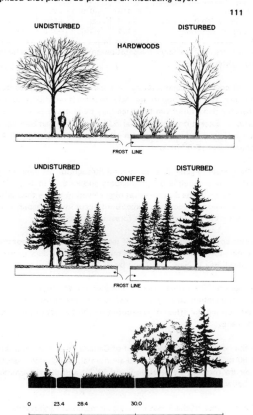

PRECIPITATION — PLANT RELATIONSHIP

112

d. Temperature Control with Vegetation

Since various references indicate that vegetation of various types affects solar radiation, the movement of wind and the percentages of precipitation and humidity in the atmosphere, it is logical and obvious that this same vegetation also affect temperatures near the ground surface. Kittredge in his chapter on air temperature has the following to say about the ability of trees and forests to modify temperature:

"The evidence . . . that a forest canopy may reduce the solar radiation to less than 1 percent of that in the open leads logically to the expectation that, because the sun is the source of the heat, the daytime temperatures where part of the sun's radiant energy is intercepted by the trees will be lower than those in the open.

"Temperature Control. Temperature control is linked directly to—and is a result of—solar radiation control, wind control, and precipitation control. Plants used for temperature control have the greatest effect by moderating temperatures near the ground.

"Shade and Absorption. Of the radiation that strikes a plant, very little will penetrate it, whether the radiation is direct or reflected. The shaded side has cooler temperatures than the radiated side."

113

	In open	Under forest	Difference
RELATIVE HUMIDITY			
	. . . . *Percentage*		
Mean annual at 6 a.m.	97.9	96.7	− 1.2
Mean annual at 2 p.m.	67.1	82.3	15.2
Difference	30.8	14.4	− 16.4
Greatest difference between mean monthly at 6 a.m. and 2 p.m.	51.8	42.9	− 8.9
Smallest difference between mean monthly at 6 a.m. and 2 p.m.	10.7	2.7	− 8.0
Difference between extreme maxima and extreme minima at 2 p.m. for whole of period recorded	77.0	72.0	− 5.0

114

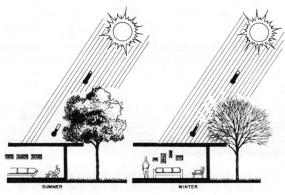

SEASONAL VARIATIONS

"The temperature of an area may be reduced by plants, even if they are not tall enough to give shade. Plants and grassy covers reduce temperatures by scattering of light and radiation and the absorption of solar radiation, and also by the evapotranspiration process.

"It is found that temperature over grassy surfaces on sunny summer days are about 10° to 14° cooler than those of exposed soil.

"Deciduous trees are good temperature control devices, in that they cool in summer and yet allow winter sun to pass through. Vines on walls or trellises are also some of nature's automatic heat control devices—cooling by evaporation and providing shade."

117

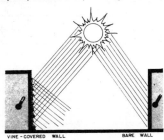

VINE - COVERED WALL   BARE WALL

118

It is found that temperatures over grassy surfaces on sunny summer days are about 10 to 14 degrees cooler than those on exposed soil. Kittredge in *Forest Influences* shows some of the ability of vegetation trees and forests to modify temperatures under a tree canopy in the following text and illustration:

"Maximum temperatures are modified by the forest also through its effect in minimizing the heating of the ground surface and thus reducing the resulting convectional currents of warm air. This influence is closely related to and acts in the same direction as the direct solar radiation. Both convection and radiation are most effective close to the surface, and thus the influence of vegetation decreased upward from the ground."

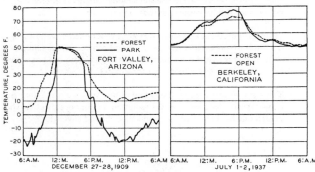

-Comparison of diurnal trends of air temperature in forest and open. (*From data of United States Forest Service.*)
123

The same author then shows another chart which indicates the temperature variation between thinned and unthinned white pine and scotch pine forests.

Type	Unthinned	Thinned
White pine	Warmer by 1.8%	Cooler by 9.2%
Scotch pine	. . . .	Cooler by 14%

The same source makes the following statements concerning the ability of vegetation to modify temperatures: 124

"Considerable evidence is available to show that a forest or other canopy of vegetation influences air temperatures significantly if not by large amounts.

"In northern Idaho during July and August, 1931, the average maximum air temperature in a clear-cut area of the western white pine type was 86.8°F (187). An adjacent uncut area showed an average maximum of 6.4°F lower."

Forest type and location	Departures, forest from open, °F				Ref.
	Mean maximum		Mean minimum		
	Jan.	July	July	Jan.	
Beech climax, N. Y.	. . . .	−8.3	+4.5	. . . .	(258)
W. white pine climax, Idaho	. . . .	−6.0	+7.0	. . . .	(187)
Hemlock, N. Y.	. . . .	−4.9	+3.9	. . . .	(258)
White pine, N. H.	. . . .	−4.8	+3.2	. . . .	(220)
W. white pine, two-thirds cut, Idaho	. . . .	−2.3	+5.2	. . . .	(187)
Douglas fir, 9,000 ft, Colo.	−3.1	−2.2	+2.5	+2.1	(34)
Jeffrey pine, 6,000 ft, S. Calif.	+0.7	−1.4	+6.7	+3.6	Munns
Jack pine, 25 yr, Neb.	−0.1	−1.3	−2.8	+1.1	(275)
Aspen–Engelmann spruce, 9,000 ft, Colo.	−1.8	−1.0	−1.0	−0.1	(37)
Douglas fir, three-quarters cut, 9,000 ft, Colo.	−1.0	−0.9	+1.1	+1.2	(34)
Ponderosa pine, 7,250 ft, Ariz.	−1.0	0	+5.0	+6.0	(282)
Chaparral, 6,000 ft, S. Calif.	+1.4	+5.0	−2.1	−1.6	Munns

125

In a summary of the effect of vegetation on air temperatures he makes the following statements:

"Forest cover in general reduces the maximum temperatures throughout the year. The reduction in monthly maxima is more pronounced in July, when the departure from the open may be as much as -8°F, and -3° in January. Exceptions are found at 6,000 ft. in the San Bernadino Mountain Jeffrey pine in January with +0.7°, and in chaparral, +0.4° in January and +5.0° in July.

"Forest cover in general raises minima throughout the year, the monthly minima by as much as 6.0° in January and by 6.7° in July. These high values occur in the Southwest at high elevation where the night radiation in the open is intense. Exceptions occur in the chaparral and to a less extent in the spruce-aspen forest in Colorado.

"Forest cover reduces the range of temperature—annual, monthly, or diurnal—the mean monthly by as much as 12.8° in July (under beech-maple climax in New York) and by 5° in January."

126

The paper on "The Influences of Forests on the Weather," in the United Nations publication on *Forest Influences* gives the following quantitative data concerning the effect of vegetation on soil temperatures under a forested area:

"In the northern coniferous forest of Canada, maximum temperatures of 50 to 70°C. on exposed sited are normal, while under canopy and litter the maxima are between 9 and 13°C., and the night minima may be below 0°C.

"The influence of the forest in reducing maximum soil temperatures results partly from the shade of the crowns and partly from the insulation of the forest floor. The effects of these two factors, combined or separated, is shown in Table 19. The shade was more effective than the floor in reducing the temperatures. The combined effect of the two factors was a reduction of 24°C. In the shade, the 30-millimeter forest floor reduced the temperature 6° as compared with bare soil. Under the forest floor, the shade reduced the temperature 10° lower than in the sunny opening.
127

EFFECT OF SHADE AND FOREST FLOOR OF *Pinus strobus* ON MEAN DAILY MAXIMUM SURFACE TEMPERATURES, IN DEGREES CENTIGRADE, IN NEW HAMPSHIRE, UNITED STATES [1]

Surface	Canopy		Departure
	opening	dense	dense from open
Bare	42	24	− 18
Covered	28	18	− 10
Departure, covered from bare	− 14	− 6	− 24

[1] Toumey and Neethling, 1924.
128

That same paper illustrated in the following chart the vertical gradients of temperature over bare dry soil and under a forest:

Height	Open	Forest
	Temperature	
Centimeters	.... Centigrade	
305	27.2	25.0
213	27.8	25.6
152	27.2	25.3
122	27.5	25.3
91	27.2	25.0
61	27.8	25.3
46	27.8	25.3
30	29.4	25.3
15	30.0	25.6
6	31.7	25.6
0.9	35.6	25.6

[1] Measurements, hitherto unpublished, supplied through the courtesy of J. S. Rothacher.

Once again this same source indicates the ability of plant materials to modify temperatures in various areas of the United States in the following words:

"The minimum temperatures in the forest are lower than in the open by about 1° in most months, and there is a maximum departure of -1.7° in May; for the cold air drainage is less rapid in the forest than in the open, where warmer air from above replaces the cold air as it flows downhill. While negative departures have also been recorded in 25-year *Pinus banksinana* in Nebraska, United States, on sand hills and in *Populus treumloides* and *Picea engelmanni* at 2,750 meters in Colorado, the usual influence of most forests is to increase the minimum temperature both in summer and in winter—sometimes by as much as 4°C., in the monthly mean in summer."

129

In the introductory remarks to that same U.N. book on *Forest Influences*, a number of references and charts are given to show these abilities of trees and forests in temperature control:

Comparison of average monthly air temperature in forest and in open land. [1]

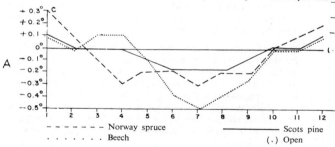

- - - - - - Norway spruce · · · · · · · Beech
———— Scots pine (.) Open

[1] Represented by horizontal line.

(after Schubert)

[3] Schubert, J. *Der jährliche Gang der Luft- und Bodentemperatur im Freien und im Waldungen* (Yearly variations of air and ground temperature in open country and in forests). Abh. des Preuss. Meteorolog. Inst., 1900-1901. 130

"According to the graph, the Scots pine causes a very slight drop in the summer temperature of the interior (2°C. less than the open), and in the winter it exerts almost no effect. Meanwhile the Norway spruce brings down the temperature of the interior appreciably in the summer and raises it in the winter (nearly 3°C. over the temperature of the open in January). The beech is the most effective of the three species in reducing the summer temperatures. On the other hand, it keeps the inside air warmer than the outside air not only in winter but also in spring; for example, it is 1°C. higher in April.

"A situation exactly similar to that in the European mesophilous forests has been noted even in other continents. In July, a drop of 6°F. was registered in a climax forest of mixed *Pinus strobus* in Idaho, while a wood of *Tsuga canadensis* in New York State showed a drop of 4.8°F. On the other hand, in a Colorado forest of mixed *Picea engelmanni* and

Populus at an altitude of 9,000 feet, the drop in the July temperature was barely 1°F. Of all the species which have been mentioned, the beech is the greatest consumer of water through transpiration.

131

Comparison of monthly temperature ranges [1]

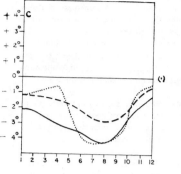

· · · · · · · Beech
- - - - - - Scots pine
———— Norway spruce
(.) Open

———— Pine stand, Migliarino, Italy
(.) Open

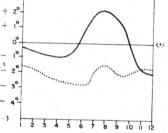

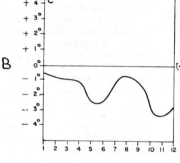

· · · · · · · Pine stand, Cecina, Italy
———— *Forteto*, Cecina, Italy
(.) Open

[1] Temperature on open land represented by horizontal line.

(after Müttrich)

132

"From data collected in Switzerland and Germany, the following percentage figures for the increase in relative humidity over the open in the summer months are available—that is, in the period of maximum transpiratory activity. For beech, the percentage figure is 9.35; for Norway spruce, it is 8.56; and for Scots pine, it is 3.87. For the larch, incidentally, the percentage increase was 7.85.

"In the Cecina pinewood, the summer and autumn temperatures at a depth of 20 centimeters are lower than in the open; but in the winter—at the same depth—the temperatures remain somewhat higher than in the open."

AIR TEMPERATURE

	In open	Under forest	Difference
	.... Centigrade		
Annual mean of maxima	32.1	28.4	− 3.7
Annual mean of minima	21.4	22.5	1.1
Difference	10.7	5.9	− 4.8
Maximum diurnal range	18.5	13.5	− 5.0
Minimum	1.0	0.3	− 0.7
Difference between extreme maxima and extreme minima for whole period	27.7	24.0	− 3.7

133

The same paper also makes references to the effect of vegetation to modify not only the air and soil under the canopy but also the air immediately above the forest canopy. These studies are mentioned in the following words:

"Italian scientists, however, have experimented on another important problem. This is the transmission of the forest microclimatic features vertically, instead of horizontally. They sent weather balloons with recording instruments up to a height of 1,000 meters. The experiments were suggested by the effects which pilots had long noted when flying over wooded country. They were also suggested by the elementary physical law that warm air currents rise and cool ones descend, while none of them work horizontally except through the action of the wind.

"Heat Transfer. Leaves absorb solar radiation, resulting in lowering the temperature in the shade of the plants. The same canopy of leaves acts to keep heat from being radiated out from under the plant, so that re-radiated heat loss during the night is reduced.

"During the day the effect of shading by trees reduces the air temperature near the ground. The amount of temperature reduction depends upon the species of trees providing shade. A. Muterich showed that in a stand of beech in July, the daily fluctuation of the air temperature at ground level is about 4.5°C; whereas, in a stand of fir, it is about 3.9°C; and in a stand of Scotch pine, about 3°C.

"On a sunny summer day the solar radiation from the sun strikes the surface of the canopy and it becomes the warmest part of the forest. Due to the absorption and reflection of heat by the canopy, the understory is cooler. Successively lower layers receive less heat and in consequence are progressively cooler. F. L. Waterhouse has observed that the highest air temperatures are usually found at the level where there is a maximum absorption of incoming radiation.

"At night the forest loses great quantities of heat by radiation into space. The canopy surface is the most effective radiating surface in the forest, for it is exposed directly to the sky. Heat radiated from the lower layers of the forest is trapped by the overlying strata. Because the canopy loses its heat most rapidly, it becomes the coolest part of the forest. As the air in the canopy cools it grows more dense and sinks toward the ground. In settling, this air is warmed only slightly, so that at night, the temperatures are nearly uniform from canopy to soil.

"Restriction of outgoing radiation by the tree crowns, from the narrow strip along a forest margin, depends on the species of forest trees and the shape of the crowns of the trees.

"If the outgoing radiation is restricted, this, with the warmer air flowing outward from the trunk space under the canopy, should theoretically produce higher night temperatures on the forest margin. However, this warmer temperature may be counteracted by the downward flow of cold air from the crowns, so that the margins of plantings may also have uniform temperature from the ground to canopy (Geiger). The diurnal amplitude of temperature fluctuations within the sheltered area under plants is directly related to the range of day and night temperatures of the air surrounding will fluctuate more than on days and nights when the temperature remains uniform.   134

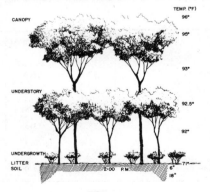

120

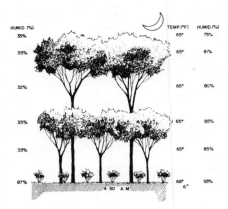

121

"Within soil under low plants without much litter, the temperature may fluctuate some, but it will still fluctuate less than will the soil temperature of bare ground. This is a barren surface during the day, heating up the soil, will be re-radiated into the atmosphere at night, warming the atmosphere and causing the soil to cool. The soil under plants absorbs little solar radiation during the day, quickly re-radiating what little heat it did absorb, thus heating the air next to the soil under the plants. Since warm air rises, a canopy of trees will tend to collect and hold warm air, retaining much of it near the ground. The capacity of the canopy to hold the warm air is directly related to the density of the foliage. The rate of heat loss by re-radiation, retarded by plant cover, results in nocturnal temperature variations of both the soil under and air within the plant cover, with temperatures not dropping as low as the soil and air temperature of adjacent areas. Thus temperature fluctuates less widely under plant cover than where the soil is bare.

"During a rain, drops of water strike virtually every exposed surface in the forest. The water absorbs heat from surfaces that are warmer than itself; and as it drips and falls, it transfers the heat to lower, cooler surfaces. Thus temperatures are equalized and in a short time after a rain starts falling, the temperature is uniform from canopy to forest floor litter under the plant. At the ground level, loose litter, dead leaves, and forest soil are efficient insulators on the forest floor. The ground temperature below the litter on a forest floor is subject to slight changes from day to day.

"In the absence of appreciable wind, cool soil of shaded areas absorbs heat from the air more rapidly than heat can be transmitted by convection or conduction from unshaded areas. Furthermore, the greater humidity of the air under plants increases the amount of heat needed to raise its temperature significantly. For these reasons, forests generally depress maximal air temperatures as well as maximal soil temperatures."

122

In an article in the April 11, 1971, issue of the *New York Times*, entitled "Green Grass That's Not So Green" by S. Elwynn Taylor and Gerald Pingel, the following statement is made concerning the ability of natural and artificial turf to control temperatures:

"It was expected that the grass would be cooler than the artificial turf because of evaporative cooling resulting from the transpiration process; this was the case. The temperature difference was also influenced by the absorption of sunlight. Measurements of reflected sun showed that grass reflected 2.94 times as much sunlight as did the artificial turf, and the asphalt 1.78 times as much. The absorptances to sunlight were: grass, 78.4 percent, asphalt 87 per cent, and artificial turf, 92.7 per cent.

"The experiment was conducted in early October, so the sun was not as intense as it sometimes is in mid-summer. The air temperature was 32°C. (90°F.) and the relative humidity was 40 percent. The wind was 50 cm sec (1.1 mph).

"The grass was near 38°C. (100°F.) and the asphalt was 60°C. (140°F.). The artificial turf heated to 72°C. (162°F.).' "

139

A short article in *Nursery Business* magazine indicated once again for popular consumption the ability of trees to modify climate in the following words:

"Trees will modify the temperature both inside and outside of your home. Bright and direct sunlight on a summer day may be as strong as 10,000 foot candles of light intensity but research shows a good shade tree will reduce that light intensity to one-tenth that or only about 1,000 candles. Trees also make the yard and garden temperatures much more bearable. Forest research shows there can be as much as 25 degrees difference between shaded area temperature and that recorded on open unprotected areas on a windless day. Shade will definitely cut down on the cost of air conditioning a home. You might want to look at your yard and home."

140

Dr. Roy Mecklenberg and his research team at Michigan State University conducted a number of on-site studies in the East Lansing area concerning the ability of vegetation to control air temperatures in the months of July and August. This team of researchers reported on their work in the following words and with the following illustrations:

"Microclimatic Investigations

Methods and Materials

Five foot air temperatures were continually recorded from 7/1/70 to 9/6/70 by Standard Weather Bureau recording thermographs placed in the Standard Weather Bureau Shelters. Two shelters were placed on the Michigan State University Forest Akers Golf Course, one at the northeastern edge of a 2 acre mature oak woods clear of secondary growth, and a second on an adjacent 2 acre grassed fairway free of woody plant growth. A third thermograph was placed on a commuter asphalt parking lot approximately 1 mile from the first two sites. These shelters also contained maximum and minimum thermometers. Statistical analysis of variance and Tukey's W procedure (H.S.D. procedure) were determined on the maximum as well as minimum daily temperatures (Steel and Torrie, 1960).

"Infrared surface temperatures were measured in a variety of sites with a battery-powered Stoll-Hardy HI-4 Radiometer (Williamston Development Co., Inc., West Concord, Mass.). Air temperatures at 3 in., 3 ft., 6 ft., and 9 ft. heights above the surface were measured with a 12 probe thermistor telethermometer Model 44T2 (Yellow Springs Instrument Co., Yellow Springs, Ohio).

"Infrared surface temperatures, as well as 3 in., 6 ft., and 9 ft. air temperatures were obtained at 2-3 hour intervals from 7:00 a.m., August 17th to 2:00 a.m., August 18, 1970, at the following sites (1) Dry Michigan State University football stadium, (2) Shaded asphalt, M.S.U. campus, (3) Irrigated grass, Washington Park, Lansing, (4) Dry grass at the Michigan-Washington Streets Mini-Park, Lansing, and (5) Unshaded asphalt, Frandor Shopping Center, Lansing, Michigan.

"In order to compare artificial turf with living grass in the same environment, infrared surface temperatures as well as 3 in., 3 ft., 6 ft., and 9 ft. air temperatures were obtained as 1-2 hour intervals, using the equipment described above at White Sox Park, Chicago, Illinois, on 8/13/70. Temperatures of the dry Astroturf infield (Monsanto Chemical Co.) were compared to the irrigated Marion Blue Grass outfield and the clay surface (Wyandotte Chemical Co.) base paths.

"The effect of irrigation on artificial turf temperatures was studied by applying 0.07, 0.13, 0.20, and 0.26 inches of water uniformly over 10ft. by 10 ft. plots of Tartan turf on August 5, 1970. One set of plots was positioned at the east edge of the playing field and a second set was posi-

tioned at mid-field. This study was made on a clear, bright day with maximum air temperatures of 89°F and wind speed up to 4 mph at 6 ft. above the artificial turf.

"Results and Discussion

Figure 1 shows the 5 ft. air temperatures from 4 p.m., June 31st to 8 a.m., July 2, 1970. The oak woods site did not get as warm during the day nor as cool at night as the grass site possibly because of the interception and re-radiation by the tree canopy of solar radiation during the d day and terrestrial radiation at night. These results in urban situations substantiate those reported earlier by Kittredge (1948) for forest sites. The asphalt site had minimum temperatures similar to the grass site and minimum temperatures similar to the oak woods side. The minimum temperatures recorded on the three sites for the months of July and August were significantly different whereas the maximum recorded temperatures were not significantly different (Table 1). As pointed out by Landsberg (1970) a heat island may develop over asphalt at night because the greater amount of energy stored in the asphalt during the day slows doen the radiative cooling process at night as compared to a grass site. Furthermore, the total radiative surface is much greater from a grass covered surface as compared to a flat asphalt surface.

"If evaporative cooling was completely responsible for the differences in the infrared surface temperatures or artificial turf and living grass it should be possible to cool artificial turf with irrigation. The application of 0.20 inches of water cooled artificial turf (105°F. as compared to 153°F. in Figure 4) for an entire day whereas 0.31 inches of water lasted approximately one-half as long. The application of 0.26 inches of water (the maximum application rate without run off) did not cool the surface the turf more than the application of 0.20 inches. All of the water application rates including 0.26 inches had only minimal cooling effects the following day (159°F. vs. 153°F.).

141

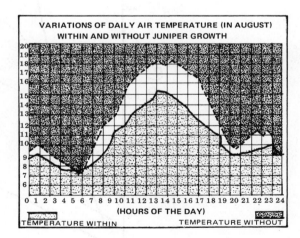

VARIATIONS OF DAILY AIR TEMPERATURE (IN AUGUST) WITHIN AND WITHOUT JUNIPER GROWTH

(HOURS OF THE DAY)

TEMPERATURE WITHIN   TEMPERATURE WITHOUT

135

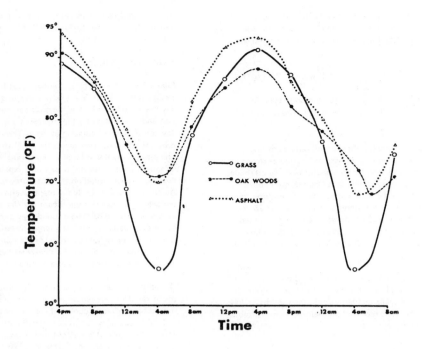

A comparison of five foot air temperatures in an oak woods site, blue grass site and asphalt parking lot site from 4:00 p.m., June 31st to 8:00 a.m., July 2, 1970.

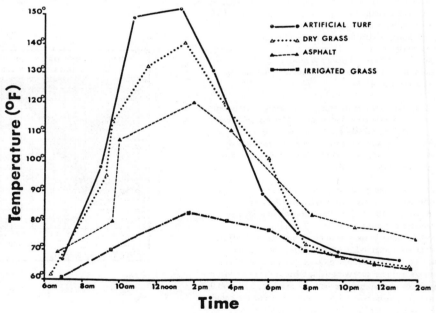

A comparison of infrared surface temperature on an irrigated grass site, asphalt parking lot, dry grass site, and artificial turf on August 18, 1970.

Mean 5 Ft. Air Temperatures for the Months of July and August.

	Oak Woods	Grass	Asphalt
Maximum	78.9°F	82.4°F	82.6°F
Minimum	60.2°F**	55.2°**	57.8°F**

H.S.D. procedure

** indicates a significance at the 1% level

Air Temperature at 3 Inches to 9 Feet Above Different Types of Urban Surfaces, from 1:30 to 2:00 p.m., August 18, 1970.

	3 Inches	3 Feet	6 Feet	9 Feet
Artificial Turf	89°F	88°F	86°F	85°F
Dry Grass	93°F	88°F	87°F	86°F
Irrigated Grass	85°F	85°F	83°F	82°F
Asphalt	85°F	85°F	84°F	83°F

Differences Between Air Temperatures Above Artificial Turf and Merion Blue Grass in White Sox Park, Chicago, Illinois, August 13, 1970 (Artificial Turf Temperature – Merion Blue Grass Temperature).

	3 Inches	3 Feet	5.5 Feet	9 Feet
9:45 a.m.	6°F	2°F	3°F	1°F
10:50 a.m.	9°F	5°F	3°F	3°F
11:50 a.m.	6°F	5°F	3°F	3°F
1:30 p.m.	3°F	–3°F	–2°F	–2°F
2:35 p.m.	6°F	–2°F	3°F	3°F
3:30 p.m.	6°F	0°F	–1°F	--
4:40 p.m.	6°F	1°F	--	--
5:30 p.m.	5°F	--	--	--
6:30 p.m.	--	--	--	--

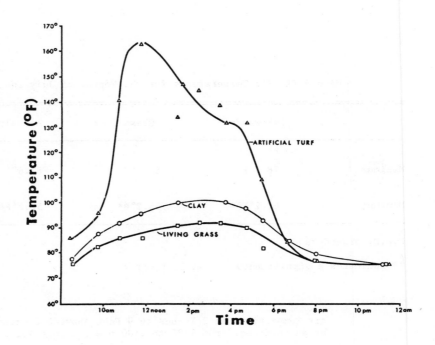

A comparison of infrared surface temperature of artificial turf, living grass and clay in White Sox Park, August 13, 1970.

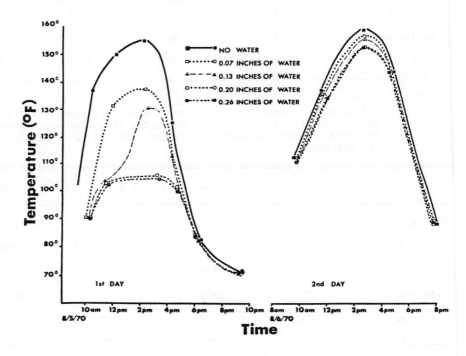

The change in infrared surface temperatures of artificial turf as the result of varying amounts of irrigation.

"Air Movement

WIND. Control of wind also controls temperature. Air movement affects human body cooling; it does not **decrease** temperature, but causes a cooling sensation due to heat loss by convection and due to increased evaporation from the body.

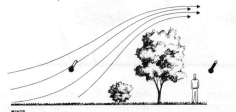

WINDS 136

"Trees generally reduce the wind velocity and product a sheltered area on the leeside and to a smaller extend on the wind side of the screen. This reduction in wind velocity brings about a lowering of the rate of thermodynamic exchanges between the air layers, with the result that protection from the wind generally permits higher temperatures to prevail in the protected zones. For instance, the temperature on the leeward side of an evergreen barrier made up of plants such as white pine, eastern arborvitae, or eastern red cedar is warmer, both during the day and at night. This concept is illustrated in the accompanying diagrams."

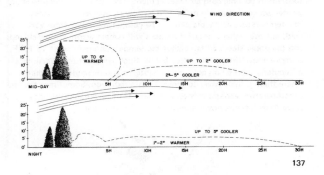

137

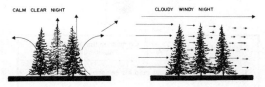

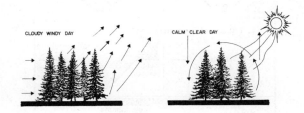

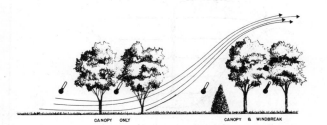

53.

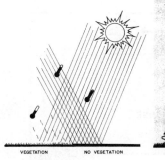

DIURNAL VARIATIONS

145

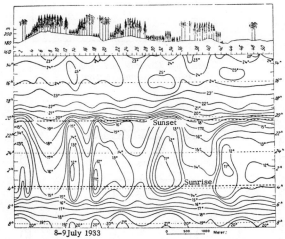

8-9 July 1933

Diurnal temperature variation within a closed forest area near Leipzig. (After H. G. Koch)

146

"Where there is free air movement, there is little or no difference between actual air temperatures in sun or shade. However, under trees flanked by shrubs reflecting air currents upward, there will be cooler temperatures in the shade.

"Dear-Air Insulation

An evergreen or row of evergreens placed next to a wall create an area of "dead" air between plants and wall. This acts much the same as the dead airspace in the wall of a house. The temperature gradient between the inside of the building and the dead airspace is reduced and held relatively constant, thus preventing the escape of heat from the building. Without the evergreens, air currents would maintain a high temperature gradient and facilitate the escape of warm air through the wall. Of necessity, such evergreens must be of a dense nature in the manner of *Arborvitae*, hemlock *(Thuja)*, or spruce *(Picea)*, and must be planted close together to form a solid wall.

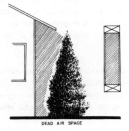

DEAD AIR SPACE

147

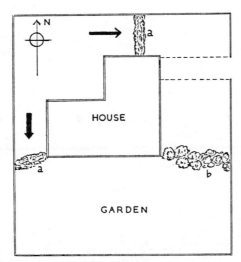

Obvious wind funnels, as at (a), should be sealed off. Checking cold northerly winds, (b), enhances a southerly aspect.

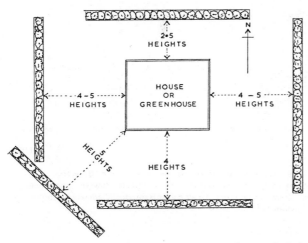

Minimum distances between a house or greenhouse and windbreaks in order to combine satisfactory shelter with least loss of light.

148

"With a 70°F. constant house temperature, the amount of fuel saved by a building protected from the wind is 22.9 percent. Victor and Aladar Olgyay estimate that with good protection on three sides of the building, the fuel savings might run as high as 30 percent.

"A by-product of such evergreen placement may be a cooling effect in summer, depending, of course, on orientation and placement. Plants slow the upward flow of warm air directly under them during the early evening hours, thus providing a tent of warmer air on cool summer evenings.

"The deciduous tree canopy combined with a coniferous windbreak planting will furnish a pleasant sheltered area under the overstory. Although shaded from the wind during the day, the air retains a certain amount of the day's warmth well into the evening, while still affording wind protection. 149

RADIATION INTERFERENCE 150

"Cold-Air Pockets. When wind disturbance is greatly reduced and air stops moving, thermal stratification and stagnation of the air may occur within a sheltered area, resulting in greater frost danger.

"Cool air is heavier than warm air and behaves somewhat like water flowing toward the lowest points. In hilly and other rough lands, a well-marked local temperature effect is often noted in valleys or depressions into which the fold air sinks during the summer nights. This flood of cold air causes "cold islands" or "cold air puddles." Accordingly any elevation that impedes the flow of air affects the distribution of the nocturnal temperature by creating a damming action; concave terrain formations become cold air lakes at night. The same phenomenon is greater when a large volume of cold airflow is involved, as in valleys. The temperature at the plateau will be cold; at the valley floor very cold; but the higher sides of the slopes will remain warm. On beechlands and the upper slopes of tablelands the temperature may be 6° to 10°F. or more above that of the valley below. Such temperature inversions are particularly marked on cold, calm nights, in arid or semi-arid regions where there is little cover to prevent rapid radiation. Sometimes the line of temperature variations is very marked. It often creates distinct air currents as it pours down mountain canyons by night.

Night cooling in a valley	Corresponding distribution of night minima
1. On the assumption that cold air behaves like cold water	
	Does not agree with actual conditions
2. Fitting the observations best	
Reservoir of heat	Cold plateau / Thermal belt / "Cold lake"
— Surface emitting radiation → Air flow	Cold ▭▭▭ warm Night minima

Development of the thermal belt.

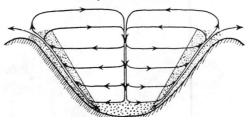

The system of upslope and upvalley winds. (After A. Wagner) 151

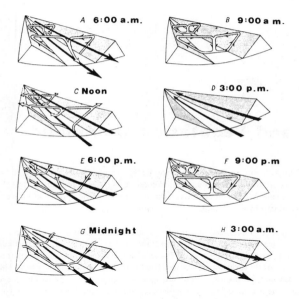

The interplay of slope and valley winds during the course of a day. (After F. Defant)

"Heated Air. During the day, hot air formed over unshaded ground rises vertically, and thus has little influence upon the temperature of the air under adjacent shade. During the nights, however, cold air formed over open ground spreads out readily to and under the adjacent cover.

"These phenomena are what Dr. Aloys Bernatzky, writing in "Anthos" magazine (1966, issue 1), is discussing when he talks about the balancing of temperatures between the green areas of the city and the built-up areas, having a beneficial effect on the climate of the city. In this article he says:

'. . . the masses of the building of a city form an artificial rock that stores up heat during the daytime. Not only with the ground surface, but with all the walls of the buildings as well and they make up a total far greater than the area itself. And as the masses of the buildings reduce the effects of air currents, the process of carrying off this stored up heat is slowed down. In addition to this, the amount of heat which on plant covered areas is absorbed by the process of assimilation and evaporation remains practically intact in the areas of the city where there is no vegetation at all. The values measured were found to be exceptionally high (60,000 K/cal/ per year/per m). During daytime the centers of the cities can have a higher temperature than that of the surrounding countryside, and the average figures for a whole year show over-temperatures of 0.5 to 1.5 centigrade, lowering the altitude of the city with respect to sea-level by 100 to 300 meters and means a change from normal climate to a more unhealthy climate. At the same time atmospheric humidity is reduced, resulting in an increase of diseases affecting the respiratory organs.'

"Manmade surfaces, then, exaggerate temperature extremes. Because plant materials absorb radiation and release it more slowly, plants are able to decrease temperature extremes. Highly reflective, manmade surfaces absorb heat to a large extent and release it rapidly, causing an overheated environment. Plant materials, on the other hand, absorb a greater amount of solar radiation during the daytime and release it slowly at night, reducing diurnal temperature variations.

"Plant materials, especially in the case of deciduous trees, interfere with solar radiation and reduce radiation reflection, causing a temperature reduction, not only in the shade of the tree, but immediately adjacent to it.

"The foliage of deciduous trees intercept solar radiation during the hottest part of the year, and their bare branches do not interfere appreciably with the winter sun, during a time when its warmth is desirable."

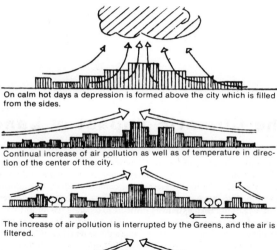

On calm hot days a depression is formed above the city which is filled from the sides.

Continual increase of air pollution as well as of temperature in direction of the center of the city.

The increase of air pollution is interrupted by the Greens, and the air is filtered.

30.3°C   33.4°C   30.3°C

Interruption of the temperature rise, cooling down of the air by the Greens, providing the built up areas with cool, salubrious air.

Diagram showing the balancing of temperatures between the Greens and the built up areas. Filtering of the falling air by trees. Cooler and cleaner air flows to the built up centers.

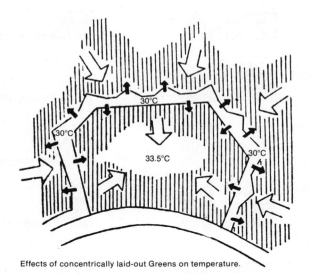

30°C

30°C   30°C

33.5°C

Effects of concentrically laid-out Greens on temperature.

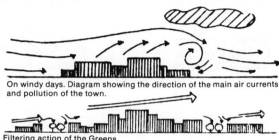

On windy days. Diagram showing the direction of the main air currents and pollution of the town.

Filtering action of the Greens.

The Climatic Impact of Landform and Vegetation Combined

Because of the embryonic nature of research concerning the environmental impact of natural elements as it relates to human environment, there has been little data-gathering in regard to the combined effect of landform on vegetation. Also, because of the developing nature and character of research in this area, it has been extremely difficult to separate the impact functions of each of these elements. At the same time it is difficult to generalize, since, in no two situations in the world do the same types of landforms and vegetation occur. The different types of landforms and vegetation combinations increase the difficulty of generalization or the development of principles or guidelines. At time, the combined inpact of landforms and vegetation on the specific environment are greater than the individual impact of either landform or the vegetation by itself. A number of studies were conducted at the University of Wisconsin in 1966 and 1967, the results of which are illustrated in the following plate:

Typical landforms were created and tested in wind tunnel situations. Typical vegetation forms were added to these scale model landforms to test the wind currents and eddy flows of the wind moving over and around the combined landform and vegetation in a variety of typical situations.

It must be emphasized in summary that both landforms and vegetation forms are dynamic and neither are static, on a particular day, in a particular season, or in a particular year. The landform is usually more solid and is a more certain wind control factor. The landform changes more slowly. The vegetation, on the other hand, changes form as it grows. The landform can raise the vegetation to a higher elevation instantly and thus add to the environmental impact of the landform by itself. Both plants and landforms of all types have the ability to alter climatic situations. They can do this by making it warmer or cooler, depending on their manipulations. Therefore, the essence of the same basic information may be used to enhance the potential of solar radiation or to provide relief from it.

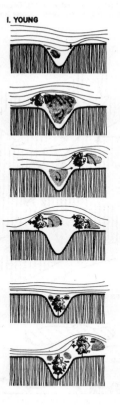
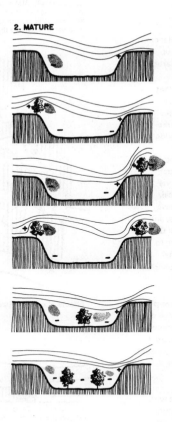
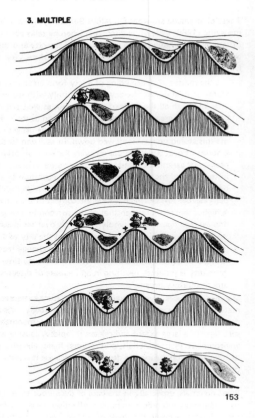

153

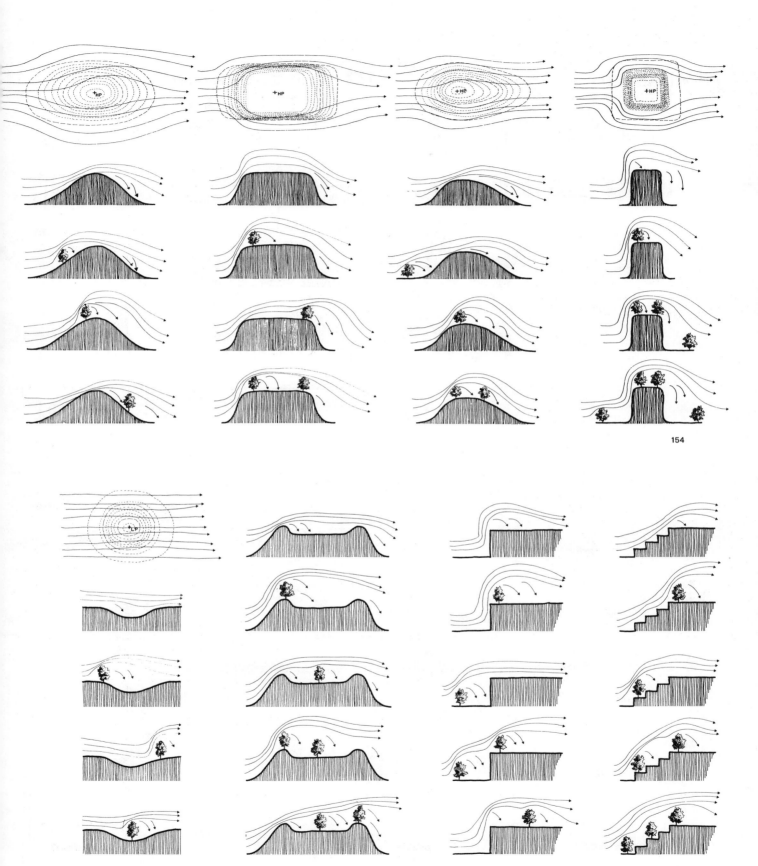

154

155

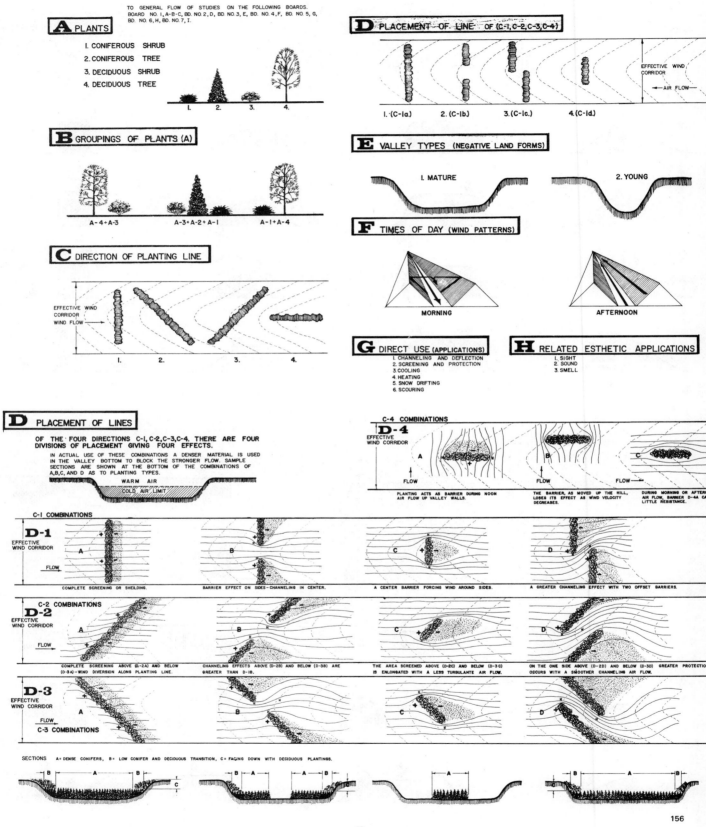

G DIRECT USE (APPLICATIONS)

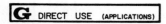

I. CHANNELING AND DEFLECTION

2. SCREENING AND PROTECTION

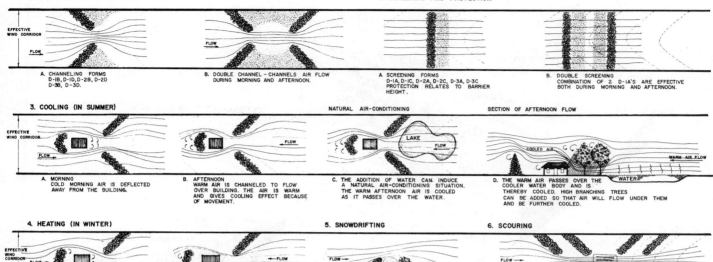

EFFECTIVE WIND CORRIDOR
FLOW

A. CHANNELING FORMS
D-1B, D-1D, D-2B, D-2D
D-3B, D-3D.

B. DOUBLE CHANNEL - CHANNELS AIR FLOW
DURING MORNING AND AFTERNOON.

A. SCREENING FORMS
D-1A, D-1C, D-2A, D-2C, D-3A, D-3C
PROTECTION RELATES TO BARRIER
HEIGHT.

B. DOUBLE SCREENING
COMBINATION OF 2 D-1A'S ARE EFFECTIVE
BOTH DURING MORNING AND AFTERNOON.

3. COOLING (IN SUMMER)

NATURAL AIR-CONDITIONING

SECTION OF AFTERNOON FLOW

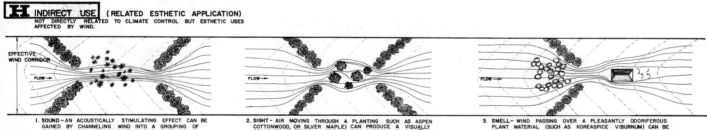

EFFECTIVE WIND CORRIDOR
FLOW

LAKE FLOW

COOLED AIR
WARM AIR FLOW
WATER

A. MORNING
COLD MORNING AIR IS DEFLECTED
AWAY FROM THE BUILDING.

B. AFTERNOON
WARM AIR IS CHANNELED TO FLOW
OVER BUILDING. THE AIR IS WARM
AND GIVES COOLING EFFECT BECAUSE
OF MOVEMENT.

C. THE ADDITION OF WATER CAN INDUCE
A NATURAL AIR-CONDITIONING SITUATION.
THE WARM AFTERNOON AIR IS COOLED
AS IT PASSES OVER THE WATER.

D. THE WARM AIR PASSES OVER THE
COOLER WATER BODY AND IS
THEREBY COOLED. HIGH BRANCHING TREES
CAN BE ADDED SO THAT AIR WILL FLOW UNDER THEM
AND BE FURTHER COOLED.

4. HEATING (IN WINTER)

5. SNOWDRIFTING

6. SCOURING

EFFECTIVE WIND CORRIDOR
FLOW

FLOW

FLOW

A. MORNING
COLD MORNING AIR IS DEFLECTED
AWAY FROM THE LIVING UNIT.

B. AFTERNOON
WARM AIR IS CONCENTRATED AND
CHANNELED AT THE LIVING UNIT
CARRYING AWAY THE COLD AIR
SETTLED AROUND IT.

IF A SKI SLOPE EXISTS ON A VALLEY
WALL, SNOW CAN BE MADE TO DRIFT ON
THE SLOPES BY ERECTING BARRIERS THAT
WOULD BE EFFECTIVE BOTH DURING MORNING
AND AFTERNOON.

THE AIR CAN BE CHANNELED TO CLEAR AN AREA ON THE VALLEY
FLOOR SUCH AS A PARKING LOT. THE DIRECTED AIR WOULD BLOW
AWAY SNOW OR LEAVES.

H INDIRECT USE (RELATED ESTHETIC APPLICATION)

NOT DIRECTLY RELATED TO CLIMATE CONTROL BUT ESTHETIC USES
AFFECTED BY WIND.

EFFECTIVE WIND CORRIDOR
FLOW

FLOW

FLOW

1. SOUND - AN ACOUSTICALLY STIMULATING EFFECT CAN BE
GAINED BY CHANNELING WIND INTO A GROUPING OF
TREES (SUCH AS PINE, SPRUCE, OR FIR) SYMBOL OF
SOUND PLANTING IS ✳. CHANNELS CAN BE MADE TO
GIVE THIS EFFECT ALL DAY OR DURING CERTAIN HOURS.

2. SIGHT - AIR MOVING THROUGH A PLANTING SUCH AS ASPEN
COTTONWOOD, OR SILVER MAPLE) CAN PRODUCE A VISUALLY
STIMULATING EFFECT. SYMBOL OF SIGHT PLANTING IS ✾

3. SMELL - WIND PASSING OVER A PLEASANTLY ODORIFEROUS
PLANT MATERIAL (SUCH AS KOREASPICE VIBURNUM) CAN BE
CHANNELED FOR A PARTICULAR PURPOSE. SYMBOL OF
AN ODORIFEROUS PLANTING IS ○.

SUMMARY OF USES-

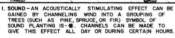

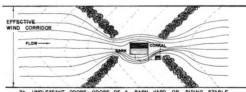

PARKING STABLE
CHALET
LOW AROMATIC PLANTING
SKI AREA

EFFECTIVE WIND CORRIDOR

EFFECTIVE WIND CORRIDOR
FLOW
BARN CORRAL

A. MORNING - COLD AIR FLOWS PAST THE CHALET AND CARRIES UNPLEASANT ODOR FROM THE
STABLES WITH IT. SOME AIR FLOWS OVER THE SKI SLOPE BARRIERS TO HELP DRIFT
SNOW. ESTHETIC FACTORS ARE USED AS SHOWN - SOUND✳, SIGHT✾, SMELL○.

3b. UNPLEASANT ODORS - ODORS OF A BARN YARD OR RIDING STABLE
CAN BE CARRIED AWAY BY CHANNELED WIND.

PARKING STABLE
CHALET LOW AROMATIC PLANTING
SKI AREA
FLOW

EFFECTIVE WIND CORRIDOR

PARKING STABLE
CHALET LOW AROMATIC PLANTING
SKI AREA
FLOW

B. NOON - WIND IS LIGHT AND VARIABLE COMING BACK UP THE VALLEY. THE VELOCITY FROM
THIS DIRECTION IS INCREASING DURING THE REST OF THE AFTERNOON.

C. AFTERNOON - THE AIR RUSHING UP THE VALLEY IS WARM HEATING THE CHALET AND SCOURING
SNOW FROM THE PARKING LOT. SNOWDRIFTING STILL TAKES PLACE ON THE SKI SLOPES
AND ESTHETIC FACTORS CONTINUE.

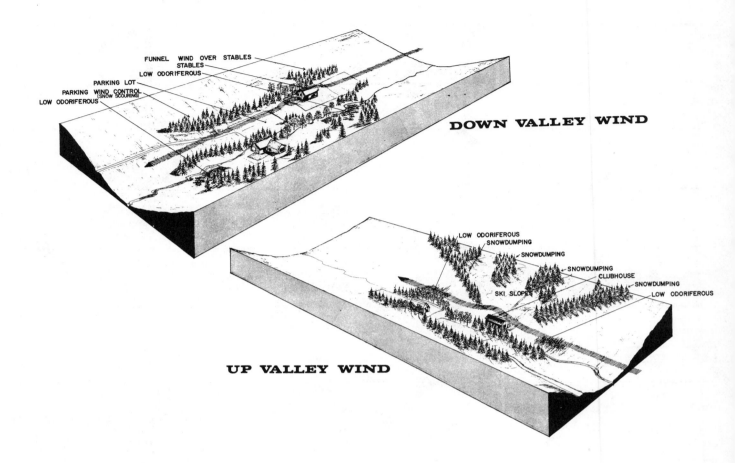

FUNNEL WIND OVER STABLES
STABLES
LOW ODORIFEROUS
PARKING LOT
PARKING WIND CONTROL
(SNOW SCOURING)
LOW ODORIFEROUS

DOWN VALLEY WIND

LOW ODORIFEROUS
SNOWDUMPING
SNOWDUMPING
SNOWDUMPING
CLUBHOUSE
SNOWDUMPING
SKI SLOPES
LOW ODORIFEROUS

UP VALLEY WIND

60.

Water

Water of all types, and in all geographic areas of the earth's surface, has a profound impact on the climate and on climate control, especially in the utilization of solar radiation and in energy conservation. Oceans are the birthplace of major climatic activities affecting areas on the adjacent land surface. At the same time, water bodies of all types modify climate and reduce its extremes. Because a large percentage of the insulation striking the water surface is stored by the water, a small percentage of the solar radiation is radiated off the water surface. At the same time, a small percentage of the solar energy striking the land surface is stored while a large percentage of the same energy is radiated back into the atmosphere. Because of this, the ocean or large water body—any large water body—has a moderating effect on temperature.

Since all water bodies radiate a smaller percentage of solar radiation than does a land mass they act as a moderating influence on abrupt temperature changes on land near the water body. The oceans are the ultimate water bodies and their more even temperatures are carried overland by air masses and help to moderate extemes of heat and cold. A large water body, such as an ocean, absorbs and stores a great percentage of the solar radiation striking its surface. Because of that the surface temperature of an ocean may vary no more than 18 degrees during the year and less than one degree from day to night. On land where much of the solar energy is radiated back into the atmosphere there may be a wide temperature fluctuation from day to night and season to season. Therefore, the more uniform ocean serves as a large air conditioning reservoir. The air moves over the warmer air in winter and moderates the cooler air on land while the cooler ocean breeze moderate the temperature of the land areas in summer.

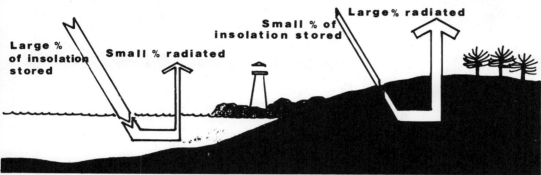

At the largest scale the ocean and ocean currents have great impact on the climate on land, particularly that closest to the edge of the water. The major ocean currents in the northern hemisphere flow in predictable patterns and may contain water which is as much as 30 degrees warmer than that of adjacent surfaces. This, obviously, has an impact on the mesoclimate of land areas which they strike. The Gulf Stream warms the southern east coast as far north as southern Virginia. The northern east coast is cooled by the Labrador current. The warmer Japan current crosses the Pacific and helps to moderate the temperature of Alaska. As that same water moves down the western side of the North American continent it becomes the California current which is cooled in the north Pacific. As this cooler current moves down the coasts of Washington, Oregon and California it brushes against the warmer land areas. As a result of differing water, air and land temperatures along this area damp fogs and heavy seasonal rains are combined with moderated temperatures. As much as 10 inches of fog drip has been recorded during a season in northern california redwood forests.

159

To a lesser extent inland lakes, especially the Great Lakes in the United States perform some of the same functions and moderate in various ways and to varying degrees. Lake Michigan moderates the climate of western Michigan and enables a large fruit growing industry to thrive because of the cool belated spring and the lengthened fall. The southern portion of the province of Ontario in Canada has a large nursery and fruit growing industry because of the climate moderated by the surrounding lakes.

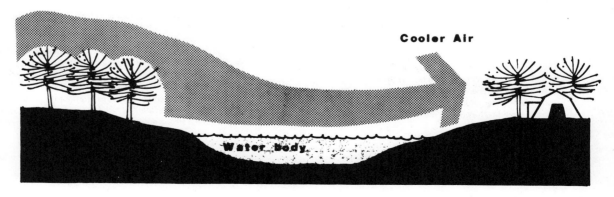

Breezes flow from the water body onto the shore during the day and off the land area onto the water body at night.

Water in the form of clouds has an impact on climate as does liquid moisture. Radiation which has been absorbed by the earth quickly and readily escapes back into the atmosphere on a clear night. On an overcast night the cloud cover inhibits this radiation loss and thus the temperatures are generally higher on an overcast night than they are on a clear night.

MORNING

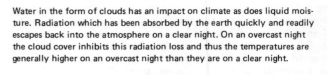

RADIATION LOSS ON CLEAR NIGHT

EVENING

RADIATION CONSERVATION BY WATER VAPOR CEILING

This natural airflow pattern may be utilized and controlled for natural ventilation and energy conservation but only if it is fully recognized and accepted.

This modifying effect of water is usually greatest near the shore.

Onshore breezes

68°   90°

160

The leeward side of the water body will always be cooler since the wind is cooled as it moves across the surface of the water body. Therefore areas or activities which need to be naturally cooler should be located on the leeward side of water bodies. Functions or areas which need extra heat or warmth should be located on the windward side of water bodies where possible because of this.

Therefore water, ranging in form from an ocean to the water particle in a cloud, is able to moderate or effect extremes of climate and to assist in energy conservation. The ocean can seldom, if ever, be modified, but its effects on microclimate can be accepted and utilized in landscape planning. Site selection and structure orientation should, on the other hand, take advantage of the distinctive climate near smaller water bodies. At the site design and materials selection scale fountains, pools and sprays may be used to lower high temperatures or affect the psychological perception of heat through the sound of falling or running water. Water, as a natural element has a strong impact in various forms and scales on the modification of climate and the conservation of energy.

Human Comfort Levels

Man's physical strength and mental activity function best within a given range of climatic conditions. Outside of this range his efficiency lessens; stress and the possibility of disease increase. Within this range he is comfortable and can function efficiently.

The comfort zone varies with people both within an area and in different areas of the world, either because of inherited or cultural differences. Most women prefer a temperature a few degrees warmer than men; young people prefer a temperature a few degrees cooler than the elderly; and Eskimos prefer cooler temperature than Africans. This is the result, in part, in variations in metabolic rates; in part, it is psychological.

It is possible, however, to define a general comfort zone which is suitable for most people of the time. Four major elements affect human comfort: air temperature, solar radiation, air movement, and humidity or precipitation. Each element tends either to offset or multiply the effects of other elements. None can be set without information about the others. The comfort zone is the product of a proper balance of all four elements.

A human receives and also gives off heat. He receives by absorption, directly from the sun or from reflective objects, and also from other radiation, by conduction. He acts as a radiator, giving off heat, and heat is conducted away from him. When a balance of heat gain and loss is present, man is comfortable, but if too great an imbalance, in either direction, occurs, he is uncomfortable.

Some writers consider sunstroke or heatstroke as the upper temperature limit for man's existence; with the freezing point as the lower limit. The ideal air temperature may be assumed to be midway between these extremes. Experiments on animals in a variable temperature tunnel at the John B. Pierce Foundation showed that animals preferred to stay at 70°F., about midway between the points calling for maximum expenditure of energy in adjustment to the environment. Therefore, some writers believe that the human being, with a body temperature averaging 98.6°F., seeks a comfortable temperature condition, an area where the temperature is about halfway between what he can tolerate in cold without being grossly uncomfortable and the point which would require real effort on the part of his circulatory and sweat secretion system in order to permit him to adapt to heat.   165

Dr. H. M. Vernon and T. Bedford, of the British Department of Scientific and Industrial Research, believe that the comfort zone is from 66.1° to 72.1°F. in winter. S. F. Markham believes the range of temperature from 60° to 76°F. is ideal for man. C. E. P. Brooks has indicated that the temperature comfort zone fir British people lies between 58° and 70°F. Others have said that the temperature comfort zone in the United States lies between 69° and 80°F.

C. A. Federer has explained that "Our temperature senses are separable. Our sense of warmth involves nerves that are triggered more rapidly as skin temperature increases. Our cold-sensing nerves pulse more rapidly as skin temperature decreases.

"A skin temperature around 32°C. seems subjectively to be thermally neutral. This temperature is maintained by a nude individual at a room temperature of about 28°C. and by a lightly clothed individual at 21°C. when there is no forced ventilation (Brengelmann and Brown, 1965). As skin temperature varies more than a few degrees from 32°C., we sense warmth or cold and our bodies start to react to increase or decrease out heat loss.

"An inactive man generated about 50 Keal. of heat per hour as a waste by-product of his metabolic processes. This heat must be dissapated to his environment. With exertion, this amount of heat increases by several times. The heat must first be transported to the skin surface, and this requires a skin temperature lower than the body or core temperature. Heat is transported partly by conduction through body tissues; but these are fairly good insulators, so most of the subsurface heat transfer is by capillary convection—mass flow of warmer blood to the surface, where it cools, and then is returned. The body has good control over this process by constrictuion and dilation of capillaries; it can alter the effective subsurface conductivity by a factor of about 10 (Brengelmann and Brown, 1965).

"The heat reaching the body surface is lost in three different ways: by convection, by radiation, and by evaporation.

"Convection or sensible heat transfer involves heating of the air next to the skin and then removal of this air and replacement by cooler air. Ventilation increases convection loss.

"Radiation loss is due to the emission of longwave or thermal radiation from the surface. Since the surface is simultaneously absorbing radiation from its surroundings, a net radiant heat loss occurs only if the surface is warm enough to emit more radiation than it gains from the environment. In a sunny environment, net radiation is a heat gain rather than a loss.

"Evaporation loss of latent heat occurs whenever water evaporates: water vapor carries away into the atmosphere about 580 cal. for each gram evaporated. For a resting man who is not perspiring, about half of the heat loss is by radiation and half is by convection, while latent heat loss is small and occurs mostly in the respiratory tract.

"Cooling of an individual can result from increasing ventilation, from a lowering of ambient temperature, from reduction of incident radiation, or—for a perspiring individual—from reduction of humidity. The corresponding reduction in surface temperature tends to increase the subsurface heat flow. The body responds in several ways. First, capillaries constrict to reduce the effective subsurface conductivity. If this is not enough, shivering starts, increasing metabolism and producing more heat. The internal temperature of the extremities is also allowed to drop, thus decreasing the heat loss from them. Humans, of course, also have the ability to add clothing. This moves the heat-exchange surface out from the skin and effectively decreases the subsurface conductivity.

"Heating of an individual can result from increasing metabolism through work or exercise, increase in ambient temperature, or increase in radiation load. The supply of radiation to the surface always includes longwave radiation and, when outdoors or near windows, solar radiation. The body can increase the loss of internal heat by capillary dilation, increasing blood flow to the skin. If this is not sufficient, sweating begins. If this is slight, all the secreted water evaporates and the skin stays dry. With further increase in secretion, the water evaporates as fast as it can, and the skin becomes wet. At this stage the evaporation rate is controlled by the ventilation rate and the humidity of the air. Evaporative cooling in hot conditions may account for 90 percent of the total heat dissipation.

"The skin temperature of an individual subject to a normal range of environmental conditions can vary by as much as 150ºC., especially at the extremities. The surface temperature of a clothed individual is obviously subject to even wider fluctuations. And the relative importance of the three modes of dissipation can also vary widely.

"An individual can thus control his feeling of comfort with temperature by internal means, and by changing clothing. He does these in response to the microclimatic variables of air temperature, wind, solar and long-wave radiation, and in warm weather, humidity. We should also recognize that much of the adverse effect of precipitation comes from the increase of evaporative cooling from the wetted skin or clothing surfaces."

166

Interior microclimates are controlled. A comfortable microclimate for humans can be created interiorly by use of heat, air conditioning, and illumination. Such positive control is not possible exteriorly, but some degree of control is needed for human comfort and use. Wind, solar radiation, precipitation of all sypes, and excessive temperature variations are factors to be considered.

A bioclimatic evaluation is necessary before designing for a climatic balance. Prevailing climatic conditions can be plotted on a chart to show what corrective measures are needed. A bioclimatic chart is an excellent way to portray temperature and humidity. The following chart shows which applies to the inhabitants of the moderate climatic zones in the United States, at elevations not in excess of 1,000 feet above sea level, with customary indoor clothing in a sedentary condition, or doing light work.

When the bioclimatic conditions—air temperature, solar radiation, wind, and humidity—have been evaluated, a determination of the type and degree of climate control necessary to provide human comfort can be made. A sun shade, a wind screen, a canopy to deflect rain or to cool, or combinations of all of these can be used to control or ameliorate climate. In addition, walls, fences, earth forms, and free-standing roofed structures may be used. Objects or elements may alter the climate positively or negatively. They may ameliorate or aggravate climatic problems—that is, they may either increase or decrease man's comfort.

MAN AND TEMPERATURE

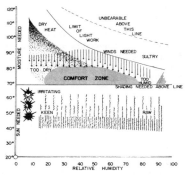

BIOCLIMATE CHART

167

MEANS BY WHICH THE BODY EXCHANGES HEAT

A. ABSORPTION FROM SUN DIRECTLY
B. ABSORPTION FROM REFLECTIVE OBJECTS
C. ABSORPTION FROM GLOWING RADIATORS
D. HEAT CONDUCTION TOWARD THE BODY
E. HEAT CONDUCTED AWAY FROM THE BODY
F. OUTWARD RADIATION TO SKY
G. OUTWARD RADIATION TO COOLER OBJECTS
H. ABSORPTION OR LOSS FROM NON-GLOWING HOT RADIATORS

HEAT EXCHANGE - MAN   168

In summary then, climate effects man both physically and emotionally and is therefore a factor of considerable importance in building and landscape design. A designer's major task is to create the best possible environment (indoors and outdoors) for the occupants' activities. The challenge of the designer is to provide total human comfort, which may be defined as the sensation of complete physical and mental well-being.

Air temperature, humidity, radiation, and air movement all affect human comfort, and must be considered simultaneously if an acceptable environment is to be provided. These factors must be considered whether or not the building being designed is heated or cooled by solar energy or fossil fuels.

To effectively design for human comfort, it is necessary to understand the basic thermal processes of the body. How the body generates and loses heat is crucial for identifying comfort zone and for designing heating, cooling and humidity control systems. Two complementary approaches for the provision of human thermal comfort have developed. One seeks to maintain thermal conditions within an established comfort zone while the other attempts to modify the comfort zone. Both approaches are used during solar dwelling design.

The body is continuously producing heat. Everyday activities such as sleeping, walking, working, and playing are all heat producing. The entire portion of the body's energy requirement is supplied by the consumption and digestion of food. The process of transforming foodstuff into usable energy is called metabolism. Of all the energy generated by the metabolic process, the body uses only 20 percent and the remaining 80 percent must be lost to the environment.

Body temperature, as contrasted to skin temperature, must be maintained at 98.6ºF (37ºC) for the body to adequately perform its functions. To maintain this constant temperature balance, all surplus heat must be dissipated to the environment. Heat gained from the environment must also be dissipated.

The body loses approximately 80 percent of its heat to the environment by convection and radiation. The remaining 20 percent of the body's surplus heat is lost by evaporation with a very small percentage of heat by conduction.

The sum total of the body's heat gain and loss should at all times equal 98.6ºF or 37ºC. If the body's heat gain is more than its corresponding heat loss, an uncomfortable feeling will occur and sweating will begin.

169

Likewise, if the body's heat loss is more than its heat gain, body temperature will drop and shivering will occur.

The body has numerous regulatory mechanisms to maintain a constant temperature. Blook circulation may increase or decrease, sweat glands may open or close, and shivering may begin to raise the body's temperature. Also, continuous exposure to similar climatic conditions, called acclimatization, can cause a change in the basal metabolism process, an increased sweat rate, or a change in quantity of blood. It is not surprising, therefore, to learn that Eskimos prefer cooler temperatures than equatorial Africans.

Like climatic regions human comfort is shaped by four major factors: air temperature, humidity, air movement, and radiation. Accordingly, the heat exchange process between the human body and its environment may be aided or impeded by these climatic variables. For example, convective heat loss is severely impeded by high air temperatures, and evaporative heat loss may be simultaneously restricted by high humidity. Different regions will have different dominant climatic features which will affect human comfort.

The four climatic variables are the primary determination of human comfort or discomfort. A number of subjective or individual factors, however, will influence thermal preferences. These include one's clothing, age, sex, body shape, state of health and skin color. Additionally, there are psychological and sociological variables which will influence thermal comfort. Whether one is happy or sad, active or confined, alone or in a group will influence thermal preferences.

64

An assessment of the impact of local climate on the body's heat dissipation process is important in providing a comfortable environment. However, to do so, four independent climatic variables must be assessed simultaneously (temperature, humidity, radiation and air movement). The difficulty of this task has led to the development of "thermal indices" or "comfort scales" which combine the effects of these four variables.

A "comfort scale" therefore is the composite of the interactions between climatic variables. Through observation and measurement in the laboratory, the characteristics of human comfort (comfort zone) have been identified and may be compared to local climatic conditions, thereby defining the need for and the type of thermal controls.

COMFORT ZONE

The comfort zone is established by analyzing the relationship between air temperature and three climatic variables: mean radiant temperature (the temperature of the surrounding surfaces), humidity, and air velocity.

The analysis will establish the range of thermal conditions (comfort zone) over which the majority of adults feels comfortable. The comfort zone is, at best, an imprecise approximation of human thermal comfort, realizing the many variations due to human preferences, physiological and psychological characteristics, and the nature of the activity being performed. However, it does provide the designer and builder with a "ball park" estimate of human thermal comfort by which climatic conditions of a locale may be evaluated so that appropriate methods of achieving a comfortable climate are chosen.

There are two ways of looking at the provision of human comfort. The first involves establishing a comfort zone based upon the users' thermal preferences and proposed activities and comparing this comfort zone to the existing or anticipated climatic conditions. In this manner, the appropriate methods for returning the climatic conditions to within the comfort zone will be established. The second viewpoint accepts the existing or anticipated climatic conditions as given, and identifies methods of altering the comfort zone to be compatible with the climate.

There are, then, certain levels of human comfort in which the temperature, humidity and air flow are more optimum depending upon the age, culture, activity and preference of different persons in the environment. There are, at the same time, limits of human adjustment or adaptation as well ad endurance. The designers responsibility is to provide, to the extent possible, the optimum human comfort levels in both the indoor and the outdoor environment. It is possible to do this to a certain degree through landscape design and with landscape elements.

The essence of all landscape development for energy conservation is to modify the aspects of air temperature, humidity, radiation or air movement in such a way as to bring existing or unpleasant conditions as closely as possible into the climatic conditions which are comfortable to specific persons on a precise site at a particular time.

The Climatic Regions and Human Adaptation

The dictionary definition of climate is "the average course or condition of the weather at a place over a period of years." The sun, wind movement, temperature and precipitation and humidity as well as many other factors shape the climate of the earth. An understanding of some of the elements of climate and climatic functions is essential in the utilization or manipulation of the landscape and landscape elements in order to conserve energy.

Global climatic factors such as the amount and intensity of solar radiation striking a portion of the earth's surface, the seasonal tilt of the earth axis in relation to the sun, the prevailing intensity and direction of air movement and the influence of water and topography determine the climatic make-up of any particular place on earth. These factors will determine the precise temperature, humidity, solar radiation, air movement, wind and sky conditions for any specific location. Regional patterns of climate will emerge from the interaction of these local climatic influences. Local climates are additionally influenced by site topography, (height, forms and orientation), ground surface, vegetation, water bodies and existing structures. The aggregation of the above factors will determine to a large degree, the need for manipulation of the site and site elements to conserve energy and to utilize more effectively existing solar energy potential. Local weather conditions and the distinctive climate of an area is directed, primarily, by the sun and is influenced by a number of other physical conditions of the earth itself. The nearness of an ocean or other large water body, abrupt changes of grade (whether mountains or valleys), the direction and intensity of prevailing winds, the type density and height of existing vegetation and similar factors will contribute to affect local climate conditions. The climates of particular localities are comparitively constant, and despite pronounced and rapid changes, have an inherent character of weather patterns which repeat themselves from year to year. Every place on the face of the earth, over an extended period of time, exhibits its particular combination of heat and cold, rain and sunshine.

In response to these differing climatic conditions, housing or building styles in one area of the country have developed, to a certain extent, which are somewhat different than those formed in other areas. This same differentiation will be particularly true, in the future, for dwellings heated or cooled by solar energy. The earth's climate, then, is shaped by thermal and gravitational forces. Regional atmospheric pressure, temperature and topographical differences influence the climatic conditions on a continental scale. The conditions of weather which shape and define local and regional climates are called the "elements of climate." The five major elements of climate are temperature, humidity, precipitation, air movement, and solar radiation. Additionally, sky conditions, vegetation, and special meteorological events or conditions are also considered elements of climate.

A designer, builder or homeowner is primarily interested in the elements of climate which affect human comfort and the design of buildings or landscape elements. The information the landscape designer, either amateur or professional, needs to know when planning or modifying the landscape for energy conservation or solar energy utilization includes: averages, changes and extremes of temperatures, the temperature changes between day and night (diurnal range), humidity, amount and type of incoming and outgoing radiation, direction and force of air movements, snowfall and its distribution, sky conditions and special conditions such as hurricane, hail or thunderstorm occurence.

Man has always existed by adapting to all types of climate or by modifying to varying degrees to specific climate of the place where he wanted or had to live. There are throughout the habitable world a variety of climatic conditions ranging from the very cold to the very hot and from the moist or damp to the dry or arid.

A commonality of climatic conditions within a geographic area will constitute a climatic region or zone. A number of systems have been proposed for classifying climatic regions within the United States. For purposes of discussing the variation in dwelling and system design which result because of different climatic conditions, only a broad description of climates of the United States is required. W. Koppen's classification of climates, based upon vegetation, has been the basis of numerous studies about housing design and climate. Using this criterion, four broad climatic zones are to be found in the United States: cool, temperate, hot-arid, and hot-humid. The areas of the United States which exhibit the general characteristics associated with each zone are shown below. The boundary between regions is not as abrupt as indicated. Each different climatic region merges gradually and almost invisibly into the next one.

The climatic characteristics of each region are not uniform. They may vary both between and within regions. In fact, it is not unusual for one region to exhibit at one time or another the characteristics associated with every other climatic region. However, each region has an inherent character of weather patterns that distinguishes it from the others. A brief description of the four climatic zones will identify the general conditions to which solar dwelling and site designs in those regions must be responsive.

COOL REGIONS: A wide range of temperature is characteristic of cool regiond. Temperatures of minus 30°F (-34.4°C) to plus 100°F (37.8°C) have been recorded. Hot summers and cold winters, with persistent winds year round, generally out of the NW and SE, are the primary identifiable traits of cool regions. Also, the northern location most often associated with cool climates receives less solar radiation than southern locations.

TEMPERATE REGIONS: An equal distribution of overheated and underheated periods is characteristic of temperate regions. Seasonal winds from the NW and S along with periods of high humidity and large amounts of precipitation are common traits of temperate regions. Intermittant periods of clear sunny days are followed by extended periods of cloudy overcast days.

HOT-ARID REGIONS: Hot-arid regions are characterized by clear sky, dry atmosphere, extended periods of overheating, and large diurnal temperature range. Wind direction is generally along an E-W axis with variations between day and evening.

HOT-HUMID REGIONS: High temperature and consistent vapor pressure are characteristic of hot-humid regions. Wind velocities and direction vary throughout the year and throughout the day. Wind velocities of up to 120 mph may accompany hurricanes which can be expected from E-SE directions.

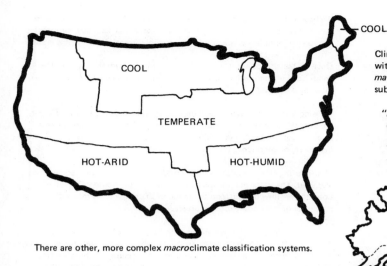

COOL

Climatic regions may be characterized as being in within a macroclimate but within those areas there are also a variety of smaller variations or *mesoclimates*. The publication *"Weather-wise Gardening"* defines these climatic subdivisions of a region in the following way.

"*Mesoclimate:* Intermediate variations in the big general weather of an entire region. Weather in San Francisco is quite different than that of Los Angeles, but both are meso-climates within the same macro-climate. The simplest answer to the variations lies in the degree of exposure to the big weather-making factors. One area may be warmer than another because it receives more solar radiation, because of its latitudinal posi-

There are other, more complex *macro*climate classification systems.

United States Macroclimates:

1. Long cold winter, hot summer, wide temperature range; northwest winter storms, scant summer wind. Look for: fall-winter, and spring sunshine; protection from storms but good secondary circulation against humidity; dry, well-drained ground.

2. Cool winter with some cold snaps; long, hot, oppressive summer; few topographic climatic differences; humid with moderate rainfall. Look for: sun protection, good air circulation.

3. Hot glaring summer, short mild winter, even temperature; humid, tropic. Look for: good air circulation, well-drained sites; protection against low western and eastern sun.

4. Cold winters, hot summer, wide temperature range; humid. Look for: storm protection, well-drained sites.

5. Cold winter, hot summer, wide temperature variation with extreme changes; low humidity, semi-arid. Look for: winter sunshine, year-round wind protection.

6. Very cold winter, warm summer; great climatic variety; low humidity and rainfall. Look for: sunshine, protection from prevailing and secondary winds.

7. Very hot summer, mild winter; very dry atmosphere; strong glaring sun. Look for: sun control, year-round wind protection.

8. Cool summer, mild winter; high humidity and precipitation, damp ground. Look for: protection against westerly winds, well-drained site.

9. Cool summer, mild winter; humid with low rainfall. Look for: south and west sun protection, protection against prevailing wind.

10. Warm summer, mild winter, humid, light to heavy precipitation; sharp changes in rainfall patterns. Look for: air circulation in wind-less exposures, well-drained sites in rainy areas.

11. Cool summers, relatively mild winters. Rainy, humid. Year-round climate moderated by warm Alaska current.

12. Very cold winters. Mild summers. Occasional hot spell up to 100°. Short, intensive growing season. Long daylight hours.

This categorization is accompanied by the following map showing the location each of those regional climatic breakdowns.

161

tion, the slope of the ground, or possibly because cloudier conditions screen out more insolation. Another location may be ranier than another because of a closer proximity to a storm track, or because it is less "shadowed" by a terrain feature. The rugged topography of the west, with high peaks adjacent to low-lying valleys, rolling hills and lofty plateaus, visually illustrates the fact that the more irregular the terrain, the more varied its climates will be.

Moderating effects of large bodies of water create more mesoclimates again more pronounced when combined with terrain features. A sea coast may be shrouded in cooling fog while a valley a few miles inland, shut off from the marine airflow by mountains, swelters in a temperature 30 degrees higher."

Not only buildings and building types have been modified or adapted but also the landscape development or landscape elements have also been moved, altered, shaped or modified in order to assist man to modify the climate to be more comfortable in a specific place or situation.

There are many, widely diverse indigenous, traditional and historical methods for accommodation of the building or building elements or modification of the site in the various climatic regions for human use, comfort or enjoyment. These methods include altered criteria for site selection, structural systems, orientation, site and differing use of the site elements. Due to the sophistication of contemporary building technologies and the development of an "international" style of architecture there has been less regional climatic modification of the architectural form and structure in recent years. Sophisticated heating, ventilating and air conditioning techniques have obviated the need for careful site planning or building design to accommodate the local climatic variations or uncomfortable conditions. These mechanical or artificial means of microclimate modification, however, are energy intensive, expensive and unrealistic in light of the costs and availability of existing energy sources. Eventually it will be necessary to more carefully study some of the methods employed historically to adapt to the distinctive climatic characteristics of each of the regions. To do that it may be necessary to update the forms and methods and to integrate these with newer technologies, materials and forms.

A brief consideration of the housing types used by indigenous peoples through North America will provide some perspective in considering human adaptation in each of the basic climatic regions. In each region, each tribe developed its housing to fit its environment, both to employ the sun and wind, and to shield itself from them, always using the materials which were at hand (which were, in turn, an expression of the climate).

The Eskimo igloo, while extreme, is a perfect expression of adaptation. It is built only of ice and snow, and a smooth ice-lining on the inside in the upper part, facing south, is made of clear, fresh-water ice. A vent in the roof faces away from the wind, as does the entrance of the shelter. The hemispherical shape deflects the wind, and maximizes the ratio of interior volume to surface area, thus minimizing heat loss. Heated only by small lamps and body heat, igloos maintain a temperature of 60° when the air outside is -50°.

Moving southward to the North Pacific coast-line, we find a less extreme climate, with more variety of building materials at hand. The primary need, however, was still for heat conservation. The houses of the Kwakiutl, in British Columbia, were built of wood plants, and joined together to reduce surface area. An outer shell was constructed around the entire group of dwellings. This provided an insulating dead-air space, and an enclosed walk-way between units. In the summer, when insulation was not needed, the outer shell was removed to allow ventilation. In other northern areas, houses were built with low-pitched roofs to support a heavy load of snow and use it as insulation.

In temperate areas, the tipi is the familiar dwelling. Built of poles supporting skins, it still faced away from wind, both at the entrance and at the upper vent. During warm weather, villages were spread out in the open. During cold weather, they were grouped together on sheltered slopes or under trees.

In the eastern woodlands, where the tipi was not used, various houses were built of wood, bark, or thatch. Several concepts, however, remained constant. Individual houses were almost always round, for most efficient heating, or dwellings were grouped together (in colder areas, usually) into long, low structures. Vents and entrances consistently faced away from the wind.

In the hot arid southwest, the predominant concern was excessive sun. The adobe pueble was built with massive walls, units atop one another, sometimes under the shelter of a cliff, all to reduce surface area. Oriented along an east-west axis, with individual units facing south, it had minimum exposure to the direct rays of the norming and evening summer sun, but maximum exposure to the southern winter sun. The massive front wall, receiving direct radiation during a winter day, would store heat into the night.
Groups which did not live communally often used partially submerged houses, providing minimum exposure in the summer to the sun and hot winds, well insulated also in the winter (but not absorbing heat significantly).

Semi-subterranean houses appear throughout the world, in areas as different as the American Southwest and Siberia, in any flat area in which drainage is not a consideration. The form seldom varies. The pit house is almost always round. Only the construction of the roof, and the total size, change.

In hot, humid areas, in the Southeastern United States and in fact throughout the world, indigenous housing has only a skeleton of a wall (in some places only poles) but usually a high, pointed, thatch roof. The roof not only protects against the rain, but also casts a large shadow. Often the floor of the house is elevated, to keep it dry and allow air circulation underneath.

Thus we see the range of conditions to which primitive man was able to respond in his architecture. Surely we should be able to respond even more accurately to the conditions present.

We have several variables to consider. While gross site selection may be predetermined, discrete site selection within an area may be made for more or less sun, more or less wind, and for microclimatic variations in temperature.

Once on the site, orientation should depend upon both the sun and the wind. Both of these will be affected by the surrounding environment, man-make or natural.

Housing forms and materials will depend upon heat characteristics sought. We may select forms which capture the sun, or those which radiate heat away, reflect sunlight, and shed water. Different materials may be reflective, insulative, or conductive.

Winds during cold periods should be intercepted, cooling breezes captured and used. The shape and location of openings in the house will depend upon the air-flow desired, and the shape of the house itself may either maximize or minimize exposed surface.

Each factor, individually, we may respond to, both in the house itself and in the site. Balance also becomes necessary, when conditions vary through the day or throughout the year. The simplest concepts are those we found when the considered the Eskimo, and the hot-humid region: complete insulation vs. no insulation; orientation away from the wind vs. maximum circulation; attempt to capture sunlight vs. maximum shade; raised from ground vs. fused into ground; sheds precipitation vs. collects it for insulation; et cetera.

In less extreme or less consistent climates, however, conflicting consideration must be taken into account. In temperate climates, some primitive peoples have changed houses or moved their houses according to the season (even the Eskimo abandons his igloo in the summer), an alternative not available as often to modern man. In the temperate and hot-arid regions, we must be able to respond adequately to constantly changing conditions, and even in the cold hot-humid regions, we should effectively recognize the changes between day and night.
Indoor temperature balance becomes the primary concern. We seek to minimize change, both from season to season and from day to night. We should create a time-lag, so that the exterior conditions do not immediately change the interior but rather are "stored up" and tapped for later use, just as man now uses the stored-up fossil fuels.

The characteristics of the four basic climatic regions, a summary of the objectives in site development, and recommendations for precise site adaptations for application in each of them are summarized in the following chart:

In each of the regions there has been an indigenous adaptation in the architecture and in the site development. By the same token, there has been, to a large extent, a contemporary architectural reflection of the region in spite of advanced technology and the ability to manufacture and produce an international architectural idiom.

The site development considerations, less able to be influenced by technological and industrial processes, have reflected where possible to a greater extent the indigenous needs and requirements of building siting, orientation, selection and use of materials and ultimate landscape architectural form.

Hopefully this publication is able to illustrate to a greater degree the principles of site design and planning for optimum solar energy utilization and energy conservation, working with and using the inherent character and indigenous materials of each of the regions.

	Temperate	Hot-Arid	Hot-Humid	Cold
Objectives	Maximize warming effects of sun in winter. Maximize shade in summer. Reduce impact of winter wind but allow air circulation in summer.	Maximize shade late morning and all afternoon. Maximize humidity. Maximize air movement in summer.	Maximize shade. Minimize wind.	Maximize warming effects of solar radiation. Reduce impact of winter wind. Avoid micro-climatic cold pockets.
Adaptations				
Position on slope	Middle-upper for radiation	Low for cool air flow	High for wind	Low for wind shelter
Orientation on slope	South to Southeast	East-southeast for P.M. shade.	South	South to Southeast
Relation to water	Close to water, but avoid coastal fog	On lee side of water	Near any water	Near large body of water.
Preferred winds	Avoid continental cold winds	Exposed to prevailing winds	Sheltered from north	Sheltered from North and West
Clustering	Around a common, sunny terrace	Along E-W axis, for shade	Open to wind	Around sun pockets
Building orientation	South to Southeast	South	South 5° toward prevailing wind	Southeast
Tree forms	Deciduous trees nearby on west. No evergreens near on south	Trees overhanging roof if possible	High canopy trees. Use deciduous trees near building	Deciduous trees near building. Evergreens for windbreaks
Road orientation	Crosswise to winter wind	Narrow; E-W axis	Broad channel, E-W axis	Crosswise to winter wind
Materials coloration	Medium	Light on exposed surfaces, dark to avoid reflection	Light, especially for roof	Medium to dark

Site Analysis Process and Techniques

Introduction

An understanding of the basic process of site selection, planning and design methodology is essential in understanding both the positive and negative effects of siting and site development on the utilization of solar radiation and in energy conservation. In the process is the key to optimum decisions which may not have to be rectified at a later time, either through site design, architecture, or the use of extensive heating, ventilating, or air conditioning equipment.

Rudolph Geiger has stated, in his book entitled *The Climate Near the Ground,* "Radiation is undoubtedly the most important of all meterologic elements. It is the source of power that drives the atmospheric circulation, the only means of exchange of energy between the earth and the rest of the universe and the basis for organizing our daily lives."

It is essential for designers, developers and administrators to understand the site analysis process and the techniques involved. Many times the process or making visible the process may be the major product of the environmental designer's input. It is necessary to develop an objective or rational approach before making any decision concerning either the site selection, building orientation, or site planning for development. It is essential to establish an objective analytical technique for gathering data or for arriving at solutions which are appropriate for the site, the climate, the architecture and the development and enhancement of the site itself after the architecture is completed.

It is necessary to ask the right questions and make the right decisions as early in the process as is possible. This, of source, encourages the use of competent environmental design professionals, such as landscape architects, at the outset of the design process, not after many of the basic decisions have been made so that it is necessary for the environmental design professional to rectify problems in the site planning or design process.

It is also possible, through a proper understanding of the site planning process itself, to establish clearly the focus or scale of control. Decisions may be made at any step along the line in the process which will influence the ability to utilize solar energy. The sooner solar radiation and climate control are recognized in the process, the more solar energy will be utilized and the more the resultant energy will be conserved. The proper decisions should be made as soon as possible in the process. In order to do that, the entire process must be comprehended by the designer, the developer, the administrator, the manager, and the user.

There are certain site-mechanical tradeoffs and it is less expensive ultimately to insure proper preliminary planning and obviate the need for a sensitive and expensive architectural or mechanical correction:

... proper gross site selection prevents problems in discrete site selection;

... proper discrete site selection prevents problems in site planning;

... proper site planning prevents problems in site design and detailing;

... proper site design prevents problems in architectural design, and

... proper architectural design prevents problems in the mechanical systems utilized in heating, ventilating and air conditioning.

The process starts at the region and ends at the individual's proximate environment inside of the structure.

The higher cost of energy entails a rethinking of the cost of preliminary planning as opposed to energy-intensive architectural and mechanical modifications to correct errors in site selection and orientation in placement planning or design. Obviously it is advisable to control the climatic factors of an individual building at as large a scale and as soon in the process as possible. Optimum siting and site planning may be more important and, in most cases, less expensive than architectural or mechanical solutions in solar radiation utilization and in energy conservation.

All of the site planning process is a matter of scale. It is a case of problem-solving objectively at a variety of scales. The same objective process is applied to specific problems at increasingly smaller scales.

Purposes

The purposes of the site analysis process are many, but need to be restated, to accentuate the importance of not only utilizing the process but making it visible and insuring its usage to fully understand the client, the site, the problem, and the potential optimum solution.

1. Natural Processes as Form Determinanants

There are, in any problem, project and site wide varieties of determinants for the final, actual form of the architecture and the site development. Among these potential form determinants are technology, economics, materials, social factors, political factors, aesthetics, and regional considerations. Depending upon the specific project or situation, some of these may have greater or lesser particular impact at the specific place and time. All of these form determinants, however, are imposed on a natural form which has inherent within itself a series of form determinants. Among these are the geology, the ecology, the climate, the soils and the inherent aesthetic characteristics of the site itself.

Louis Kahn, the late well-known architect, developed a philosophy of "volitional will." This concept basically had to do with analysis to find what a particular thing "wanted to be." Inherent in the concept is the idea that within the problem, the site or the client is the germ of the optimum solution in a particular situation. This, of course, requires careful and studious analysis to find very specifically what a particular site wants to be. There is, at the same time, an extremely high environmental price in ignoring what a particular thing or particular site wants to be and how it can and will allow itself to be altered or amended.

2. Natural Constraints or Barriers

There are, in any particular site, a series of natural barriers or constraints to what we want to do or think should be done. Among these constraints may be climate, geology, soils, vegetations, or a special ecological system. To arbitrarily destroy or disrupy any part of this complex system without fully understanding it may be not only extremely dangerous, but inordinately expensive in a particular situation. In effect, it is not wise to "fight Mother Nature" and to do it may, in effect, be "shovelling against the tide" and may require a great deal of remedial expense and energy at a later time for rectification through architectural or mechanical systems.

3. Cost/Benefit Site Selection and Planning

The cost of proper initial planning at the appropriate stage may be less expensive than the remedial or retrofit activity needed to correct the oversight for lack of adequate preliminary analysis. What appears to be less expensive initially may cost more in the long run. Various tradeoffs may be discovered in the process of site or problem analysis which may enable the environmental designer, the owner, or manager or administrator to understand more fully the various costs and benefits of a variety of options in decisions, options and hidden costs and benefits deriving from a variety of decisions.

4. Site Selection, Orientation and Activity Placement

Proper site and problem analysis will insure the right thing or the right activity will be placed eventually in the right specific location on a particular site. There is a certain logic as to where each item or activity should be placed which can be discovered in the course of a proper analysis process. At the same time, the logic of how much environmental control is needed on a particular site may be ascertained in the course of detailed analysis. The parameters and the framework may be objectively determined in the course of the analysis which would insure the selection of the proper site and the proper locations for activities on a precise site as well as the orientation for the building for outdoor activities and areas. It is possible to interpret the basic analytic information in a variety of ways. One of them is obviously to weight heavily the utilization of solar radiation. Obviously, another possible area which could be weighted heavily is the development of the site for energy conservation.

5. Life Cycle Costing

It is possible to use site analysis data as a basis for life cycle costing studies. The basic question which may be asked is: "if you do this in this precise way, what will be the total cost of the project twenty years from now?" Front end planning costs may have to be more with $11.00 per barrel oil than it was with $3 per barrel oil. It may be necessary to spend more time planning to utilize the site and its elements to the fullest for solar energy utilization and for energy conservation. The increased costs of planning and study as measured against the savings in energy or the ability to use solar energy coupled with the increased initial costs of roads and utilities or construction may be measured against the lower costs of heating or cooling over the life cycle of the building itself. These total costs of the life cycle of a specific building and its site element may be carefully, objectively and dispassionately ascertained in a careful analytical process.

Summary

A careful, rational, analytical process in the selection, design and development of a site is extremely important:

> . . . it ensures omplete input;

> . . . it ensures objectivity;

> . . . it requires data gathering;

> . . . it allows tradeoffs and evaluations to be made;

> . . . it gives a framework for decisionmaking; and

> . . . it is a skeleton for form giving.

The environmental designer must ask what is given and what he can control and objectively understand the process in order to know he can recommend a variety of optional decisions and to what extent each of these decisions may at various times control the process, the cost and the effectiveness of the site and the eventual building.

Techniques

There are certain basic techniques which are involved in data collection and data depiction as a part of the overall process or methodology of site design for any particular purpose. The need for detailed and precise data collection is even more important in gathering and illustrating basic information in regard to site planning for solar energy utilization and for energy conservation. This is true since so often the entire process of information collection and illustration is taken for granted by the site planner and insufficient care or understanding is shown in the distinctive problems of site planning for solar radiation utilization or energy conservation. It is all the

more important to re-emphasize the need for utilization of new resources as well as for the collection of new types of data and of newer and more efficient methods of depiction or illustration of this data for fuller utilization of solar energy.

1. Data Collection

Data collection may be of two types. These are: (1) *primary*—where the material is gathered directly by the user of the information and (2) *secondary*—where it is necessary to gather information data or material from other primary agencies, companies and individuals who collect or gather the basic raw primary data.

Much of the data collection necessary in site planning for solar energy utilization is being done by a great variety of agencies and companies. The material which is available can be used for any number of interests for a variety of purposes. However, certain techniques, information or materials have more relevance or pertinence to knowing, feeling, and understanding the site.

There are three basic methods of data collection. These are two forms of remote sensing and one of direct data gathering. Much information is gathered not by the person who ultimately will use it but by others for a variety of reasons. This information and data is then stored and is regathered and interpreted by the site planner or the environmental designer wishing to deal with a specific site or project. Some of the newer technologies of remote sensing have supplemented the old surveying and mapping techniques which have traditionally been used for the gathering of information, material or data. The earliest and most elementary means of data collection is, of course, on-site gathering of information or materials concerning the site, for its potential users.

a. Remote Sensing

A contemporary technology of remote sensing has assisted the environmental designer in a variety of ways by providing new techniques and methodologies for gathering data from afar without walking or touching a particular site. Much of this technology is less expensive, provides wider coverage, and faster means of collecting greater depth of material and information.

(1) *Air Photo Interpretation:* This is substantially the most rudimentary means of remote sensing or data gathering. Currently there is much extensive data or information available at a variety of scales and prices. The advent of stereo pairs, and more sophisticated single frame photography, has assisted in providing more accuracy and greater detail in remote sensing information. Aerial photography is the least generalized and sophisticated of the contemporary remote sensing techniques. Often, a particular site may be covered with clouds or with vegetation, thus handicapping a full view of a particular site or area at a specific time of year. The advantages of aerial photographs are that they show vegetation, cultural features and landforms.

(2) *Radar Imagery:* Of relatively more sophistication is radar imagery which provides the view of a specific site or specific area by cutting through cloud cover and other impediments to reveal the essential geological and geographical elements of a specific land area. The radar imagery data does not always show vegetation to the optimum degree. It does, however, have certain advantages which can be explored not to the exclusion of other methods of remote sensing, but to supplement data gathered from other sources.

(3) *Infrared Photography:*

a. *Black and White*—black and white infrared photography provides a photography with distorted value system. It does reveal certain features not otherwise observable by cutting through cloud cover, smog, fog, and haze. It also reveals vegetation during the growing season in a very light color with land areas showing much darker, though in a clearly differentiated way.

b. *Color Photography*—this provides an aerial photography with a distorted color spectrum. Color infrared photography reveals sources of heat, temperature variations, as well as a variety of forms of vegetation and cultural elements.

(4) *Multi Spectral Scanning:* In this technique of remote sensing for data gathering it is possible to have a variety of photographs on different visual bands, all taken at one particular time. Therefore, it is possible, through interpretation, to extrapolate a wide range of information concerning a specific area or site at the same time.

(5) *Satellite Technology:* Contemporary remote sensing has benefitted greatly from the earth resources satellite program under the ERTS and the EROS (Earth Resources Orbiting Satellite) programs, developed jointly by NASA and the U.S. Geological Survey of the Department of the Interior. This satellite technology is able to provide greater sophistication at a variety of scales of specific areas or sites. At the same time sensor satellites are continually orbiting the earth, it is possible to provide sequential and up-to-date information showing changes in vegetation patterns, land use, climate, or other factors relating to specific areas under review.

b. Utilization of Map and Survey Information

There is usually, for any given site, or for any specific area, a variety of levels of map information. The most commonly available map, of course, shows topography as well as cultural features and is usually a U.S. Geological Survey Map. Quite often as well local communities have topography maps of the particular area. Specific sites may have more detailed mapping information which has been prepared either from computer mapping operations or from satellite data. It is also usually advisable or necessary to have very detailed site or site surveys for each of the potential sites being con~idered so that the environmental designer, client and administrator involved in the decision-making may have available the latest and the most accurate types and degrees of information.

c. On-Site Survey

It may be necessary, in addition, to have very specific site surveys conducted by civil engineering or surveying firms for a specific site or specific sites if they are not already available. In addition, it is advisable for the individual designer to conduct detailed on-site review and analysis. There are a variety of reasons for this, but usually it is for the designer himself to discover the *genius loci.* As mentioned previously, each site or specific area has a "volitional will" of its own. There is something which it "wants to be" and there are places where things, activities or structures want to go on a specific site. It is necessary in such a visual survey to review the slope, the orientation, the vegetation, the ecology, the soils and the geology to the extent possible. It is also necessary for checking and ascertaining the reliability of other gathered material. The information contained on the topographic and vegetation surveys can be reviewed in greater detail by the site designer. The on-site physical survey by the designer is not necessarily the best nor the least expensive way of obtaining information, but it is a good check on information obtained from other sources and does enable the designer to gain a better grasp of the site and its inherent characteristics.

2. Data Depiction

It is necessary in the analytical process to extrapolate from all of the available data that information or material which pertains or relates to a specific site or series of sites under consideration. It is also necessary to make visible information concerning the client, the site and the specific project under consideration. This is desirable, advisable and necessary for a variety of reasons. First of all, when it is depicted in a form suitable for easy comprehension, it is also much more easily interpreted, manipulated and communicated. Obviously, all of the mass of basic information relating to the specific project or problem under consideration must be selectively culled in order to provide, in a rather uniform and graphic format, the basic information on which future decisions will be made. There are, of course, a wide variety of ways of depicting such information graphically. The environmental designer must, in each situation, evaluate the cost/benefits of various ways of communicating this sort of information. He must evaluate his resources, time frame, and the availability of technology or manpower necessary in order to weigh the advantages or disadvantages of any specific or particular means of data depiction. The time, the cost, the readability and the resources available will determine to a large extent the ultimate need of conveying information and the degree of detail of information graphically portrayed or illustrated. Various techniques are well known to most envir-

onmental designers. It seems appropriate to review the most commonly used methods of analytical data depiction.

a. Manual Mapping

This is usually the use of pencil, magic marker, pen and ink or some other graphic means to depict or display information, ranging all the way from contour or topography maps to depiction of soils, vegetation, slope, climatic factors, cultural elements. This is a very traditional method, easily understood by a wide variety of viewers. It is also extremely labor-intensive, but does require very simple technology, simple tools, and is able to be done by the designer or his staff. Information storage, on the other hand, is very difficult since it requires the storage of plans, or maps, in a raw form.

b. Overlay Technique

A familiar technique developed and utilized to a greater extent during the late 1950's, early 1960's, was the depiction of each major factor in the site or problem analysis, depicted on a separate sheet, preferably of clear or frosted acetate sheets. In this way, each analytic factor, item of information or area of concern is isolated. As these overlay sheets are superimposed one upon the other, the composite is easily done and shown. It is necessary to use either various colors or texture patterns for each specific subject or area. This methodology is extremely labor intensive; it also provides difficult storage problems and entails a certain amount of difficulty to explain and understand for a lay audience.

c. Computer Modeling

With the increasing sophistication of computer technology, the computer printer or plotter has been used to depict certain basic information or data utilized by the environmental designer. This may be through the use of generally accepted numbers and letters or it may be by utilization of a drum or flat-bed plotters. At the same time, the basic or raw information is stored in the computer itself. There are certain advantages of a system of this type, having primarily to do with speed of depiction and ease of storage. It is also possible to manipulate the stored information in a variety of ways and utilizing the basic raw data it is possible to do predictive modeling based on various optional directions. At the same time, one of the disadvantages of use of this system has to do with the psychological distrust of the computer by many of the lay public, though it is widely used and accepted by other design professionals. It requires a high degree of technology and the necessary hardware to be readily available inexpensively for use by the environmental designer. It also requires the continued use of the printer and the terminal for the storage of the information in a readily retrievable format.

d. Three-Dimensional Modeling

Oftentimes it is possible to depict the basic information concerning a specific site or a series of optional sites through the use of a three-dimensional chipboard or styrofoam model. These models used for illustrating the basic information have the advantages of being extremely visual in a three-dimensional way. They also have the ability to be used to project with the slide projector or other mechanical device some of the basic inventory data onto the model itself. Information presented on a three-dimensional model is often quite easily understood, explained and shown. It is, on the other hand, extremely difficult to store or transform into other materials since it is bulky. It also has the disadvantage of being rather rigid and not subject to easy or inexpensive manipulation of the data on the model itself.

There are, of course, a variety of other methods of data depiction, ranging from the use of the cathode ray tube to video tapes, movies and slide presentations, requiring varying degrees of technology, machinery, equipment and expertise. These four mentioned in some detail, however, illustrate the most commonly used techniques and methods for analytical or design data depiction.

Process or Methodology

There is a basic, rational objective method for approaching or organizing design problems. This process is no different in the development of the optimum site arrangement for the utilization of solar energy and for energy conservation. This design process is always important in solar energy utilization since an improper organization or weighting of factors may change significantly the cost of the entire project during its useable lifetime. The designer's analytic process has, as its purpose, the finding, in effect, of the specific distinctiveness of a particular client, site, or problem. The same process will be utilized at a variety of scales for various purposes, but a different weighting of the input factors will be made, depending upon the purpose or the end product desired or needed. Basically, all design methodology is oriented around the particular scale of design problem being observed. It is problem solving objectivity at a variety of scales. The same objective process may be applied to a wide variety of scales. When energy was cheap, any mistakes in preliminary planning could be offset, either architecturally or mechanically, at a relatively minor cost. When energy is expensive, to rectify the mistakes in planning is prohibitive; therefore, the cost of initial planning may be more expensive and certainly more important.

The suggested steps in a site analysis of either one or a series of sites which might be considered for energy conservation or solar energy utilization purposes would include many traditional steps with new emphasis but should cover the following traditional steps:

1. Establishment of Program

At the outset of any design process it is imperative to clearly formulate and state the needs, wants, and constraints of a particular client, problem, site, or project. It should also be determined when each of the elements are needed and where—who needs what information and in what form—what is it that the client and the designer want to do—when do they want to do it—why—and where? This is the stage of development of a problem statement, a problem synthesization, distillation and explanation of what should be done.

2. Research

It is often misunderstood by environmental designers that in the design methodology, research may be any number of efforts of gathering background data; included in this would be information and material that was specific to the precise client to site as well as information which is also tangentially related to the problem being undertaken. Included in the research in this context would be:

    a. data collection;

    b. resource inventory:
      1) pertinent people with expertise in the particular subject;
      2) professional offices engaged in related activity;
      3) agencies involved in similar or related projects;
    c. literature search for published material on the subject;

    d. review of previous related studies; and

    e. structuring of gathered or available information.

The bibliography at the end of this book gives a number of sources of information and material concerning site planning for solar energy utilization. Undoubtedly, in the next few years, there will be many more documents, publications, reports and studies available on this particular subject.

3. Program Analysis

It is necessary to analyze the client and his particular needs for architecture and site development prior to formulation of either a program or a final design. It is necessary to elaborate the needs and the desires of the client, we whether this is an individual or a corporate client, and to seperate these into the fixed and the variable needs with an eye toward maximization of the site and the building for solar energy utilization. In the program analysis, it is advisable to isolate and sepe
is advisable to isolate and separate the functions and the interactions desired on the site as they relate to the areas inside of the future architecture. It is also desirable to provide in the initial program analysis for growth and expansion of the architecture and for various functions on the site. It is also helpful to analyze carefully the necessary circulation which will take place on the site. This will include the primary entrance to the site and the provision for service facilities. By diagramatically dealing with the various activities to take place in the building and on the site, it is possible for the design-

er to deal with the givens of function, interaction and circulation in the program analysis. This diagram can then be related to the potentials and constraints discovered during the course of the site analysis. It could also be of help, during the program analysis, to isolate the activities, elements or areas which would make requirements upon the site or on various areas of the site and their relative size and optimum configuration. These, then, could be placed on the site conceptually before actually relating them to the site consider
conditions after an optimum site analysis.

4. Site Analysis

It is essential to understand fully and completely the site itself and the factors acting upon it prior to disturbing it or placing objects or accoutrements into its delicate natural balance and framework. During the program analysis stage, a review should have been made of the functions which are to take place on the proposed site. Within the site itself there are a variety of form determinants. As mentioned previously, the geology, soils, ecology, vegetation, and orientation are all form determinants. The designer must find what will enhance the environment the most and cause the least disruption of the delicate natural balance between soil, climate, vegetation and water while utilizing to the maximum the climatic potential of the site. In some cases, climatic factors will act as locational or relational constraints. The site analysis is an integral part in the overall design methodology. It entails a step-by-step analysis of the site, its climate and existing cultural elements with an increased emphasis on optimum solar energy utilization and minimum energy loss or dissipation. Various facets or aspects which normally are utilized in site analysis are as follows:

a. Slope

It is necessary to use air photos, contour maps or other sources to determine the angle, degree or steepness of slope throughout the site. It is also necessary to analyze carefully the direction and orientation of each of the slopes. All of this is done with an eye toward the constraints embodied in the slopes on the site. It is also done to determine the optimum and marginal uses of various slopes on the site as they now exist.

b. Geology

It is essential to analyze the site to determine the underlying rock masses as well as the geological history of the site, including the cycles of upheaval and deposition. It is advisable to determine the layers of rock beneath the site and the depth of each of these layers as well as the history of the soil formation and the types of soild, at least as they relate to the geological history of the area. Of a great deal of use and interest in any site analysis is an enumeration of any geological constraints to building in a specific area on a site as well as the potential materials already located on the site which may be used in the architectural or landscape architectural construction. This survey is appropriately done during this phase of a site analysis.

c. Ecology

On nearly every site, if possible, it is essential to study the indigenous and disrupted ecology before removing any vegetation from any part of the site. It is essential to know and understand where vegetation on that particular part of the site stands in the succession cycle as the vegetation moves toward climax. It is also essential to understand the dominance of either the indigenous or the introduced vegetational varieties on the site. Careful mapping of the vegetational zones is extremely helpful before the removal or the replacement of any vegetation on the site for any purpose. A review of the effects of clearing and the development of optimum "band-aid" planting to heal scars if the natural vegetation is removed from that specific area, is also essential. Understanding, analyzing, and completely revealing the natural ecology is vital for the full utilization of the site for solar energy exploitation and conservation.

d. Soil

Site analysis must cover the history of the development of the soil as well as an overview of the usage, the destruction and the depletion of the soil on the particular site. It is also desirable to develop maps showing optimum use for zones, types or areas of soils on the site. Both the engineering and agricultural potentials of the soild should be mapped and indexed.

e. Vegetation

It is essential to analyze, index and map the existing vegetation, both by areas and by individual specimens, if possible. The age, condition and utility of the existing vegetation should be known before determining those particular areas or plants to retain and those which are expendable. An understanding of the relationship between native and introduced vegetation is essential in expressing the ecology and in suggesting possible replacements for existing trees, shrubs and ground covers.

f. Climate

Obviously, climatic analysis is of extreme importance in solar energy utilization and energy conservation programs. Therefore, it is essential to analyze very carefully the precise climate of the site, the area, and the region. Each site will be located in some part of one of the major climatic regions. These regions, obviously, are very generalized, no matter how many divisions into which they divide the nation. The macroclimate of a particular site and the region must be clearly characterized and understood in any site analysis during the preliminary stages of climatic analysis. Next a review of the mesoclimatic characteristics of the site within the larger region will reveal some of the subleties of climatic conditions distinctive to this part of the larger region or area. Finally detailed microclimatic analysis of a specific site or a series of sites may provide insight into the optimum site, location or orientation or functions or activities from a climatic, energy conservation or solar energy utilization standpoint.

Each site has a precise microclimate; each section or area of a larger site has, as well, a more defined microclimate which must be analyzed carefully by the environmental designer. As indicated elsewhere in this publication, each slope, valley, and exposed hill of any particular site has a precise, discrete and finite microclimate which will affect greatly the ability to utilize various portions of site for architectural or landscape architectural purposes. Included in the climatic analysis is the review of the sun movement and passage over the site. The development of the sun chart for a specific site or a series of alternate sites may be advisable. These sun charts would show the direction of the sun at various seasons of the year and times of the day. Also in such an analysis would be the review of the amount of sunlight received and the degree or percent of cloud cover at various seasons of the year on specific sites. The degree of exposure of various slopes and areas of the site as well as the existence of natural sun pockets should also be made visible during this stage of the analysis. The precipitation reaching the site in amount, type and season should also be reviewed, since it has an important effect on the microclimate of a site. The resultant temperatures in various areas and at precise points on the site should also be reviewed in such an analysis.

A landscape designer, either amateur or professional is primarily interested in the elements of climate which affect human comfort and the use of either the buildings or the parts of the landscape. This is especially true if one of the primary aims of the site development is either energy conservation or greater solar energy utilization. The information a landscape designer needs or would like to know includes averages, changes and extremes of temperature, the temperature differences between day and night (diurnal range), humidity, amount and type of incoming and outgoing radiation, direction and force of air movements, snowfall and its distribution, sky conditions and special conditions such as extreme pollution levels, temperature inversions, hurricanes or severe storm directions, patterns or frequency.

The following is a cursory outline of the climatic factors which should be formed during this stage of the site analysis.

Climatically protected areas on the site

- areas protected at certain times of the day or year
- areas protected by topography
- areas protected by vegetation

Climatically exposed locations on the site

- areas exposed to sun or wind

- areas exposed primarily in winter
- areas exposed primarily in summer
- areas exposed all seasons of the year

Solar radiation patterns on the site

- daily and monthly
- seasonal
- impediments (e.g., vegetation which may cover the site or shadow certain areas on the site)

Wind patterns on the site

- daily and monthly
- seasonal
- impediments (e.g., thick vegetation or underbrush which may block air movement on or through the site)

Precipitation patterns on the site

- fog movement, collection or propensity patterns
- snow drift and collection patterns
- frost "pockets"

Temperature patterns on the site

- daily and monthly
- seasonal
- warm areas
- cold areas

Water or drainage patterns on or across the site

- seasonal air or water flow patterns
- daily air or water flow patterns
- existing or natural impediments to air or water flow patterns

Climatic data is gathered for major factors such as amount of sunshine, air temperature, humidity, wind direction and intensity and precipitation (type and amount) at airports and meteorlogical stations by the National Weather Service. The information is generally not collected specifically for use by landscape designers and may omit some data relevant to efficient energy conservation and solar energy utilization. Therefore, it is often necessary to supplement published data with information obtained directly from the meteorological station. In many cases, however, the most frequently used climatic data is organized into helpful design manuals such as the *Handbook of Fundamentals* published by the American Society of Heating, Refrigeration, and Air Conditioning Engineers, the National Association of Home Builders *Institutional Manuals* or the U.S. Climatic Atlas published by the Government Printing Office. The use of these and other climatic design tools eliminates the need for painstaking analysis of unedited climatic data. The careful analysis of climatic elements and factors will identify the local climatic features which are potentially beneficial or detrimental to human comfort and energy conservative site design.

An excellent example of a climatic analysis undertaken for a specific project is that undertaken by the Office of Wallace, McHarg, Roberts and Todd, for the Woodland New Community Site north of Houston, Texas. The brochure concerning this information spoke of the site in the following words:

"The site lies within the humid sub-tropical belt which extends northward from the Gulf of Mexico. Rainfall is abundant and fairly evenly distributed throughout the year. On an average, total monthly rainfall exceeds four inches. Storm intensities of 4.3" rainfall in an hour can be expected to occur. Snowfall is rare and none has been recorded for several winters. Because of nearness to the Gulf, the prevailing winds are from the southeast and south. In January, frequent passages of high pressure areas bring invasions of polar air. Prevailing winds during this period are from the north. The average wind speed ranges from 8 to 15 mph through the year, with daily fast wind speeds of 30-40 mph. During the fall storm season daily wind speeds of 55 mph are not uncommon.

The area is subject to intensive storm activity during the fall season. Extensive flooding results, which is often aggravated by obstruction of natural drainage channels. There is great disparity between normal runoff and storm flood, and storms are generally accompanied by high winds. Thinning of the forest will increase the wind throw hazard, to which shallow rooted trees are especially vulnerable. Whenever possible forests should be left in solid stands or thinned selectively after detailed examination of the existing trees

Refrigerated air conditioning can alleviate the discomfort caused by hot humid summer days. This high cost can be reduced if care is taken in locating development and controlling its form. Clearings in the forest will

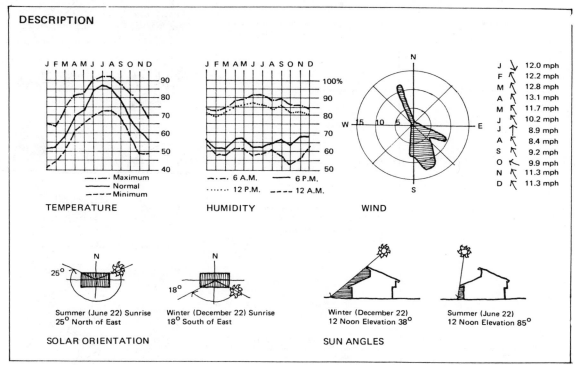

DESCRIPTION

TEMPERATURE

— · — Maximum
——— Normal
- - - Minimum

HUMIDITY

— · — 6 A.M.   ——— 6 P.M.
······ 12 P.M.   - - - 12 A.M.

WIND

J	12.0 mph
F	12.2 mph
M	12.8 mph
A	13.1 mph
M	11.7 mph
J	10.2 mph
J	8.9 mph
A	8.4 mph
S	9.2 mph
O	9.9 mph
N	11.3 mph
D	11.3 mph

Summer (June 22) Sunrise
25° North of East

Winter (December 22) Sunrise
18° South of East

SOLAR ORIENTATION

Winter (December 22)
12 Noon Elevation 38°

Summer (June 22)
12 Noon Elevation 85°

SUN ANGLES

184

The region within which the site is located favors the development of both ground and advective fogs. On an average of 16 days of heavy fog and 62 days of light fog can be expected every year. The short spells of freezing temperature occur between December 11th and February 15th, producing an average growing season of 270 days.

increase both the wind speed and the incidence of solar radiation. Alternative choices for forest clearings of variable shape, size, and orientation have been tested. The results show what the expected increase in wind intensity of 20% is negated by higher solar insolation resulting from removal of forest. Selected clearing and retention of some shade, however, is desirable."

OBJECTIVES	ADAPTATIONS	
Reduce heat absorption in summer.	Orientation along east-west axis minimizes overheating.	
	Hip roof reduces heat load and provides insulating space.	
	A roof moderately pitched toward the north intercepts less radiation than a flat roof or dome.	
	A double roof reduces reradiation to the interior.	
	The ground cover adjacent to the structure should have a low albedo and relatively high conductivity to prevent reradiation and reflective heating.	paving planting

185

The same report included the following climatic summary, graphically depicted:

SUMMARY: CLIMATE

OBJECTIVES

Maximize shade in summer.

Maximize sun in winter.

ADAPTATIONS

Locate structure in deciduous woods or plant deciduous trees for summer shade and winter sun.

In summer the structure should be shaded in the late morning and afternoon.

morning   afternoon

Frequently used outdoor spaces should be shaded from late morning and afternoon sun in summer.

Pedestrian circulation routes should be shaded by vegetation, canopies, pergolas, or arcades.

Deciduous trees should be located on the south side of pedestrian paths to allow winter sun.

Increase comfort and reduce humidity by maximizing summer ventilation.

Protect from storm winds.

Orient structures and outdoor spaces to capture summer breezes and protect them from storm winds.

Deciduous trees with open understory permit the passage of cooling breezes.

Evergreen trees with dense understory offer protection from storm winds.

A preferred location for development is N—NW of a predominantly hardwood forest with open understory and S—SE of a dense predominantly pine forest.

SUMMARY: CLIMATE

OBJECTIVES	ADAPTATIONS	
Increase comfort and reduce humidity by maximizing summer ventilation. Protect from storm winds.	Clearings oriented NE to SW should have a cross section 5-10 times the canopy height.	
	Orient cleared or planted corridors of vegetation to increase wind velocity.	
	Roadways can act as wind corridors, and if heavily shaded can be used to cool residential areas.	
	Hedges should be placed at a distance from buildings so as not to block air movement or used as a wind funnel.	
	Capture of breezes over shaded lawns should be maximized. Parking areas should be located downwind from houses or recreation areas.	
	Siting of housing clusters should respond to desired air movement patterns.	
	Rectangular and circular buildings are preferred over irregular forms with greater wind exposure.	
	Slightly elevated structures are preferred to avoid dampness.	
	Internal organization of structure should allow free cross ventilation.	
	Overhangs can be used to channel breezes into structure.	
	Adjustable louvres and screens will block solar radiation and storm winds and allow passage of cooling breezes.	

OBJECTIVES

ADAPTATIONS

Increase comfort and reduce humidity by maximizing summer ventilation.

Protect from storm winds.

The optimum position for ventilation openings is the lower half of the wall on the windward side. The leeward vent location can be variable.

If recreation areas or paths are located adjacent to trees with an open understory and next to a heat generating land use, air movement will be increased as cool air replaces the rising warm air.

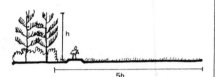

To protect paths from storm winds, they should be located on the leeward side of the forest edge within a quiet zone equal to five times the height of the canopy.

The recreation system should include "breeze towers" that enable the user to appreciate the site while enjoying the more intense air movement at upper elevations.

When larger buildings or groups of buildings are required—as in community facilities and recreation areas—they should be open and loosely connected with free passage for breezes and shaded pedestrian walks.

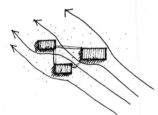

The internal organization of uses should respond to the time sequence of activities in relation to solar radiation and to factors of wind direction and humidity.

Non-living areas of the house (garage, storage areas, laundry) can be used to buffer living areas from NNW storm winds or undesired solar radiation from S and SW.

Heat and moisture producing areas of the house should be ventilated and separated from the rest of the structure.

	N	S	E	W
Living		●	●	
Dining		●	●	
Kitchen			●	●
Sleep				
Storage	●			
Laundry	●			●
Garage	●			●
Outdoor Space	●		●	●

Two drawings developed by Richard P. Browne Associates of Columbia, Maryland show the "winter thermal regions" and the "winter winds, site exposure" for a specific property in Montgomery County, Maryland. These show in a very clear way those areas of the site which are exposed or protected and the warm and cold slopes, plateaus and valleys on this particular tract of land.

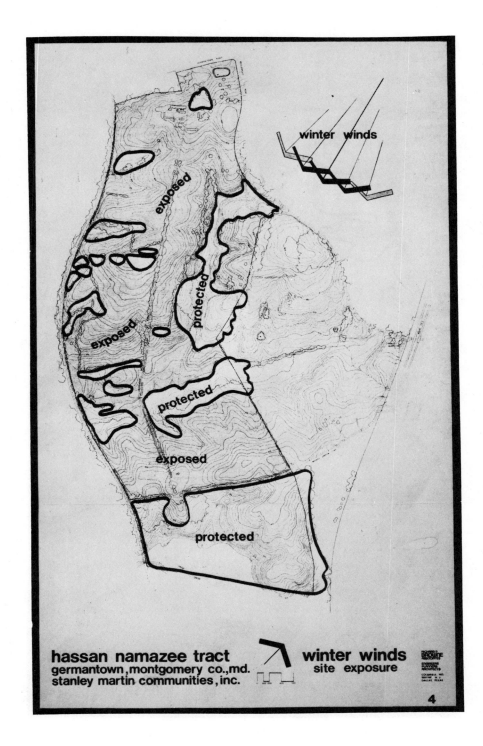

hassan namazee tract
germantown, montgomery co., md.
stanley martin communities, inc.

winter winds
site exposure

4

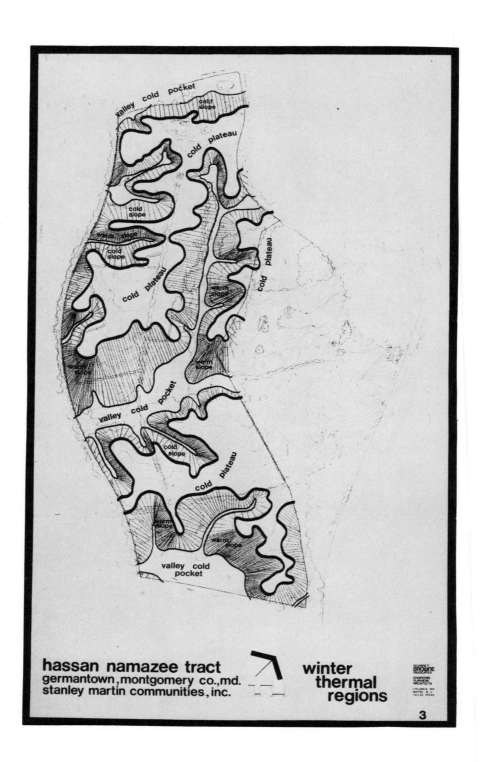

hassan namazee tract
germantown,montgomery co.,md.
stanley martin communities, inc.

winter thermal regions

3

189

The publication "Environmental Characteristics Planning; An Alternative Approach to Physical Planning", prepared by Mr. Jacob Kaminsky for the Regional Planning Council of Baltimore, Maryland utilizes a unique technique to indicate the various slope orientations on a particular site and the climatic implications of these orientations. The following test and illustrations explain more fully the analysis process and what is able to be indicated graphically.

The planning area was divided into eight categories on the basis of slope orientation. In the physiographic summary for the planting area, the eight groups were combined into two categories; warm slopes, including eastern, southeastern, southern and southwestern; and cold slopes, including northern, northeastern, western and northwestern. The map opposite shows slope orientation in the case study area.

The micro-climate of areas with different slope orientations differ depending upon the effects of solar radiation and wind direction. Eastern and southern slopes provide better habitats for people and plants since they receive more solar heat in the winter and cooler breezes in the summer. Northern and western slopes, on the other hand, receive less solar heat and more cold wind in the winter. Development plans prepared on the basis of the variations in micro-climate of different slopes could result in significant reductions in heating and cooling costs, as well as conserving energy. Energy conservation is a critical need today and will be even more so in the future, therefore, every effort should be made to conserve energy. For these reasons it is suggested that the higher intensity development be located on the southern slopes and less intensive development on northern slopes.

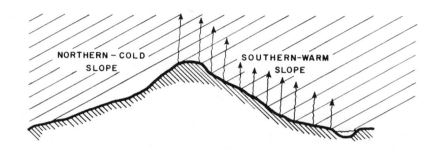

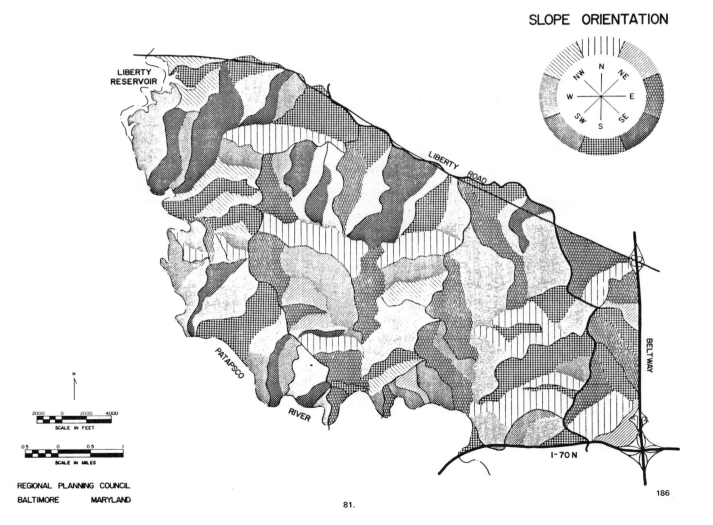

SLOPE ORIENTATION

REGIONAL PLANNING COUNCIL
BALTIMORE   MARYLAND

186

Site analysis information may be depicted or shown graphically in a variety of ways ranging from a very pictoral delineation of optimum areas or locations on an aerial perspective of a specific site to a stylized or realistic indication on a plan of the site or sites. The following illustration from "Planning the Garden" by Robert B. Deering of the University of California at Davis appeared in *The Art of Home Landscaping,* by Garrett Eckbo shows in some detail the considerations given to local variations in climate on two different areas.

Consider direction of wind.

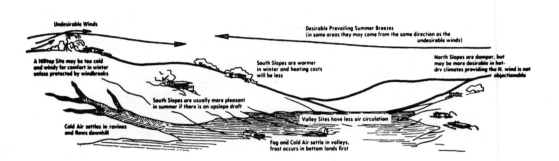

Best location for an inland home site is generally part way up a slope that faces the desirable direction. This will depend upon prevailing wind, sun, and the views.

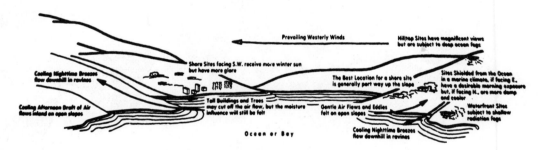

The combination of topography and exposure may determine the desirability of a coastal site. Generally, best location for a shore site is part way up a slope.

176

The process of site analysis and site planning applied to a residential property is illustrated in some detail in *The Art of Home Landscaping,* by Garrett Eckbo in the following drawings.

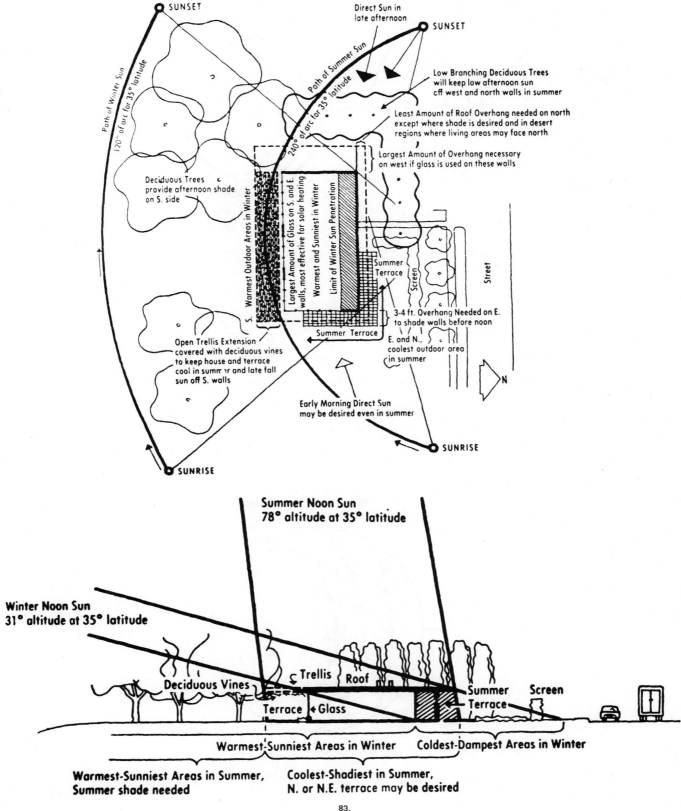

The solar angles for a specific site may be calculated and shown in a variety of ways. At least three of these, which are relatively easy to depict, are shown in the publication *Solar Dwelling Design Concepts* as follows:

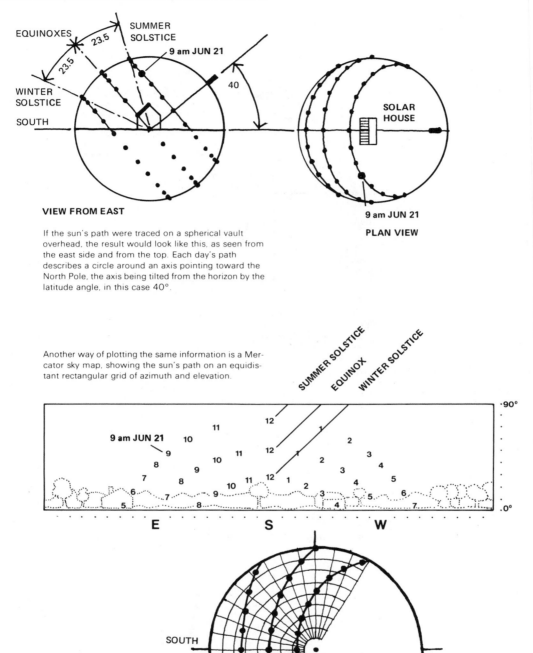

VIEW FROM EAST

PLAN VIEW

9 am JUN 21

If the sun's path were traced on a spherical vault overhead, the result would look like this, as seen from the east side and from the top. Each day's path describes a circle around an axis pointing toward the North Pole, the axis being tilted from the horizon by the latitude angle, in this case 40°.

Another way of plotting the same information is a Mercator sky map, showing the sun's path on an equidistant rectangular grid of azimuth and elevation.

A Sun Path Diagram, such as this one for latitude 40 North, is an equidistant circular graphic "map" of the azimuth (compass direction) and elevation (angle from the horizon) of the sun at all hours of the day for each day of the year.

84.

An effective, though time consuming method of showing solar angles is
through the use of a perspective drawing of a particular site with the seasonal and daily sun pattern shown as indicated on the following drawing
from the Jack Kramer book on greenhouse development. In this way it is
possible to calculate, to see and to show where the sun will shine and where
it will cast a shadow, depending upon obstructions, anywhere on the site.

SOLAR ANGLES

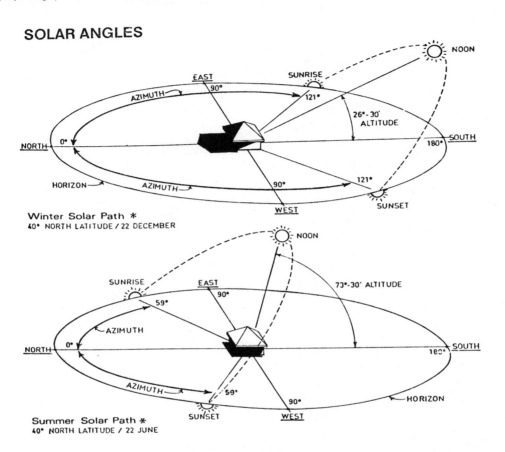

Winter Solar Path ✳
40° NORTH LATITUDE / 22 DECEMBER

Summer Solar Path ✳
40° NORTH LATITUDE / 22 JUNE

LATITUDE	SEASON	SUNRISE	SUNSET	AZIMUTH	ALTITUDE
50°	WINTER	8:00	4:00	128°-30'	16°-30'
	SUMMER	4:00	8:00	51°-30'	63°-30'
45°	WINTER	7:40	4:20	124°-30'	21°-30'
	SUMMER	4:20	7:40	55°-30'	68°-30'
✳ 40°	WINTER	7:30	4:30	121°-0'	26°-30'
	SUMMER	4:30	7:30	59°-0'	73°-30'

LATITUDE	SEASON	SUNRISE	SUNSET	AZIMUTH	ALTITUDE
35°	WINTER	7:10	4:50	119°-0'	31°-30'
	SUMMER	4:50	7:10	61°-30'	78°-30'
30°	WINTER	7:00	5:00	117°-30'	36°-30'
	SUMMER	5:00	7:00	62°-30'	83°-30'
25°	WINTER	6:50	5:10	116°-30'	41°-30'
	SUMMER	5:10	6:50	88°-30'	63°-30'

NOTE : THESE LATITUDES COVER THE CONTINENTAL UNITED STATES. HOURS INDICATED ARE STANDARD TIME.
AZIMUTHS ARE AT SUNRISE AND SUNSET, NOON AZIMUTHS ARE ALWAYS 180°. NOON ALTITUDES
ARE GIVEN, ALTITUDES AT SUNRISE AND SUNSET ARE ALWAYS 0°.

190

The site analysis process which results in site planning for energy conservation is illustrated in the following drawings by Prof. Craig Johnson of the Department of Landscape Architecture at Utah State University. The following site analysis shows the problems which exist on a site and the potential areas of climatic modification which are desired and needed.

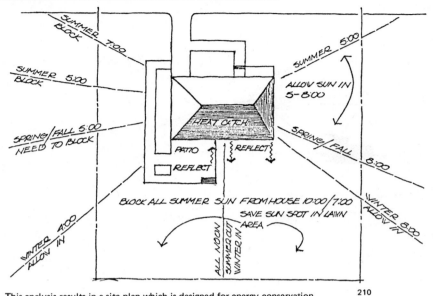

210

This analysis results in a site plan which is designed for energy conservation and maximum solar energy utilization as shown on the following drawing.

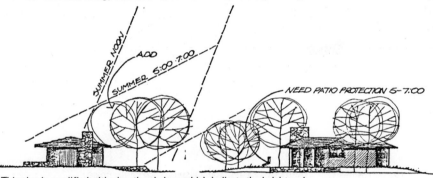

This plan is amplified with elevational views which indicate the height and the canopy characteristics of the shade trees used for climate control on this particular site.

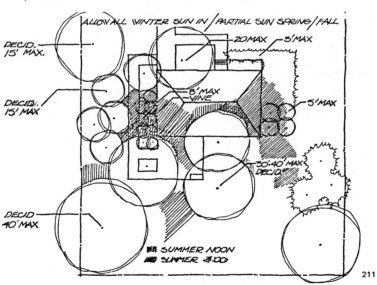

211

g. Aquifers

There are, on any site, natural drainage patterns and there is, within each larger region, a series of underground recharge basins where moisture is stored after falling through the atmosphere and flowing through the soil. These must be carefully located in a site analysis to determine what constraints to building or location of elements they may provide on a specific or precise site.

h. Infrastructural Elements

Complete analysis or survey of roads, buildings, and any overhead or underground utility lines and other man-made infrastructional elements must be catalogued on a particular site or on alternate sites to indicate what constraints to building or functional locations they may provide. These should also be evaluated and examined with an eye toward which of these should be removed and which should be retained. There are many other elements which might be reviewed or analyzed on specific sites or in specific problem situations. There is extensive literature on the techniques and methodology of site analysis which should be examined in order to provide the best possible extraction of information concerning the site itself.

5. Site Survey

It may be necessary for the environmental designer to conduct or have conducted for him a very careful site survey locating significant areas or pieces of vegetation and other natural and man-made elements on the site. This survey should consist of precise dimensions, topography and the specific location of all elements germane to the use of the site or sites to be considered. This site survey will assist in legal and in decision-making problems encountered during later stages of the process.

6. Synthesis

During the synthesis stage of site design methodology, all of the diverse elements collected during previous stages should be pulled together. All of the wants, needs, and desires of the client should be incorporated, assimilated and synthesized into a minimum number of drawings or displays. All of the positive and negative potential elements on the site or sites should be distilled and depicted in a way that would make it easy for decision-making purposes. All of the constraints of the site, of the client or of the program should be extrapolated in such a way that they can be measured against the wants, needs and desires and the potentials. Synthesis should be expressed graphically for easy manipulation and comprehension.

7. Solution Communication

An actual, proposed solution, either for the selection of a site or for the utilization of one or more of the sites being considered. The basic solution communication stage answers the question about what needs to be done, how it needs to be done, and a cycling or sequencing of when each part of it should be done. The solution should be communicated both conceptually and philosophically and as possible be supplemented with a hard, defined, nuts-and-bolts implementational plan. The solution communication may consist of drawings, plans, details, specifications, and estimating data. This step is to this point the culmination of the design process to provide solutions and answers to specific problems for a precise client about a particular site or optional or alternative sites.

8. Policy Development

In some cases it is not enough merely to present a plan or a proposed solution since it is not possible to implement some solutions in a short period of time or by a single entity decision making. Therefore, it is essential to develop an overall policy that can guide and direct future decisions beyond the individual single plan. If it cannot be done now, this stage is, in essence, a long-range extended program which can be implemented over a longer period of time. It is essential to supplement the precise solution with a longer range policy development and program which contains the essence of the solution for eliciting broader support and longer range realization.

9. Evaluation

Increasingly the design professions are understanding and accepting the need for an evaluation of completed projects. In essence, this is a feedback mechanism to test how a specific plan, proposal or configuration works, or has worked. This evaluation stage will increasingly be built into projects in the future. It will be used to build a literature and a resource of experience and information. The relative effectiveness of various site planning concepts, approaches, or designs, will be tested and evaluated in coming years. The effectiveness of a particular design solution or a specific material in a discrete location will hopefully be evaluated to a larger extent and should be considered in any site planning program for solar energy utilization.

Application

The design methodology and process as well as the information developed as part of the analysis can be applied to a wide variety of types of problems and projects, depending on what the needs of client, of society and of the other environmental design professions are. It is necessary, in any application of either the process or the basic developed information to establish a listing of hierarchies of priorities, as to what is most and least important. Of primary importance to the current study is the application of this material to at least the following four areas:

1. Solar Energy Utilization

As energy collecting apparatus in residential, commercial, or industrial buildings will either be on the site or on the architecture, it would be necessary for the site development to supplement and augment the work of the engineering devices incorporated into the architectural framework. It would also be necessary to structure the rest of the site to maximize solar energy utilization. In order to do this, it will be necessary to know the full potential of the site and the site elements.

2. Climatic Utilization and Control

This has been done traditionally, but will be done to a greater degree in time to come because of the data being gathered and the interest being generated in solar energy utilization and in energy conservation. It will be necessary to the greatest extent possible, in an energy-short nation, to utilize natural climatic factors, conditions, and elements. In a time of less expensive heating or air conditioning this was of less importance—in the future, it will be of much more importance and the weighting of analysis factors will show that by indicating the needs or inadequacies of specific or particular sites or site development programs.

3. Energy Conservation

Certain environmental problems can most efficiently and effectively be dealt with in the site selection, orientation, planning and design phases. Energy conservation is one of those. Through the use of earth berms, proper building orientation, well selected and well placed planting, the integration of building site and the use of indigenous materials, it will be possible to conserve a great deal of energy through proper site planning and design.

4. Options, Tradeoffs and Interchanges

By careful analyses and adherence to an objective methodology, it is possible to carefully and rationally evaluate various alternatives and their long range cost and impact. Examples of such options, tradeoffs and interchanges would include the following:

— the flat site with its relative ease of construction might be evaluated against a sloping site which would offer better opportunities for solar radiation collection, but increase cost of initial construction;

— the relative values of building winter sports facilities on the north or south slope—the south slope would, of source, allow for better solar energy utilization whereas the north slope would allow for better skiing and snow retention possibilities;

— building on a heavily wooded site or on a bare site—the bare site would enable solar collection devices to be placed in any number of locations on the site but would require a great deal of expenditure on site development, whereas the heavily wooded site would require very limited planting but very careful thinning and clearing of the vegetation, to allow for optimum solar radiation collection in the heavily vegetated or wooded site;

— the evaluation of scenic sites in the valley with its attendant predictable wind patterns or the precise location on lake-front property, knowing full well the windflow characteristics from land to water and water to land in the day and evening hours.

To understand more fully the characteristics of the site and of the climatic factors on specific sites, means understanding more fully the costs and benefits in a variety of tradeoffs; by conducting careful research, it is possible to index fully the site and climatic aspects of various options or alternatives.

The book *Energy Conservation in Building Design* previously developed by the American Institute of Architects Research Corporation has the following to say concerning site analysis in its relationship to building architectural construction:

"A building design that ignores the impact of the natural environment will almost always have to use energy in the form of mechanical, structural, or material interventions to compensate for the resulting discomforts and inconveniences of adverse natural conditions. Clearly, then, a building project should start with a thorough analysis of the assigned site or potential site alternatives. An architect should understand and anticipate the effects of a particular site or climate on the energy flow of a building if his design is to use the environment to advantage.

Man can deal with the natural environment in several ways. Where it is hostile to his living or working needs, he can build shelters or structural enclosures to separate himself from the outdoors and so minimize its undesirable effects. He can also develop the site to minimize and economize his structural needs. In either case, his main natural concerns are temperature, precipitation, humidity, air quality, sun, wind, and neighboring developments, animals, and people.

Weather

The intensity and duration of seasons, the wet and dry bulb temperatures, the frequency and amount of precipitation, storm patterns, and wind and sun characteristics of a site are vital to determining the energy needs of a building project. Computerized weather tapes, containing hour-by-hour data for the various regions of the country are available from the National Climatic Center, National Oceanographic and Atmospheric Administration (NOAA) and can serve as the basis for analysis of a given project. The tapes use statistics based on 10-year averages and contain information on temperatures, wind, cloud cover, etc. (However, it has been reported that much of the data on clouds and sun are inaccurate, partially because many statistics were drawn from rudimentary observation and partially because the instruments were faulty.)

Such statistics make it reasonably easy to calculate peak loads for HVAC, lighting, and other systems. Usually, however, engineers and architects use block load calculations, and these can be off by as much as 30 percent. A better method of calculation is to divide a building into energy zones, instead of calculating by gross energy requirements, and to measure these in terms of diurnal cycle scales. The National Bureau of Standards (NBS) has such a computer program. The architectural firm of Caudill, Rowlett, Scott is among the professional firms using computers to analyze buildings by energy criteria. Called EAP, or the Energy Analysis Program, it assesses weather data; heating and cooling loads; energy requirements; and various short- and long-term cost factors.

In any case, the requirements of a particular building must be related to the particular site conditions, and the latter often require a far more complete and specific weather analysis than gross regional statistics can provide.

Sun

Where energy conservation is a major goal of a building design, sun is perhaps the single most important natural element to consider. It affects virtually every portion of a building's design from siting and orientation to its envelope and glazing, HVAC, lighting systems, and operating and maintenance policies.

In the Northern Hemisphere, the sun factor is most important to the design of the south, east, west, and north sides of a building, in that order. But the precise effects of the sun will vary according to the season of the year and the time of day. Accurate calculations of solar loads, therefore, should differentiate among the sides of a building, different height levels, and time periods. Air quality can also determine the amount and components of sunlight that will reach a building. (In New York City, for example, air pollution reduces the amount of sunlight reaching building surfaces by as much at 25 percent.) Ultraviolet light can affect not only the longevity of some building material, such as vinyl, but also the growth of bacteria, which may necessitate an increase in air-filtering requirements.

The intensity, direction, swing, and duration of sunlight, and the effects of its direct penetration into a building or structure, are the first calculation. The second is to determine whether and how this energy is to be controlled or collected. Solar controls, such as internal or external shading devices for glazed areas or cooling ponds or sprays for roof areas, can help achieve maximum energy savings. But sunlight may be collected for heating, cooling, and domestic hot water needs of the building, and eventually for power generation as well. . . .

Wind

The velocity and prevailing directions of wind on a given site should affect the shape and orientation of a building, as well as the components of its envelope. Wind affects infiltration (air leakage) and transmission (thermal conductance) over the entire skin of a building, and of glazed or windowed portions particularly. In the Northern Hemisphere, the north and west sides of a building are most exposed to wind loads.

Winds can decrease the exterior film of still air that usually surrounds a building and so increase the thermal vulnerability of roof and wall elements. This can increase heating and cooling loads. Wind can carry heat built up by solar radiation away from a building and evaporate moisture on wet surfaces, thus cooling the skin to temperatures lower than the ambient air. By removing water vapors, wind increases humidifying loads.

Trees

Trees comprise a special category because they serve as wind and light breaks that can have a significant effect on the design of low buildings. They can substantially alter the effects of arctic winds, the wind direction off arid land or large bodies of water, the effects of solar radiation, and the flow of air. They affect natural ventilation, air pressure, surface temperature, and humidity levels. They shade the paved areas surrounding a building as well as the building itself.

The natural landscape of a site should be considered in designing and orienting a building, since it takes energy to destroy, remove, or replant trees and other plants. But installed landscaping also can be based on energy considerations.

In the colder regions of the Northern Hemisphere, deciduous trees (which lose their leaves in winter) should be planted on the south side of a building. They will provide sunshade during hot months, and yet allow maximum sun penetration in winter. Evergreens can be planted on the northern side, where there are no cold-weather solar gains. Evergreens also may be planted to protect building entrances and windows from prevailing wind conditions.

In general, the relative advantages and disadvantages of trees in terms of energy should be weighed in relation to the whole building system. However, no building design should rely entirely on their presence since trees may be destroyed through disease or accident, and it may take many years for replacements to fill their roles.

Knowing the direction of prevailing winds may determine where entrances and exits should be placed, and whether or not they should be shielded. It may determine the effects and desirability of natural versus mechanical ventilation systems.

Wind is a dynamic force and must be analyzed as such in calculating its effects on a building. Its natural consequences must be weighed against the energy costs of mechanical or structural compensation.

Topography

The dimensions and shape of a site often determine much of the building envelope's design. Both can have a serious impact on the potential energy consumption of that building. For example, a rectangular site can impose a similar shape on a building design. Whether the long sides face east-west or north-south, determines the effects of solar and wind loads. Building orientation affects heating and air conditioning energy requirements. A rectangular building with a 2.5 length/width ratio absorbs considerably less solar heat if its long axis is aligned in an east-west instead of a north-south direction. (The sun bakes east and west walls longer and with more intensity than even a south wall, which intercepts solar rays at less direct angles.) A triangular site, with its base side facing north, could be a tremendous asset in a hot climate by eliminating a southern wall and so the area of greatest heat gains. In the north, however, this could be a serious loss because the solar gains on a southern facade can lighten heating loads.

Whether a site is flat or hilly can influence energy requirements. For example, there are large night and day temperature differentials at the foot of a hill. Cold air tends to collect in low areas. Wind directions, velocities, and pressure zones can be altered by hills, valley, trees, and adjacent buildings. Also, the sun warms the east, south, and west slopes of a hill, but the northern slopes remain shaded and therefore cooler. A building on a northern slope would therefore experience heavy heating loads in cold weather, but its warm-weather air conditioning systems could be designed for lighter loads.

Available water and drainage patterns also can affect building systems design. Well or river water can be used as a heat sink for cooling, or as a heat source for heating systems. If building development can be planned so water run-off is maintained at predevelopment levels, storm drainage systems can anticipate lower capacities than otherwise. Retention ponds can help reduce run-off and then supply irrigation systems or a closed waste-water recovery system as a made-up supply. Leaving enough site exposed to absorb run-off also leads to energy savings. For example, not covering a site with a parking lot or building means less storm drainage and consequently less horsepower devoted to pumping storm water.

The planted, earthen, or other exposed natural surroundings of a building can reduce the temperature gains of the structural masonry or other surfaces, and also can reduce any reradiation of that heat into the building. (Concrete paving, on the other hand could increase heat radiation.) The quality of soil and subsoil, and what can grow there, is therefore important to landscaping plans.

Off-Site Development

No discussion of site analysis can be complete without mention of related development, and proximity to desired services and utilities. In terms of energy, there are some advantages to degrees of concentrated, or urban, development.

It takes energy to move people, goods, services, garbage, etc. It takes energy to install sewer lines, electricity, phone, and other services. Centralized electrical generating plants lose efficiency when energy must be delivered over long distances. Likewise, steam heat and hot water lose usable energy if they have to be transported too far.

The proximity of a site to available transportation, to fuel supplies, to service personnel, and to other supply systems, immediately affects energy consumption of a given building, though generally this consumption is not charged off to the building. While very large cities sometimes suffer from uncontrolled congestion and extremely high per capita costs

for the delivery of services and materials, the densities of smaller, medium medium-sized urban areas do offer some advantage in this regard. (Smaller cities avoid severe congestion and its attendant cost, yet offer convenience and proximity. This is, however, a generality and must be analyzed according to particular project proposals.)

Central utility plants usually can be operated more economically than individual units. High-rise, cluster, and common-wall structures use less energy than free-standing structures of comparable size because they increase the ratio of enclosed space to surface volume. Surrounding buildings and structures can have beneficial or adverse effects on wind velocity, wind direction, and solar loads. They should be considered accordingly. The reflective surfaces of other buildings, pavements, etc., may lessen heating loads or increase cooling loads. (Environmental noise pollution, on the other hand, may lead to extra insulation or other accoustical controls, and hence more energy consumption.)"

Site Selection for Energy Conservation

Proper site selection may do far more than planting or grading to assist in energy conservation and in optimum solar energy utilization. If site selection is possible, there are certain criteria to be established and considered from a climate control and energy conservation standpoint. Some degree of site selection is a factor in almost any landscape design and development process. Therefore it is essential to know the energy conserving options available at each stage and to make the decision as early as possible in the development process. A certain degree of site selection is possible in almost any situation, either in gross site selection, discrete site selection or in activity or functional site selection. The site analysis process must be applied to a specific site for proper site selection at all scales for energy conservation. Therefore it is essential to understand the climate of an area and the microclimate of a site or an area before selecting the site or the location for a building, a function or activity.

The selection, either of the optimum site from the number of alternatives, or the precise siting of a specific building on a much larger site, is one of the most important roles in the environmental design process. If there are a number of options, the best site from a climatic standpoint must be chosen since the problems created in a less than optimum site will have to be overcome by either architectural or engineering means, both of which may be more expensive and less satisfactory from an energy life cycle costing standpoint. It is impossible and superficial to generalize with a series of formulas or platitudes about how to select specific sites for various functions and purposes. It is necessary to know what is going to happen on the site and what are the inherent characteristics of a particular site or alternate sites. There is no such thing, for instance, as a right slope or a wrong slope on which to place a particular function or piece of architecture. There are, however, bad slopes, less bad slopes, good slopes, and better slopes and best slopes. In discussing site selection, the analytical data either for the single site or for a variety of alternative sites must be reviewed with an eye toward

- existing vegetation
- angle of slope
- orientation of slope
- surrounding geography
- buildable areas
- access routes
- protected points
- exposed location
- existing underlying geology
- existing soil classification from engineering and agricultural view points.

Site selection may be at least three levels. *Gross site selection* may deal with the choice of the optimum site or sites within a region or area. The rating of sites could be done according to solar energy potential. *Discrete site selection* deals with the selection of a location for a specific activity or a particular building within a specific site. *Activity or functional site selection* will usually deal with where each of the activities or functions go on a particular site. In many cases a building may already be placed and built, but site activities and elements may be adjusted, placed or moved in such a way as to conserve energy.

As mentioned previously, the entire analytical data base should be reviewed once again with the specific aim in mind of selecting either among a number of sites or the possible alternatives on one particular site. Included in site selection process would be the following:

Development of Criteria

In this stage, it would be necessary to answer the following questions:

- What do you need on the site?
- How much of it do you need?
- How important is it?
- Where on the site or on the sites are the suitable locations for structures, elements or activities?
- What are the optimum relationships for functions?
- Which functions take precedence?
- What are the criteria for this project?
- What are the climate criteria for maximum use of solar radiation and energy conservation? It is possible to apply the basic criteria for a specific project or client to a variety of sites.
- How do you develop criteria?
- What are the criteria for this particular client?

It is possible and it is necessary to know the specific region in which the project is to be built. It is also necessary to know the functions and the client and enumerate them in considerable detail. This is a very complex question to answer and it cannot be done this simply. The following is a perfunctory listing of some of the elements which might be included for housing in the Washington, D.C., area. Some of the case studies contain more complete data for their respective areas.

- *Orientation:* South-facing slopes are most desirable for the location of solar collectors; west-facing slopes are next best, since they are warmer in the afternoon. East-facing slopes are less good since they are warmest in the morning. North slopes should be avoided or overcome in some way, either through the alteration of the architecture or the collector itself. The optimum site would be on a slope of approximately 20 percent. The optimum location on the slope would be part way, and, if possible, over halfway up the slope, but not too near the brow of the hill. The topography should not be in a fog belt nor on a particularly rocky site. Vegetation should be thin, or in distinct ecological communities. New planting should consist of evergreens to screen from the northwest winter winds and consist of lower deciduous plants on the south and in no case should either the existing nor the introduced vegetation be so high as to interfere with the sun's rays and the work of the solar collection device.

- *Location:* The location should be on the windward side of a water body and protected on the west, if possible, by topography from the cooling winter winds. The enumeration of the activity areas should include any winter sports facilities which should be over the brow of the hill on the north slope. Tennis courts should be protected from the wind and with shade nearby, but with no shade on the courts themselves. Paving when included on the south side of the building, should have low deciduous shading over the paved area. A swimming pool, if utilized, should be on the southwest side of the site and should be protected on the west and on the north by a tightly planted coniferous hedge. Dual patios or terraces should be developed on the east and west sides of the house. The east patio could be utilized most effectively in the mornings and in early spring, while the terrace or patio on the west side of the house could be utilized later into the fall, and in late afternoons.

A complete checklist should be made by the environmental designer for the criteria for a specific site project or client. Under optimum conditions, this should be in the form of a hierarchical continuum from suitability to unsuitability. If possible, the environmental designer should deal with the criteria in an abstract, generic way; either dealing with single or multiple family housing or with industrial, commercial, or recreational areas and buildings, or with specific building types.

1. Suitability

The environmental designer should arrange a list of what is suitable for a particular project or for a specific client.

A suitability hierarchy or continuum should be developed, answering the questions:

— what do you want for this client?
— what do you want for this project?
— do you want it or feel it is absolutely necessary?
— how important is it?
— what will you give up for it?

2. Unsuitability

A similar list of what is unsuitable should be arranged in a hierarchical continuum. In the unsuitable category the following questions should be asked:

— what don't you want?
— do you feel it is absolutely necessary not to have it?
— to what extent do you not want it?
— what are you willing to give up not to have it? (An example of this would be increased construction costs or cost of transporting materials to a site with optimum orientation.)

3. Tradeoffs

What are the general and specific tradeoffs possible? What are the techniques for measuring and comparing the various tradeoffs? Some examples of these have been mentioned previously and others will be covered in case studies. The various aspects which must be evaluated in site selection tradeoffs have to do with short-term costs as compared to long-term costs or locational costs as compared to construction costs. The entire matter of higher construction costs as compared to higher energy costs will come under increasing scrutiny in years to come. The cost of preserving existing vegetation as opposed to replacing it with more appropriate vegetation in a better location has to do with the measurement of the relative cost and time requirements. A comparison of the more expensive construction materials must be made with more expensive maintenance cost in each area of a specific site or on a number of potential sites.

4. Site Selection Criteria

Many times we tend to think of a specific climate as a certain condition uniformly distributed over a large area. This is particularly true since weather or climatic data is collected where "undisturbed conditions" prevail, and to a degree because large scale climate maps show equal mean temperatures by means of a few smooth lines. However, in reality at ground level a variety of distinct and different minute exist next to each other, varying a great degree with the elevation of a few feet and within the distance of a mile or less. This is demonstrated dramatically in late winter by the differences in melting snow cover patterns and in early spring when the north slopes of hills may be frozen and grown while the south facing slopes of some hills may be turning green with emerging vegetation plant materials of favorable climatic conditions. This effect is well known to farmers and

orchardists, who prefer southern slopes for growing grapes or cultivating orchards. The difference between the kind of plants which would grow on either side of a hill, if nature were allowed to select, would be as great as the difference between locales a hundred miles north or south of one another. In addition, every change in elevation, every change in the character of the land cover, every surface covered with water, every structure and major masses of vegetation will induce variations in local climate. These effects within the large scale "macroclimate" form a small-scale pattern of "mesoclimates" and even smaller "microclimates." These minute and discrete deviations in climate play an important part in land use and land planning. Obviously, in site selection, the most favorable or desirable locations from a climatic standpoint should be considered and utilized for building construction and landscape development for optimum energy conservation and solar energy utilization. Secondly, a less favorable site may be able to be improved through the use of vegetation, land forms, architectural extension or paving of surrounding surfaces to induce an advantageous reaction to temperature and radiation impacts.

There is an inverse relationship between temperature and altitude. Air temperature decreases with increased altitude. In fact, the temperature in a mountainous area may approximate 1^oF for each 330 foot rise in altitude during summer months and for each 400 foot during the winter period. This effect is especially in the hot arid or hot humid regions where temperatures become more favorable at higher altitudes. As larger land forms, such as mountain ranges affect the macroclimate, smaller variations in terrain can affect inordinately large modifications in the microclimate. Cool air is heavier than warm, and at night the outgoing radiation causes a colder air layer to form near the ground surface. This colder air acts in much the same way as a liquid flowing toward the settling into the points of lowest elevation. This movement or flow of cool air is the cause of cold islands or puddles. By the same token land forms or changes of grade which serve to impede the flow of warmer or cooler air and affects the distribution of night time temperatures through the damming action. Due to this concave terrain areas which do not have a lower outlet become, in effect, cold-air lakes during the night time. This same phenomenon is exaggerated when a larger volume of cold air flow is involved, as in valleys. The plateau, valley walls, and bottom valley surface cool off at night. This cooler air flows toward the valley floor. On the valley slopes a series of smaller circulation systems mix with the nearby warm air, causing intermediate temperature conditions. The temperatures on the plateau will thus be cold, that on the valley floor will be colder, but the higher sides on each side of the valley (especially the south, east and west facing slope) will remain warmer. This area which often contains much of the vegetation and is referred to as the warm slope on this thermal belt is the best location for a structure. In certain instances, care must be taken not to place a structure or activities too high on the slope since it may be exposed to crest winds which may offset the higher temperatures.

The area on the leeward side of a medium size hill, especially if it is on the south, east, or west slopes, will also have some of the same characteristics though such sites may need more wind protection from either side since the wind may flow around the hill to a greater extent than it would over the brow of a slope.

Obviously the quantity of solar radiation striking a particular site also has a dominant effect on the microclimate of that site. Any hillsides receive varying degrees of radiation impact depending on the angle of inclination, the direction of the slope, the season of the year and the degree of cloudiness. In selecting the optimum location on a sloping site, from the standpoint of available solar radiation, it is advisable to see an inclined surface which receives larger amounts of radiation during underheated periods, and less at overheated times, than a horizontal site. Slopes steeper than 200° are generally considered unsuitable for ordinary building purposes. Due to the angle of inclination solar radiation intensities received on south facing slopes will be received on level sites a few weeks later. Thus a sloping site receiving 40% more winter radiation will, by the same token, be 3½ weeks ahead of the horizontal site. Bearing in mind that morning ground temperatures are cooler and afternoons warmer, preference might be given to a site on the east in a warmer mesoclimate and to a site on the western slope in a cooler climatic region.

A hill, or any change in elevation, has a modifying effect on the distribution of both wind and precipitation. Victor and Aladar Olgyay in their book *Design with Climate* discuss this in the following words, "a wind flow is diverted by a hill in both its horizontal and vertical stream patterns, causing higher speeds near the hilltop on the windward side and less turbulent wind conditions on the lee slope. The resultant wind distribution on a hill creates high velocity areas below and at the sides on the crest; the lowest speeds are near the bottom of the hill in the wind "shadow."

Precipitation on the windward side is carried over a hill by the wind which strikes the slope, and falls on the lee side, where irregular weak air movements prevail. However, high mountains cause exectly reversed precipitation conditions. When air is forced to ascend on the windward side, this produces adiabatic processes of condensation and precipitation. This pattern of rainfall shapes the climatic character of the California coast. The water-laden Pacific Ocean winds bring about 20 inches of rainfall to the seaside valley. The rising air on the slope of the Sierras causes a deposit of more than 50 inches. The descending air at the eastern slope, compressed and warmed by the drop, sucks up moisture instead of releasing it. Thus the arid character of the Nevada side, where Reno receives only 6 inches of rain yearly. A similar effect prevails on the Riviera, where the protecting Alps shut out the cold north winds; and the descending air, heated by compression, provides mild winters.

Natural and Built-up Surroundings

Water, having a higher specific heat than land, is normally warmer in winter and cooler in summer, and usually cooler during the day and warmer at night, than the terrain. Accordingly, the proximity of bodies of water moderates extreme temperature variations, and in winter raises the minimums, in summer lowers the heat peaks. In the Great Lakes region this effect raises the average January temperature about 10°F., and the annual minima about 15°F. Average July temperature is decreased about 3°F, and the annual absolute maximum is depressed about 5°F. In the diurnal temperature variations, when the land is warmer than the water, low cool air moves over the land to replace the updraft. During the day, such offshore breeze may have a cooling effect of 10°. At night the direction is reversed. The effects depend on the size of the water body, and are more effective along the lee side.

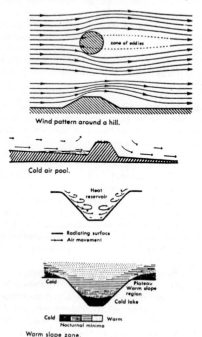

Wind pattern around a hill.

Cold air pool.

Warm slope zone.

191

The natural cover of the terrain tends to moderate extreme temperatures and stabilize conditions through the reflective qualities of various surfaces. Plant and grassy covers reduce temperature by absorption of insolation, and cool by evaporation. This reduction can amount to 1,500 Btu/sq ft/season. It is generally found that temperatures over grass surfaces on sunny summer days are about 10° to 14° cooler than those of exposed soil. Other verdure may further reduce high temperatures; temperature under a tree at midday was observed to be 5° lower than in the unshaded environment.

Conversely, cities and man-made surfaces tend to elevate temperatures, as the materials used are usually of absorptive character. Landsberg refers to observations, where asphalt surfaces reached 124°F in 98° air temperatures. He also measured the temperature distribution on a bright summer day in Washington, D.C., which varied 8°F within horizontal distances of a few miles. At night the differences in temperature were even larger, some suburban territories had temperatures 11°F lower than those downtown. A schematic drawing illustrates the effect of the "city climate." Note how closely the temperature lines follow built-up areas.

If one considers that a 9° difference in average temperature occurs in the United States roughly over a 9° latitude distance change, the importance of carefully selected sites becomes evident. Therefore, zoning should be differentiated according to the desirability in living conditions, based on microclimatic survey. Sites can be further improved by layout, windbreaks, and shade-tree arrangements . . .

Criteria for Site Selection

In various environments, according to the specific bioclimatic needs of a region, different topographic exposures will be desirable for habitation, and other human activities.

In the cool zone, where heat conservation is the main objective, protected sites are preferable. The lower part of the "thermal belt," on slopes placed in "wind shadow" areas but well exposed to winter insolation, offers advantageous positions. Orientation somewhat east of south secures balanced heat distribution. Accordingly, sites about halfway up a slope located in a SSE direction would offer the best location for desirable cool-zone habitation.

In the temperate zone, location requirements are not so strict as in the cold zone; however, they are broader in scope, inasmuch as needs for both over- and underheated periods must be correlated. Desirable site exposure tends to move farther east of south, as does the orientation index. The cool-air-flow effect is less important, allowing the utilization of lower portions of a slope. The upper topographical locations of a "warm slope" become advantageous provided there is adequate windbreak sheltering. Breeze utilization in warm periods grows in importance. This need not conflict with winter wind protection, as prevailing seasonal wind directions often do not coincide. In the temperate zone the varying needs of sun-heat gain and shade protection should be carefully considered.

In the hot-arid zone desirability of heat loss overrules the demands of the cool periods. Lower hillside locations, benefiting from cool air flow, are preferable if arrangements are made to avoid the flow during underheated times by "dam" action. A "courtyard" type of solution coincides with the need of capturing the air of the immediate surroundings that is cooled by the outgoing night radiation. Wind effects have relatively small importance. The larger daily temperature range makes easterly exposures desirable for daily heat balance. In a large portion of the year afternoon shade is required; accordingly sites with ESE exposure are preferred in the hot-arid zone.

In hot-humid areas air movement constitutes the main comfort-restoring element. Sites offset from the prevailing wind direction, but exposed to high air-stream areas near the crest of a hill, or high elevations on the windward side near a ridge, are preferable. East and west sides of a hill receive more radiation than other orientations where the sun rays come in a more oblique angle. Therefore southern and northern slope directions are more desirable. However, wind-flow effects will remain the dominating consideration, as shading might be provided by other means."

There is no way that a finite and clear cut and perfectly ordered site selection process is able to be spelled out in enough detail to insure that the optimum site for energy conservation will be chosen in all circumstances and under all conditions. There are, of course, certain guidelines to be followed in any site selection process. These may be modified somewhat to emphasize those characteristics which would emphasize greater energy conservation and solar energy utilization. The information and material discovered, assimilated and organized during the site analysis phase is able to be utilized to assist in site selection. In the broadest possible terms, it is essential to locate, within any region, those sites which have south, southeast or southwest slopes since they will be the warmest and the ones best able to provide maximum **solar** utilization.

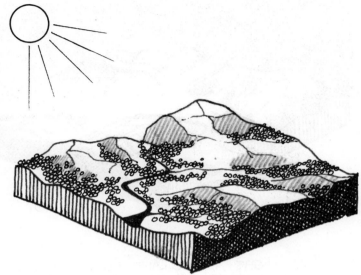

 INVENTORY THE TYPE AND EXTENT OF VEGETATION COVERAGE ON THE SLOPES IDENTIFIED AS HAVING GOOD EXPOSURES FOR UTILIZING SOLAR ENERGY.

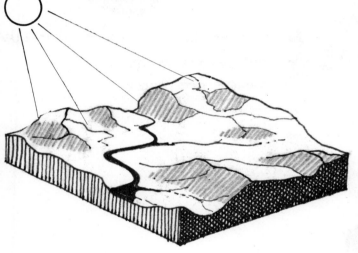

Next it would be important to trace the patterns of colder airflow during the evening hours. Since colder air is heavier it settles into lower areas and flows down valleys unless impeded by land forms or massive blocks of vegetation. Those sites, either not in the direct path of this colder air or located in the valleys themselves would obviously make the best sites both for energy conservation and also for solar energy utilization.

INVENTORY SLOPES IN REGION HAVING SOUTHERN EXPOSURE.

Next an inventory should be made of the type and extent of vegetation coverage which exist on the slopes identified as having the best exposures for utilizing solar energy. Heavy vegetation cover will provide dense shade which might have to be partially cleared before development could take place on it.

All of the analytical data is assimilated in such a way as to show those potential sites which are unsuitable to one degree or another as well as those sites which are either acceptable or desirable for development or conserving or using natural energy sources. This shows graphically those sites within a region which are less desirable as well as the optimum sites using the analytical data and the criteria of optimum energy utilization. One of those specific sites is chosen in the process of gross site selection.

GOOD SITES ARE IDENTIFIED THRU ANALYSIS AND A PROCESS OF ELIMINATION.

AT NIGHT, AS THE LAND BEGINS TO COOL, THE COLDER AIR WILL SINK AND COLLECT IN LOW POCKETS AND VALLEY FLOORS, AND WILL FLOW DOWNSTREAM, MUCH LIKE A RIVER, THROUGH THE REGION. THE IDENTIFICATION OF THESE COOL AIR MASSES AND FLOWS WILL HELP IN CHOSING THE BEST SITES FOR ENERGY CONSERVATION.

⬇ OPTIMUM SITES FOR SOLAR RADIATION USAGE

⇩ LESS DESIRABLE SITES

◯ EXAMPLE SITE BELOW

The prevailing winds, both in winter and summer move across a large site or land area in relatively predictable ways. Earth forms, of various sizes are able to block, deflect or channel the windflow pattern in relatively predictable ways. In order to select the optimum site for conserving existing energy and for using more of the sun's radiation, these wind patterns must be analyzed and those sites which receive cooling summer breezes and which are protected from colder winter winds must be identified as optimum potential building or development sites.

WINDS BECOME CHANNELIZED, THEY DEFLECT AND DISPERSE; THEY CONSTRICT AND SPREAD OUT, AND INCREASE AND DECREASE VELOCITY AS THEY MOVE OVER AND THROUGH LAND FORMS.

This site is then analyzed in greater detail to find on it the best possible location for the structures and activities. Vegetation patterns, slope angle, orientation and variation, natural wind flow patterns, precipitation amount, type and location as well as solar radiation interception by the site should all be carefully analyzed and depicted in order to make an accurate discrete site selection.

GOOD SITES WILL HAVE CORRECT ORIENTATION, LIMITED VEGETATION COVERAGE, PROTECTION FROM WINTER WINDS, FEW COLD AIR POCKETS.

As a result of this process all structures, functions and activities may be located on a particular site so as to place each element where it will use the least possible energy and be able to use the most possible natural heating, cooling and lighting resources. The site selection process is sequential, logical and rational. In order to use it to select specific sites to save more expensive and scarce energy and to utilize more effectively relatively inexpensive natural resources different priorities and values have to be introduced into the process and used for guidance and direction. Obviously not all of these options are open to all designers at all times, however, when the options are available they should be carefully evaluated and the same methodology applied for optimum site selection of energy conservation.

IN DEPTH ANALYSIS OF THE SELECTED SITE, IS NECESSARY TO DETERMINE THE OPTIMUM LOCATION FOR FACILITIES AND ACCESS WAYS.

Siting and Orientation for Energy Conservation

The actual siting and orientation of the buildings, activities and functions to take place on the site is one of the more important factors in the entire site selection, siting, orientation, planning and design process. This stage requires a great deal of time and study since it is a very sensitive link in the preservation of the inherent characteristics of the site and in the accommodation of the basic activities or functions to take place on the site.

There is, for each given site, client and situation probably one best solution, given all the tradeoffs and a proper weighting of all of the factors. It is necessary, in a building siting orientation study, to develop general as well as specific guidelines. The elemental decision is where to put the functions and the infrastructural elements after the exact site is selected. This may be either when the architecture is given or where the architectural structure may be modified and adapted. Obviously, the siting and orientation process should take place in conjunction with the development of the architectural form, so that the two may fit together in the best possible way.

There is a basic optimal orientation for each region and for each function that takes place on the site, not just for the building itself. Within each of the regions on specific sites there are a great many subtleties and variations. The preliminary siting study should require the minimum fitting and adaptation of the site to accommodate the architectural and landscape architectural activities and functions. A series of careful studies should be made so as not to disrupt the site more than is absolutely necessary in order to accommodate the building, access to it and the activities taking place on the site.

The following is an overview of the process of building siting and orientation. The selection of the precise appropriate site will, hopefully, have been made prior to beginning of this stage. Therefore, the specific location of each of the various functions on a specific site is the end purpose of this particular stage in the overall process. Initially, the problem should be dealt with in an abstract way, answering such questions as: "where should each function go in relation to each other function?"; "what is the interaction and interrelationship between the various functions?" Secondly, "where should each function go in relation to the climatic situation of the precise site?" Suggestions should be made in an abstract way as to the optimum building configuration in the indoor/outdoor relationship or functions which chould also be coordinated with the climatic situation on the precise site. This may lead to an alteration of the building mass and an enumeration of the tradeoffs of alteration of the site or of the architecture. At the same time, the relative cost/benefit of the various adjustments to the site should be considered and evaluated against possible options or alternatives. The morphology of the building envelope may be developed in cooperation between the landscape architect and the architects and other environmental designers. It may be that certain site alterations will be essential or will be required to obtain optimum orientation. After the best of possible orientation is developed, its relative cost may be prohibitive and therefore compromise orientation studies might be able to remain.

The use of solar collection devices, either on the building or on the site, may require some additional disruption of the site which will lead to a lessening of the deleterious environmental impact of any mechanical structure. Various energy conservation devices may be incorporated into the building and into the site which will decrease the life cycle costing of the building, its planning and construction, and may lower the energy cost over a 20 to 40 year life cycle. The use, preservation or replacement of vegetation and landforms should be considered as should the use of architectural extensions to rectify deficiencies resulting from the ultimate decision.

A review of the topography with an eye toward the runoff and hence the water pollution caused by various siting orientation alternatives, the orientation of parts of the building and various activities to catch the sun or the wind or to remove precipitation or enhance precipitation should be studied. The review of various orientations or locations to utilize or diminish onshore or off-shore winds or upslope/downslope winds should also be considered in this phase. At the same time, the relative means and techniques for damming, channeling or controlling valley winds as well as side valley slope winds should be considered and utilized by the environmental designer in orientation studies.

1. Functions

The basic elements which are found on most housing sites include:
- the building itself;
- the entrances to the site and to the building (both primary and secondary);
- an area for service with both entrance and sufficient storage space for service, supplies and materials;
- an area for outdoor living as well as
- areas for recreation such as play areas, swimming pools, tennis courts.

2. Region and the Functions

The following chart shows the suggested optimum orientation for each of the basic functions taking place in the single or multi-family housing site for each of the four regions under consideration.

It must be remembered that not all of the solar radiation leaving the surface of the sun ever reached the surface of the earth. Some of it is diffused throughout the earth's atmosphere, some of it is reflected off the earth and its atmosphere, some of it is dissipated in the earth's atmosphere and over the earth's surface. A very small portion of this energy and radiation is ever directly available on a particular site even on the clearest of days. This is shown on the following chart.

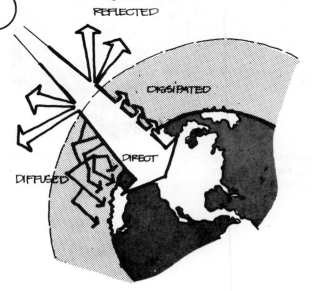

REGIONAL SITE FUNCTION ORIENTATION MATRIX

1. Quadrant location on the site
2. Facing which direction
3. Protection from which direction

ACTIVITY	Temperate			Hot-Humid			Hot-Arid			Cool		
ARCHITECTURE	1	2	3	1	2	3	1	2	3	1	2	3
Housing Unit	SSW-ESE	S	N-W		S	S-E	ESE	E-W	S-W	S-W	S	N-W
Auto Storage	E-N	N	N-W	N	E-W	S-W	N	E-W	S-W	N-E-W	E-N	N-W
Storage	E-N	N	S-N-W	E-N	N	S-E	S-W	N	S-W	N	N	N-W
SITE												
Access												
Primary	E-W	N	N-W	N-W	E-W	N-E	N	N-E	S-W	E-W	N	N-W
Secondary	N	N	N-W	N	N	S-W	E-W	N	S	E-W	E-W	N-W
Service												
Entrance	N	N	N-W	N	N	E	E-W	N	S	E-W	E-W	N-W
Storage	E-W	N	N-W	N	N	E	N	N	S-W	N	E-W	N-W
Parking	E-W		N-W	N	E-W	S	N	N	S-W	E-N-W	S	N-W
Outdoor Living	E-S-W	S	N-W	N-E	N	S-E	N-E	N	S-W	E-S-W	S	N
RECREATION												
Tennis	E-W	N-S	N-W	N	N-S	E-W	N	N-S	W-N	S-W	N-S	N-W
Pool	S	S	N-W	S	S	N	S	S	E-N	E-S-N	S	N-W
Play Areas	S-E	S-E	N-W	N-E-W	N-E	S-E	N-E	E-W	S-W	S-E-W	S	N-W
Passive Activities	E-W	E-W	N-W	E-W	N	S	N	E-W	S-W	E-S-W	S	N-E-W

The sun's energy which does not strike the earth is measured in "langleys" and the amount of solar radiation striking different sections of the United States varies from under 100 to over 300 langleys during January to from under 500 to over 700 langleys during July. The following maps show the different amounts of solar radiation in different sections of the country in July and January and a more discreet delineation of the four major climatic regions.

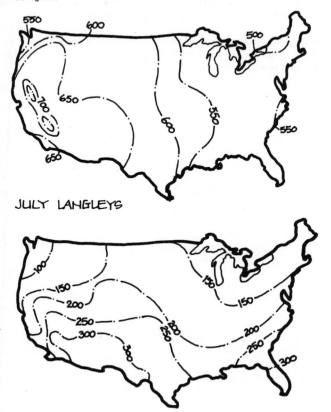

JULY LANGLEYS

JANUARY LANGLEYS

In order to conserve the maximum energy it is imperative to orient both buildings and site related activities and areas in the best possible direction so as to maintain temperature and humidity levels near the comfort zone. By orienting buildings and outdoor areas in certain directions it is possible to maximize the impact and effects of both sun and wind to raise or lower the temperature or humidity and thus to increase to comfort levels naturally and decrease the use and dependence on scarce and expensive energy to heat or cool a building or an area.

The relative importance and need for added solar radiation and the heat provided by the sun will vary from region to region and from season to season. In a cold region or season additional solar radiation and thus heat will be welcomed and needed. Therefore a building or a section of a building or outdoor living areas desiring such natural warming will be positioned to receive as much radiation as possible. In a warmer region or season, on the other hand, the orientation of the same structure or functions should be such as to decrease the undesirable or unwanted solar radiation. The best possible orientation on a particular site would be one which would provide for maximum radiation and heat during the colder period and which would, at the same time, reduce excessive radiation and heat during warmer periods.

In the cool or the temperate regions where winters are longer and the air is cool there is a greater need to utilize the latent solar energy and warmth. Therefore, structures and outdoor living functions should be oriented in such a way that they receive the maximum amount of solar radiation, particularly during the colder periods of the year. The same building on a similar site in the hot arid or the hot humid areas, where excessive heat is a problem, should have a different orientation. In this case the buildings or the functions should be located to avoid excessive solar radiation and to pick up, to the extent possible, any prevailing cooling winds or breezes. Even though it is desirable, it is not always possible to choose one orientation which maximizes both incoming natural heat or warmth and directs natural cooling wind patterns. Different building configurations and envelopes will, of course, react in a variety of ways to varying orientations in the different regions. Buildings which are open or have glass on one side and a wall on the other will react differently than will a building with a central wall and with glass on either side of the structure in each of the regions. The following charts show the most desirable locations on a site in each of the four major climatic zones in the United States.

In order for any building to be most energy effective it should be oriented and placed on the site to use, to the maximum extent possible, the natural heating of the sun and the cooling effect of natural winds and breezes which prevail on a specific site or in a region. In addition, outdoor terraces, decks and patios, swimming pools, tennis courts, play areas and other outdoor living areas will be able to be used earlier and later in the season and require less artificial heating, cooling or cleaning if they are properly located in relation to surrounding and existing topography, architecture, sun and wind patterns and vegetation. In this way these areas may be heated and cooled naturally in order to make them not only pleasant but usable for a greater period of time during the year.

The chart given at the beginning of this section on orientation gives suggestions as to these site related functions in each of the major climatic regions.

The above charts are from the book *Design with Climate* by Victor and Aladar Olgyay and indicate the optimum orientation for various building configurations in each of the four major climatic regions in the United States.

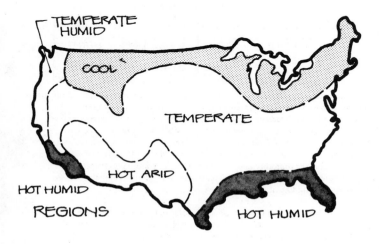

Architecture and Landforms

It seems somewhat platitudinous to indicate that in the building, siting and orientation it may be necessary and advisable to sculpt and mold the land forms to supplement the design intent of the architecture. It is also desirable to request this same process to accommodate the buildings and to use the architectural design to reflect and supplement the landforms which exist on a particular site. In siting and orientation of the building, the natural terrain should be used to optimum advantage for climatic purposes. In some cases, this may mean that the solar collectors (if placed on the site rather than on the building) should be integrated into the landform and into the topography rather than be alienated and isolated from both the individual landforms and the overall topography of the site.

Architecture and Vegetation

In the siting and orientation of any building on which there is existing vegetation of any significance, there are at least two aspects to the particular problem. The first of these is to retain as much vegetation as possible in the right location. The second is to select and supplement existing plant material with new plants that are both in harmony with the existing ecology, with the architecture and with the intent of the site development. It may be necessary to initially clear a certain amount of vegetation for placement and proper orientation of the collector. It also may be possible to use planting to supplement the solar collector and channel the sun's rays toward the solar collector without undue dissipation. It may also be necessary to use both plants and some landforms to screen the engineering and architectural structure of the solar collector, whether it's on the building or integrated into the site itself. On the one hand, it is advisable in site orientation to use vegetation to supplement the architecture; on the other hand, it is only fitting to design the architecture to reflect the existing vegetation.

Regional Variations

Obviously, building, siting and orientation, as well as the character of the architecture and the site development will depend to a large extent upon the landforms and vegetation on the site as well as the indigenous architectural forms. The solar collector should be integrated with the architecture and the site elements should reinforce the architecture toward the end of solar energy collection and energy conservation. This, of course, may entail completely different orientation, site development and elements in each of the major geographic regions as mentioned previously.

Single or Multiple Building

It is less difficult, obviously, to orient a single building on a specific site for maximum solar energy utilization and for energy conservation. It is much more difficult to orient entire groups or clusters of buildings on all the same site to provide both optimum orientation for solar radiation utilization and energy conservation and also to provide the necessary utilities, services, and roads in the most efficient and economical manner.

Increased costs for utilities and roads may be justified in order to better orient and enhance a building's ability to conserve energy or to achieve solar energy in siting or orienting planned unit developments, trailer parks, office complexes, or other multi-building unit systems. It is imperative not to use a rigid formula for building layout on diverse sites. It is advisable to use looser and more dispersed layouts in which individual units are each oriented properly in order to make use of the microclimate on the site. Each of the units should be adjusted to the site, to the topography, to the vegetation, and to the climate. There will, of course, be regional variations, with those multi-building units in cold areas possibly clustered closely together to minimize heat loss and share exterior walls. On the other hand, similar multi-building units in the humid or hot arid regions may wish to be separated in order that the cooler air may flow between the building units.

The following illustrations show examples of single or multi-residential building units oriented and sited with major emphasis on solar energy utilization and energy conservation.

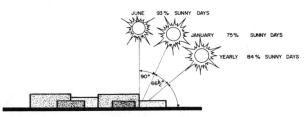

JUNE 93% SUNNY DAYS
JANUARY 75% SUNNY DAYS
YEARLY 84% SUNNY DAYS

90° 66½°

SOLAR CHARACTER

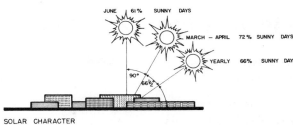

JUNE 61% SUNNY DAYS
MARCH – APRIL 72% SUNNY DAYS
YEARLY 66% SUNNY DAYS

90° 66½°

SOLAR CHARACTER

OPTIMUM ORIENTATION - 25° E. OF SOUTH
EXPOSURES SOUTH TO 35° E. OF SOUTH ARE GOOD

OPTIMUM SOLAR ORIENTATION

OPTIMUM ORIENTATION 5° E. OF SOUTH

OPTIMUM SOLAR ORIENTATION

OBJECTIVES

1. BLOCK HOT DRY DUSTY WIND
2. PROVIDE MAXIMUM SHADE
3. CREATE COOL, DARK MICROCLIMATES
4. VEGETATION IS DESIRABLE AS A RADIATION ABSORBENT SURFACE AND FOR IT'S EVAPORATIVE AND SHADE GIVING PROPERTIES.

VEGETATIONAL CONTROL

HOT-ARID REGION

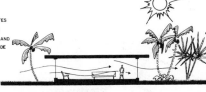

OBJECTIVES

1. PROVIDE YEAR AROUND SHADE
2. CUT SKY GLARE
3. SHADE TREES SHOULD BE HIGH BRANCHING SO AS NOT TO OBSTRUCT AIR MOVEMENT.
4. LOW VEGETATION AWAY FROM STRUCTURE TO PREVENT DAMPNESS AND NOT BLOCK BREEZES.
5. BREEZE ACROSS SHADED LAWNS ARE DESIRABLE.
6. EAST AND WEST WALLS NEED SHADE.

VEGETATIONAL CONTROL

HOT-HUMID REGION

SEPTEMBER 68% SUNNY DAYS MAXIMUM
JANUARY 53% SUNNY DAYS MINIMUM

SOLAR CHARACTER

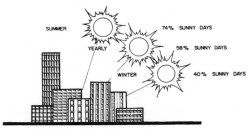

SUMMER 74% SUNNY DAYS
YEARLY 58% SUNNY DAYS
WINTER 40% SUNNY DAYS

SOLAR CHARACTER

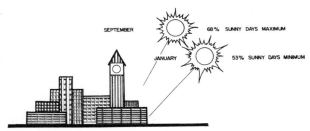

17½° EAST OF SOUTH

OPTIMUM SOLAR ORIENTATION

12° EAST OF SOUTH

OPTIMUM SOLAR ORIENTATION

OBJECTIVES

1. DECIDUOUS SUMMER SHADE ON EAST AND WEST
2. TURF FOR RADIATION ABSORPTION
3. SUMMER BREEZES UNOBSTRUCTED
4. ADMIT WINTER SUN

VEGETATIONAL CONTROL

TEMPERATE REGION

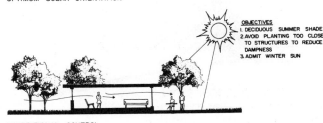

OBJECTIVES

1. DECIDUOUS SUMMER SHADE
2. AVOID PLANTING TOO CLOSE TO STRUCTURES TO REDUCE DAMPNESS
3. ADMIT WINTER SUN

VEGETATIONAL CONTROL

COOL REGION

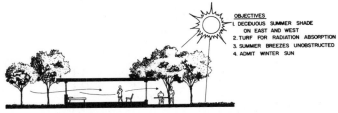

Temperate Regions

Objectives:

Maximize the warming effect of solar radiation in winter months and maximize shade in the summer months.

- Utilize deciduous trees for summer shade and winter warmth
- Orient active living spaces to the south for winter warmth
- Design building overhangs to shield the high summer sun and expose the area to the lower winter sun.

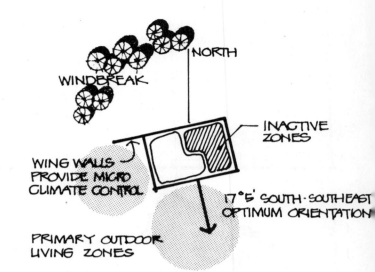

WINDBREAK

NORTH

WING WALLS PROVIDE MICRO CLIMATE CONTROL

INACTIVE ZONES

17°5' SOUTH-SOUTHEAST OPTIMUM ORIENTATION

PRIMARY OUTDOOR LIVING ZONES

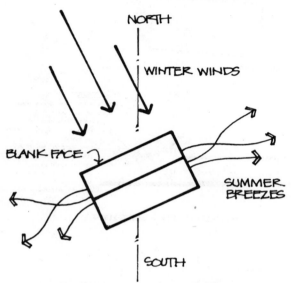

NORTH

WINTER WINDS

BLANK FACE

SUMMER BREEZES

SOUTH

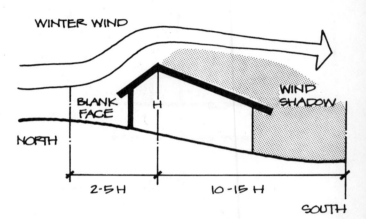

WINTER WIND

BLANK FACE

H

WIND SHADOW

NORTH

2-5 H

10-15 H

SOUTH

Reduce the impact of winter winds but maximize summer breezes.

- Steeply pitched roofs on the windward side deflect wind and reduce the roof area effected by the winds
- Blank walls, garages, or storage uses on north exposures
- Protect north entrances with earth mounds, evergreens, and walls or fences
- Allow for natural ventilation with prevailing summer breezes

Cool or Temperate Regions

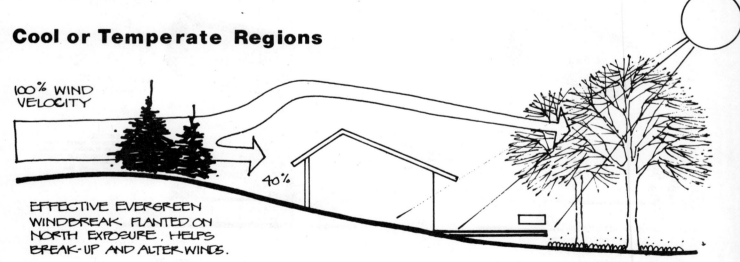

100% WIND VELOCITY

40%

EFFECTIVE EVERGREEN WINDBREAK PLANTED ON NORTH EXPOSURE, HELPS BREAK-UP AND ALTER WINDS.

DECIDUOUS PLANT MATERIAL ALLOWS SUNLIGHT TO PENETRATE DURING WINTER MONTHS AND PROVIDES SHADE FOR THE SAME SPACES DURING SUMMER.

PLANTING CONCEPT FOR TEMPERATE REGIONS

Cool Regions

Objectives:
Maximize the warming effects of solar radiation

- Utilize south to south-west facing slopes as much as possible
- Orient active living areas to the south to take full advantage of the winter sun
- Utilize exterior walls and fences to capture the winter sun and reflect warmth into living zones
- Create protected sun pockets
- Utilize darker colors which absorb radiation

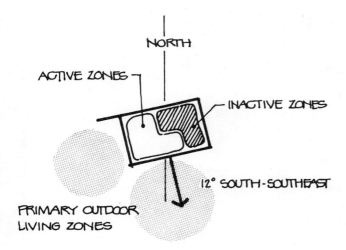

Hot-Arid Regions

Objectives

Maximize shade for late morning and afternoon solar radiation.

- Orient active living areas to the south east to collect early morning sun
- Glass areas should face south with properly designed overhangs
- East and west windows should be avoided to minimize radiation with low sun angles
- Cluster buildings and utilize solar panels for shade

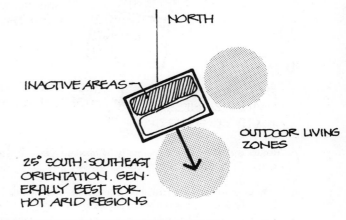

MAXIMIZE DESIRABLE AIR MOVEMENTS

Maximize the humidity and cooling effects of evaporation across water bodies.

Utilize the lower hillsides to benefit from cool natural air movements in early evenings and warm air movements in early morning

Reduce the impact of cold winter winds

- Locate buildings on the lee side of hills in the "wind shadow"
- Utilize evergreens, earth mounds, and exterior walls to protect the northern exposures
- Flat or shallow pitched roofs collect and hold snow for added insulation
- Structures can be built into hillsides or partially covered with earth and planting for natural insulation

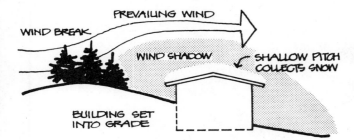

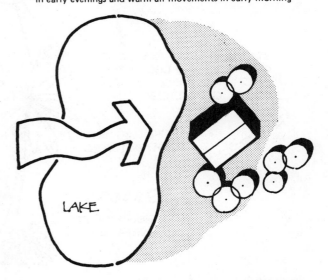

Hot-Humid Regions

Objectives:
Maximize shape throughout the day

- Orient active living zones to the south with properly designed overhangs, trellis, or other sun control
- East or west window should be avoided to minimize radiation with low sun angles
- Minimize energy intensive paving and building materials

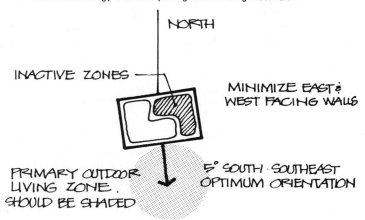

Reduce the effects of high humidity by maximum exposure to air movements

- Orient streets and structures to maximize cool breezes prevailing wind vary with regions and micro climates
- Utilize the psychological effects of falling water of large water bodies but minimize the humidity of small water ponds and low areas

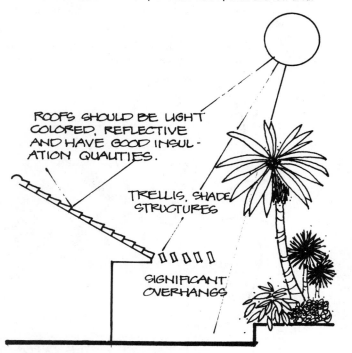

102.

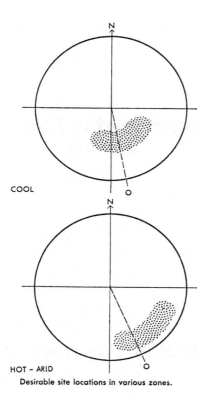

COOL

TEMPERATE

HOT - ARID

HOT - HUMID

Desirable site locations in various zones.

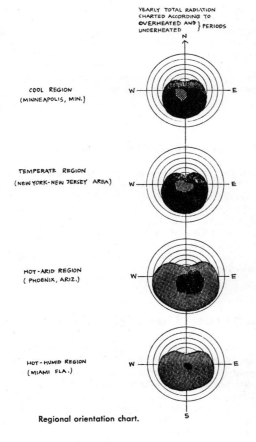

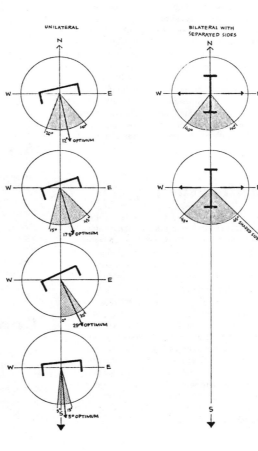

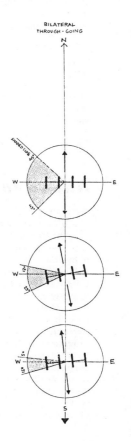

YEARLY TOTAL RADIATION
CHARTED ACCORDING TO
OVERHEATED AND } PERIODS
UNDERHEATED

UNILATERAL

BILATERAL WITH
SEPARATED SIDES

BILATERAL
THROUGH - GOING

COOL REGION
(MINNEAPOLIS, MIN.)

TEMPERATE REGION
(NEW YORK - NEW JERSEY AREA)

HOT - ARID REGION
(PHOENIX, ARIZ.)

HOT - HUMID REGION
(MIAMI FLA.)

Regional orientation chart.

Integration of Building and Site for Energy Conservation

The optimum integration of the building with the site on which it is located for maximum microclimate modification for energy conservation and solar energy utilization.

This integration may either be accomplished through the adaptation and accommodation of the structure or through the adaptation and accommodation of the site and site elements. The architecture may be integrated through the use of adaptive architecture, indigenous forms or materials, through architectural extensions, through construction considerations in form or materials, or through developing or following ecological guidelines adapted to the geology, geography or the forms or materials suggested or directed by the ecology of the site or the region. The site may be modified by the preservation, addition, subtraction or adaptation of the landforms, the vegetation. In order to do that the inherent energy conservation potential of a particular site must be analyzed, accepted or recognized, before it is changed in any way to integrate any building or building type.

Ideally, each piece of architecture is designed for the specific site on which it is to be placed. The site is then designed to supplement the basic concepts of the distinctive piece of architecture in the precise location. Realistically, however, building plans, especially in multi-building unit types, may be replicated a number of times on varying sites or pieces of geography. Site solutions, because of the distinctness of the region, the geography, the geology and the ecology are not easily repeatable and therefore must be done for every piece of architecture on each specific site. Quite often, the site designer is responsible for the integration of the building with the site. It is possible, however, on a regional climatic basis, to provide guidelines for site designers and architects to utilize in integrating more fully the building with the site. By doing this, it will be possible to utilize the potential solar energy in each region more effectively and at the same time assist in conserving energy in utilizing local forms and materials. Historically, there have been and are a number of ways in which buildings and sites have been integrated. Among these are the following:

Adaptive Architecture

The concept behind adaptive architecture is that the building itself consists of two shells. The outer shall is adapted to the region or the climate and takes the form appropriate to that specific area. The inner shall is adapted to the uses, perceptions, needs and wishes of the persons using the interior of the structure. The inside and the outside shell of such pieces of adaptive architecture may never meet or touch. Therefore, the internal form of the building does not dictate the outer form, nor vice versa. In addition, there have been a number of architects who have dealt historically with the adaptive architecture of various types and various degrees of innovation and integration. Paolo Soleri and his "Archology" practices a type of adaptive architecture. Frank Lloyd Wright with his Taliesen or "shining brow" used a southern orientation just over the brow of a projecting hill for the location of his building which was usually well integrated with the site throughout material selection, siting and orientation. Malcolm Wells, the contemporary New Jersey architect, with his underground architecture of course has one approach to the problem of energy conservation and orientation. Grillo, in his book *What is Design?*, has also shown a number of illustrations of various proposed forms of adaptive architecture.

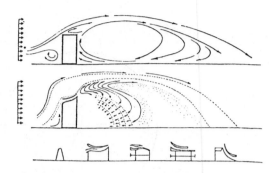

203

Indigenous Architecture

Bernard Rudofsky, in the early 1960's, put together an exhibit which was later published in book form entitled *Architecture Without Architects*. There is historically in every region of the United States and of the world an indigenous architecture that is usually less contrived and less architecturally sophisticated. On the other hand, it is more carefully integrated with the site in which it is found. Must of this sort of information could be found as a part of any research phase on any specific region. Much of this indigenous architecture and landscape architecture has been developed as a result of observation, sensitivity and adaptation to the climate, the geography and the geology. Local alteration of roofs, walls, and of the building mass, the use of indigenous architectural forms, materials and relationships should be studied to understand possible ways of integrating building and site.

Architectural Extensions

To remove any arbitrary lines between architecture and the site entails a specific way of thinking and teamwork between the architect and the site planner. It is possible, through the extension of parts of the site into buildings, to achieve a high degree of integration. It is also possible through extension of the architectural units such as walls, fences and canopies to integrate architecture to a great degree with any specific landscape types in any region. The use of the same materials, the same forms, and the same modules in the site and in the building allows for a great deal of integration between the building and the site.

Constuction Considerations

Through the use of materials and forms found either in the architecture or on the site it is possible to achieve a great deal of integration between building and site. Through the use of some of these same materials and forms within the architectural extensions, it is further possible to achieve even a greater degree of integration under ideal circumstances. The site planner should, if possible, attempt to make the solar collector part of the site and site design. It may be necessary to alter the site to incorporate or hide the solar collector and to use various materials or elements found on the site for energy conservation purposes.

Vegetation, whether trees, shrubs or ground cover, is able to be used in conjunction with earth forms or inadequately to assist in integrating building and and site for energy conservation and solar energy utilization. Vegetation may be added to provide wind breaks, shade, wind or breeze channels, to raise or lower the humidity or site adaptation for integration.

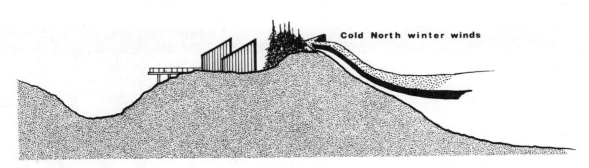

Cold North winter winds

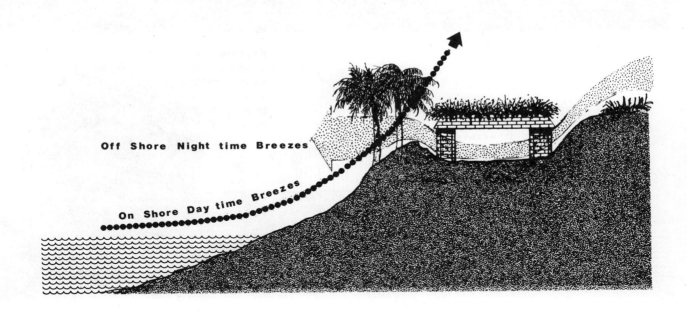

Off Shore Night time Breezes

On Shore Day time Breezes

When the earth is brought up "around" the structure it may mean the use, movement and modification of existing earth forms or the introduction of fill from off the site to fully integrate building and the site. Such earth forms can be used to block cold north winter winds, channel cooling summer breezes, divert up valley or down valley winds or to control on or off shore breezes either during the day or at night.

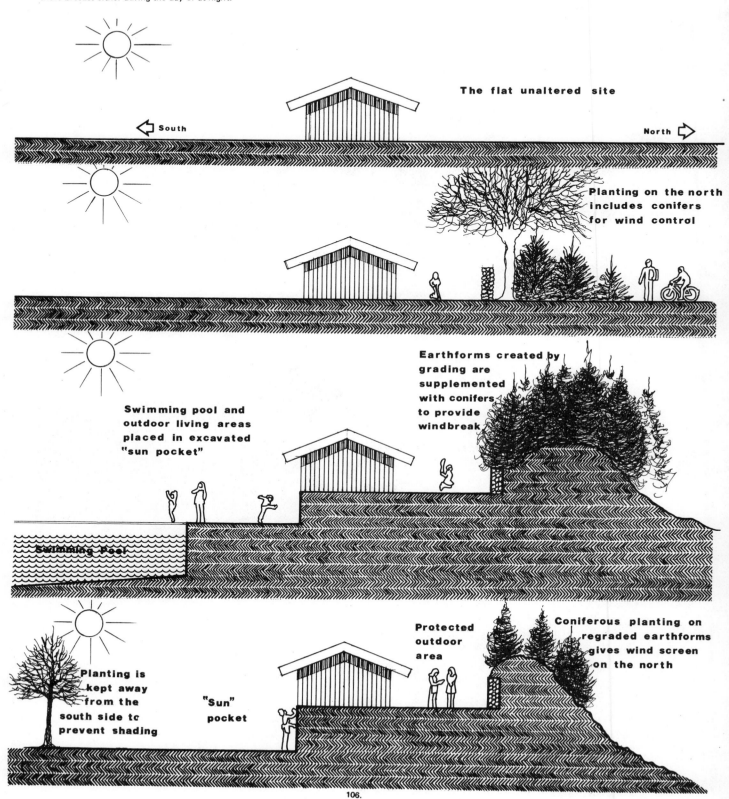

South

North

The flat unaltered site

Planting on the north includes conifers for wind control

Swimming pool and outdoor living areas placed in excavated "sun pocket"

Earthforms created by grading are supplemented with conifers to provide windbreak

Swimming Pool

Planting is kept away from the south side to prevent shading

"Sun" pocket

Protected outdoor area

Coniferous planting on regraded earthforms gives wind screen on the north

106.

Ecological Guidelines

Each geographic area of the United States has a distinct series of ecological communities associated with the geography and the geology. The site designer should be aware of the design characteristics of the ecological community and then use their essential characteristics to supplement, complement and replicate those characteristics in the site after the structure and solar collector are completed. At times, the ecological communities may dictate availability of materials, such as the use of wood products in the northwest; the use of masonry in the southwest; etc. In other situations the ecological communities found in the geographic area may dictate the use of forms of the architecture of the region. The stilt or platform houses of Florida and the Gulf Coast are cases in point. These ecological guidelines should be understood and used as fully as possible by the site designer.

If the site is not chosen properly, if the building is not sited or oriented properly, if the architecture is not appropriate in form or materials and the building is not designed to be fully integrated, then the site design and designer must bear the brunt of the burden and be modified or altered to insure optimum integration. The site is able to be altered, either through addition or subtraction, more easily and less expensively than is the architecture, or the siting or the orientation of the structure or structures.

There are, of course, a number of things which can be done to or with the site and site elements to insure the integration of building and site. Among these are the raising and lowering of earth levels, through grading and earth movement on the site, the addition or subtraction of vegetation of various sizes and shapes, the preservation or introduction of water on to the site, and the use of construction forms, materials and surfaces which may assist in integrating building and the site on which it is or is to be located.

In the manipulation or movement of earth there are limitless degrees of movement or alteration of the ground plane which are possible. All of these, however, may be categorized either as depressing the building *down* into the land or by bringing the earth *up* around the building.

Altering the Ground Plane for Optimum Building-Site Integration

The level site

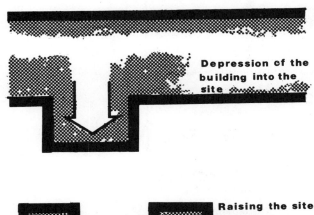

Depression of the building into the site

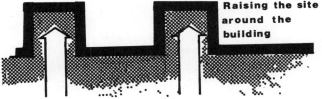

Raising the site around the building

Any building or structure is able to be "tucked into" a site in most instances. This may range from depressing one floor below ground as in a basement, or half a floor, as in an exposed basement to a complete burial of the building with openings for only light and ventilation. In a conference on "Alternatives in Energy Conservation: The Use of Earth Covered Buildings" Kenneth Labs of Washington University in St. Louis, Missouri makes reference to a simple taxonomy of terratectural types with the following illustration:

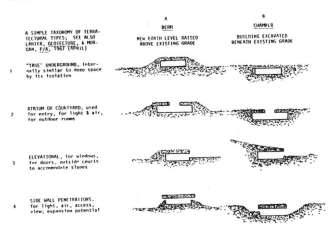

At the same conference, the proceedings of which have been published by the National Science Foundation (NSF RA-76006) through the Superintendent of Documents, U.S. Government Printing Office, Mr. Malcolm Wells the New Jersey architect who has specialized in recent years in designing underground architecture presented the following drawings to show some of his concepts for using the earth for integrating the building and site.

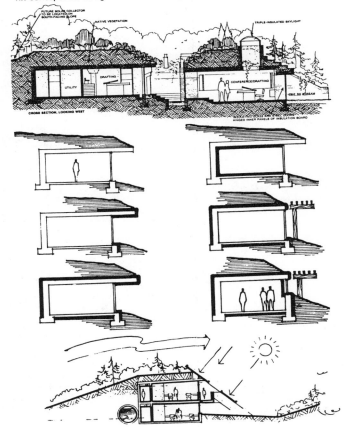

Site Planning for Energy Conservation

The purpose of the site planning and design process is to give form to the basic concepts which are appropriate to the client and to the site, with appropriate elements at appropriate locations and of the appropriate size, shape, and material. The site planning stage in this entire process has to do with planning and organizing the site for optimum human use and enjoyment and for solar energy utilization and accommodation of the necessary equipment and machinery. It must be understood that all of the landscape elements utilized on any specific site have to do largely with providing suitable surfaces for specific activities or for providing either dividers, definers or articulators for the various functions taking place on specific portions of the site.

The site planning process deals with the planning and organizing of the site on a theoretical, conceptual basis. It is just that; planning and organizing the site and site activities. It deals with the distribution of specific activities in a general way, and organizing and locating functions which the client and the designer wish to take place on the site. In site planning, the designer provides for the interaction, the optimum relationship and the optimum location in relation to other activities on the specific site, and in surrounding areas. After locating the functions or activities, it is necessary to specify and locate in a general way the separators, the dividers, the definers and delineators, which may be any of the traditional landscape elements. The following is a suggestion for the site planning process for greater solar radiation utilization and for energy conservation on prototypical sites.

Location of Functions

In order to locate the functions to take place on a specific site, it is necessary to first of all enumerate or list the basic activities which are normally found on single and multiple-family sites. This listing is shown on the following chart.

ENUMERATION AND LISTING OF POSSIBLE FUNCTIONS AND ACTIVITIES ON RESIDENTIAL SITES		
	Scale of Residence	
	Single Family	Multi Family
FUNCTION		
BUILDINGS	Residence Garage or carport	Residential units Garage or carports Storage facilities
ACCESS	To the site To the buildings To storage areas	To the site Throughout the site To each building To storage and service areas
SERVICE	Firetrucks and ambulance Garbage or trash trucks Delivery vehicles Drying yards (optional)	Firetrucks Ambulances Garbage trucks Trash removal vehicles Delivery vehicles
STORAGE	Autos Wastes Tools and supplies Maintenance equipment Pets Recreation equipment	Autos Wastes Maintenance equipment
RECREATION	Play areas Tennis courts Swimming pool Court games	Central play areas Tennis courts Gold courses Swimming pools Outdoor sitting areas Communal gardening areas
OUTDOOR LIVING	Outdoor sitting Entertaining Private living	Community gathering areas Private living areas
CIRCULATION	Walks Driveways	Paths Trails Congregating or gathering areas
SOLAR RADIATION DEVICES	On the residences On the site	On the residential units On the site (individual units) (collective units)

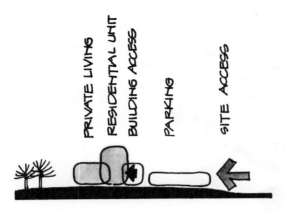

There are normal relationships between various activities that take place on a site. The following diagram shows the relative location of some of those functions on a typical multi-family housing situation and their relationship to one another.

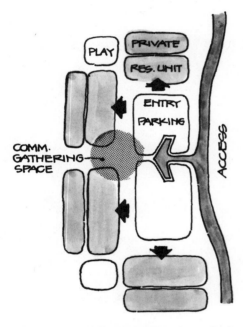

It is necessary, early in the site planning phase, to organize functions to take place on the specific sites and to show the relationships, the interaction and the location of each of these functions. It is also necessary to show the optimum location or possible locations of solar collection devices on the site in this particular phase. The site planning phase is used as a basis for the detailed site design which entails a selection of specific landscape elements which should be used in order to make the site more effective in solar energy utilization or in energy conservation. The functions taking place on the site may be located in different configuration in order to be better able to assist in this function.

There are obviously some functions which are more closely related to one another than are others. Access to the building site and to the individual units on a multi-building complex must be provided. At the same time, parking and storage must be in close proximity to the entrances to the buildings. There is an optimum location for the function in regard to the site, in regard to the client, and in regard to the other functions or activities taking place on the site. These need to be very carefully and graphically depicted by the environmental designer since they dictate to a large degree the location of the landscape elements and use them to define, delineate or separate these functions.

109.

Location of Elements

The traditional or typical landscape elements such as paving, vegetation, fences, walls, site planning, canopies, et cetera, all serve to define, delineate or articulate the area in which various functions take place. The following illustration shows a relationship between the functions indicated on the previous diagram and the elements used to define various functions.

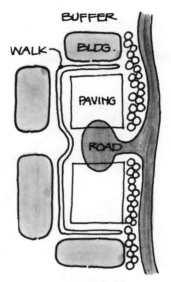

The following illustrations show some of the dangers in the use of vegetation around residential units. It is possible to either deflect or dam cold air flow around the individual structures. Planting should be arranged to deflect rather than dam the air flow, either through or around the individual structure. On the other hand, fences should not be designed or located to restrict cold air drainage. Adequate provision should be made for cold air in the small residential yard to be drained out of the area, thus making it more habitable.

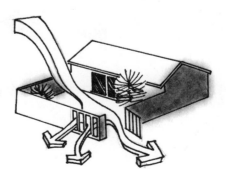

PROVIDE COLD AIR DRAINAGE

VENTURI EFFECT BY PLANTING

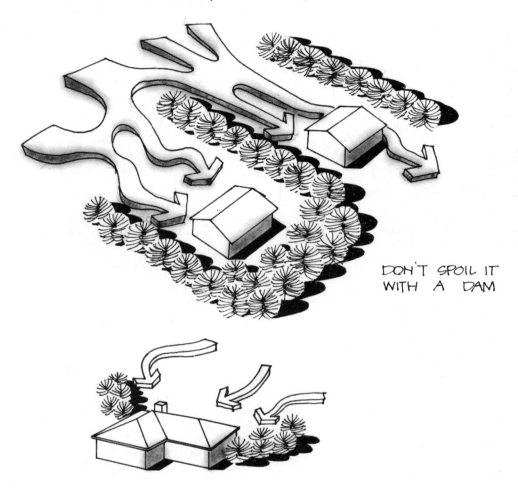

DON'T SPOIL IT
WITH A DAM

DEFLECT RATHER THAN DAM AIR FLOW

The following illustrations show a number of optimum relationships between the various functions taking place on a single- and multi-family site.

The following illustrations show the methodology utilized in siting and orienting various functions or activities on a prototypical site. This illustrates in greater detail the process utilized in siting and orienting architectural and landscape architectural construction.

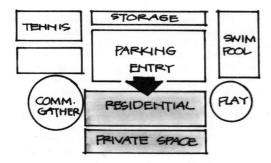

Conceptual site functions and optimum relationships.

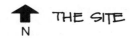
THE SITE
N

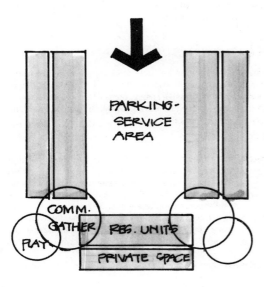

Multi-family site activities and optimum relationships

Optimum site activity relationships modified by site constraints.

Individual site planning elements should be developed so as to take advantage not only of the primary wind currents, but also of secondary currents, either located on the site or in the neighborhood.

The placement of site elements such as paving, fences, canopies and vegetation should be planned so as to provide the optimum relationship and yet division between the functions taking place on the site and more climatic control, specifically for solar radiation and energy conservation.

At the location of functions in the site planning stage of the design process, a major new area of emphasis should be placed on the emerging need for greater solar energy utilization and energy conservation. In order to do that, traditional site elements will probably be placed in new locations because of their new emphasis. The south side of any structure receives the maximum solar radiation. Therefore, various functions taking place on the site should be rearranged in order not to screen the south side of the building arbitrarily but should be reoriented in order to allow the south side of a structure to receive maximum solar radiation.

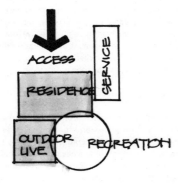

Single family site activities and optimum relationships

The east side of most structures will receive maximum morning solar radiation. If this is desirable, the traditional functions which may be in a particular instance, site or for a specific plan which will be located on the east side of a structure possibly should be reconsidered in light of the need for optimum solar radiation from this particular direction. In some instances, activities which would take place in the early morning time period may possibly be moved on the site to the east side of a particular structure.

The west side of any structure will receive maximum afternoon solar radiation and in most areas of the United States, prevailing winds are also from the west. Therefore, this specific area of the site will receive maximum afternoon sun and will need a certain amount of wind protection.

The north side of most buildings in the United States receives little or no sun except in the very early morning or early afternoon. At the same time, the prevailing winter winds throughout much of North America are from the north and northwest and therefore the maximum wind protection should be provided from this direction. The north side of the building then should be used for activities requiring shade and the shade of the building itself should be used to the maximum extent so as not to require other natural or man-made shade elements to be developed or to be erected on the site. Earth forms, vegetation, fences and walls may always be used to provide wind protection in such situations. Wind control devices, when located on the north side of the building, should, where possible, be made of as light a colored material as possible on the south side so that they will reflect solar radiation into the shaded area on the north side of the building. Paving materials located on the north side of the building should also be made out of light colored material so that they will radiate or reflect the maximum so solar energy reaching them. On the other hand, paving materials, if they must be located on the south or west, should be of a dark material with as coarse a texture as possible. This is so that they will minimize the solar reflection or radiation into any windows located on the south or west side of the particular structure.

Functions which will take place in the late evening hour should never be placed at the bottom of a slope unless adequate air drainage is provided. Care should be taken in locating functions in regard to adjacent water bodies that may have some climatic effect. Since the prevailing winds are from the water onto the land during the day, and from the land back onto the water in the evening, functions which take place either in the early morning or in the late evening should be located so as to utilize these winds rather than to be located in opposition to them.

The traditional elements used to articulate, divide or separate basic functions to take place on the site for maximum solar energy utilization and energy conservation may be the same elements normally used in landscape development or in any landscape design. They may, however, be located in new ways as a result of current and emerging interest in solar energy utilization and in energy conservation to the site development. Many of these traditional elements should be oriented and located in order to take advantage of the sun and to protect against unwanted or undesired wind at specific seasons of the year.

Streets and roads should be located, where possible, on the north side of the structures and not be placed in prime solar radiation areas on the south side of the buildings. Oftentimes it is advisable or possible to screen various areas or activities. Such screening should occur as often as possible on their north side. In this way, the screening device will not block the solar radiation and will provide, at the same time, a wind screen. If the screening is located on the south side, it will, of course, provide a visual screen but it will also block solar radiation. This will put the street or road on the north side of the screening in the shade; this may, in the cool or temperate regions, provide icy conditions in the wintertime. Site planning for solar energy utilization and energy conservation requires the *optimum* dual usage of all elements. This will mean much more careful study of each of the traditional elements used for circulation, articulation or delineation. Walks, trails and paths should be very carefully located as part of the circulation system on the site in order to insure not only optimum orientation but the most desirable cross-sectional configuration.

If walks, trails, or paths are located improperly their configuration may be altered, making them unpleasant areas through which or on which to walk.

Paved areas, such as parking lots, terraces or patios, should be located very carefully in order that the warmth and heat from the solar radiation may be achieved without the undesired glare or reflection from the paved surface which is required for the function of parking or outdoor living.

Oftentimes it is not possible to alter the location of the function or the element; therefore, some tradeoffs will have to be made later in the process in the actual selection of material. The degree of reflectivity of various surfaces should be kept in mind in their location on the south side of buildings. Water, of course, is extremely reflective and would provide undesired glare, whereas asphalt, with a much darker surface, might provide the heat and warmth and yet not provide the unwanted reflection into windows.

Vegetation should be chosen very carefully, understanding fully the total spectrum of the functional abilities of plant materials, not only for climate control but for architectural, for engineering, and for aesthetic uses. Vegetation, many times, is added in the design process near the end; because of that it can assist materially in correcting or rectifying mistakes which are found to still exist by the site designer wishing to fully utilize solar radiation and to conserve energy.

The manipulation of landforms should be very carefully accomplished for climatic effect, and to achieve optimum orientation and location for solar radiation utilization and to block undesired wind in those seasons of the year during which it is not wanted.

Relationships

The basic relationships between functions to be placed on the site during the site planning process may be arranged differently in order to fully utilize solar radiation and to conserve energy. Some of the suggested ways in which these relationships might be changed in light of the new needs, demands and criteria might be as follows.

1. Circulation

It may be necessary to have longer or more extensive means of circulation in order to get the best possible orientation for all of the various activities and functions to take place on the site. At the same time, paved surfaces radiate much more the heat from the sun and should be used and controlled very carefully to direct that radiation where it is needed and wanted at various times of the day and seasons of the year. Walks, as mentioned previously, should be exposed to the sun on the south and should be located so that screening takes place on the north sides of the walks where possible.

Cross-sectional views of walks should not provide for a wind tunnel configuration.

It may be possible to study more carefully the grouping or clustering of buildings to reduce the lines of circulation and to make dual usage of some of the traditional methods of circulation. For instance, extra-wide pathways may be used more often as occasional service drives. Proper facilities should be made on the individual site to allow for use in parking of bicycles, smaller cars and motorcycles as low energy consuming means of circulation. Driveways should be planned so as not to become wind tunnels nor screened so as to become ice covered in cold winter weather. The design of driveways and the screening of these means of circulation should be done so as to provide the maximum solar radiation on the paved surface and low energy consuming usability throughout the entire year.

2. Storage and Service

Where possible, the north side of buildings and other relatively cold and unusable areas should be used for service and for storage. Screening around such storage or service areas should be carefully planned so that the resulting means of screening does not provide a device for shading the paved surface of the storage or service area itself. Parking areas should be lightly screened and shaded with at least a deciduous canopy tree. Careful analysis of the site conditions should be made so as to place the actual storage and service areas in otherwise unpleasant climatic areas, or on areas of the site which are not easily or readily as usable for solar energy utilization or energy conservation.

3. Activities

Some traditional activities which take place on the site, such as recreation, outdoor living, gathering, should be located very carefully in prime areas on the site to make them usable for the maximum amount of time during the year. At the same time, the traditional site elements should be used carefully to ameliorate any resultant unpleasant climatic conditions at the specific or particular season of the year. In the spring or fall, these areas may be more usable if they are located in a protected area on the south side of the structure, while mid-summer usage might be possible and pleasant if these same areas are screened or canopied in some way with deciduous vegetation. Care should be taken that these areas require as limited lines of circulation as possible on the site. Such outdoor activities should be clustered as closely as possible to provide for dual usage of outdoor paved areas and to provide for multiple uses of the site elements, used either for screening, paving or sheltering.

Site Design for Energy Conservation

Site design entails a finer degree of analysis and decision making than have the previous steps in the design process or methodology. In the site design stage the site planner deals with solutions in an abstract way, but he does begin to deal in specifics with such decisions as exact size, configuration, and materials. The site design phase begins the finely honed detailing of the elements to be included on the site. In the future, the site design will be done with solar energy utilization, and energy conservation in mind to a greater degree. The traditional design methodology will continue to be used, with new emphasis on the use of the sun and other climatic factors, and their manipulation through site design decisions.

A. GRADING AND LAND MATERIALS

As mentioned previously, the moving of the earth to increase, decrease or direct climatic factors is probably the least expensive method or technique for site manipulation. The grading plan is eventually the basis for all else that is done in the site design phase. Therefore, a finely honed grading plan supplemented by cross sections, elevations and spot grades should be developed during this particular stage. Earth forms may be manipulated in a great number of ways in order to provide protection from the wind, to create sun pockets and to provide inexpensive screening as well as improving the orientation and location of either architectural, landscape architectural or solar collection structures. Early in the site design phase, very careful calculation on where and how to move and adjust the earth should be made.

After preliminary decisions are made, it is advisable for the site designer to go back over his basic decisions concerning the use of grading and land materials to see that he has not, in fact, created problems in his primary design. If these do exist, a review should be made as to whether they could be corrected by other landscape materials to be added later in the program or whether this basic grading plan needs to be altered so as to rectify at this stage these particular problems.

The angle of slope of the ground surface has a significant effect on the micro-climate of a specific area. The orientation of the earth's surface determines the amount of solar radiation which will strike a particular site. The topography and orientation of an earth form also determines, to a certain extent, the movement of air and thus the amount and type of wind which will strike a site. Orientation, especially in relation to the sun is most important in the temperate region, since quite often in the cool region the sun is lower in the sky and even the summers are somewhat cloudy and at times north slopes and south are illuminated uniformly. In the hot humid and hot arid regions the higher summer angle of the sun has a tendency to lessen the differences between the orientation of either the north or south slope. The maximum direct solar radiation is received by the surface which is perpendicular to the direction of the sun as it strikes a site. This direction and angle will depend upon the latitude in which the site is located and on the season of the year and hour of the day. Because of this a south slope will receive more solar radiation than will a flat site. For instance, a site which slopes 10 percent forward will receive as much solar impact, and will, in essence have the same climate, as will a level site 6 degrees closer to the equator or the difference in latitude between New York City and Memphis, Tennessee. According to Kevin Lynch in his book, *Site Planning,* "On a cloudless day at 40°N. latitude, the total direct and diffuse radiation on a 10 degree (17½ percent) slope attains the following approximate percentages of the possible maximum, depending on season and the orientation of the slope:

Slope Direction	Mid-summer	Equinox	Mid-winter
North	95%-	55%	15%
East or West	100	60	25
South	100	70	35

The same data for a perpendicular wall (where the possible maximum is about one-half of that in the first table) is

Wall Faces	Mid-summer	Equinox	Mid-winter
North	40%	15%	5%
East or West	90	70	25
South	50	95	100

Land forms also have an effect on winds, breezes and wind movement, as mentioned in the section on the climatic impact of natural elements. In essence, wind speeds at the crest of a hill are greater than those on level land while wind speeds generally are slower on the lee side than on the weather side of a hill depending upon the relative slope angles of either side of the hill. Air flow is also generally faster through an opening in a land form. Gen-Generally cold air is heavier than warm air, and because of this warm air rises and cold air sinks. The layer of air near the ground is cooled by the earth at night. The earth is cooling at night through the loss into the night sky of heat stored during the day. Because of this down hill cold air flow is a nightly occurance on open slopes. It is a thin sheet which gathers into a stream in open valleys. It also may form cold air pools where its downhill flow is blocked by a topographical, vegetational or man-made damming object. Therefore areas at the foot of long open slopes are usually cold and damp. Low hollows with no outlet thus become frost pockets. At the site design scale this colder air must be released or accepted in order to make full night usage of outdoor areas. Earth forms are also responsible for the characteristic day/night onshore/offshore wind movement patterns characteristic near the edges of large water bodies. Larger scale land forms near the ocean are responsible for temperature inversions and smog situations.

B. USE, SELECTION AND PLACEMENT OF VEGETATION

The primary purpose of site design is to give areas of comfort and maximum usage on the site itself. Vegetation may be selected, placed and specified to enhance solar radiation utilization, to be used not to interfere with solar radiation utilization, and to conserve the maximum amount of energy. Beyond climatic control vegetation has a full range of other functional uses which may be utilized in the site design phase. Some of these are shown on the accompanying illustration.

Space Articulation

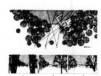
Progressive Realization

Privacy Control

Screening Objectionable Views

Architectural Uses

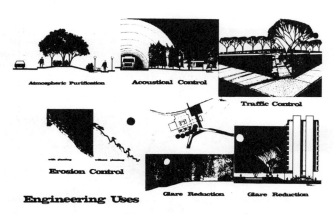

Engineering Uses

Atmospheric Purification
Acoustical Control
Traffic Control
Erosion Control
Glare Reduction
Glare Reduction

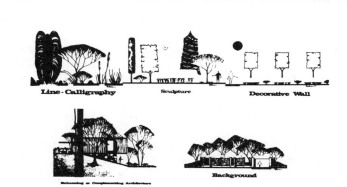

Line - Calligraphy
Sculpture
Decorative Wall
Enhancing or Complementing Architecture
Background

Esthetic Uses

In the site design phase, the site planner should be concerned with preserving, repairing and replanting the vegetation on the site to assist in the energy conservation and solar utilization program of the rest of the entire project. Vegetation may direct, obstruct, or otherwise control the sun's rays, the wind, precipitation and humidity and the temperature. Precise decisions should be made during the site design phase as to possible form, the height, the spread and the shape of the plant to be used in each precise place on the site. A number of possible plants should be suggested and evaluated for each specific location during the site design phase.

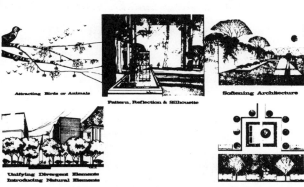

Attracting Birds or Animals
Pattern, Reflection & Silhouette
Softening Architecture
Unifying Divergent Elements Introducing Natural Elements
Enframement

Esthetic Uses

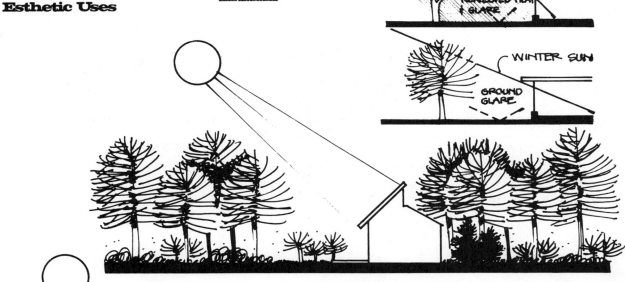

SUMMER SUN

REFLECTED HEAT & GLARE

WINTER SUN

GROUND GLARE

SELECTIVE THINNING TO ENHANCE SOLAR COLLECTION

In some instances vegetation may have to be thinned on a particular site to a certain extent in order to accommodate not only any architectural structure but also a clear path to a solar collector at all seasons of the year and times of the day. It may be that some planting on a heavily wooded site may be damaged and have to be removed because of reflection of the sun's rays off a solar collection device.

At other times it may be necessary or advisable to selectively thin vegetation on a particular site in order to increase air circulation and thus to cool the site naturally.

In still other instances multiple layers of planting may be developed in order to provide a multiple braking effect beyond that which is possible on a site with a single row or layer of planting no matter how dense. By slowing the flow of cold air, it is possible to make the site feel warmer than it would with a faster moving prevailing wind or breeze.

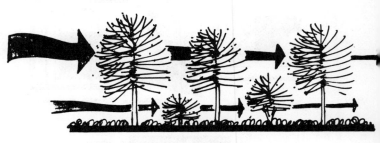

MULTIPLE BRAKING EFFECT

INCREASED AIR CIRCULATION

Obviously the protective zone leeward of windbreak planting is larger than that which is created on the windward. It may be 5 times as large as shown on the following chart, but in addition there is a dead air space, of maximum protection from unwanted breezes or wind, immediately in front of and behind such wind break plantings.

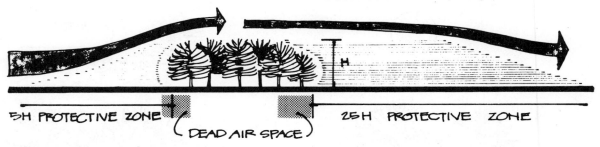

5H PROTECTIVE ZONE   (DEAD AIR SPACE)   25H PROTECTIVE ZONE

Plants and land forms may be used in conjunction, as mentioned previously, to give even greater protection than is possible from either element used separately. Planting in front of a structure on the windward side of a hill provide a short zone of protection while the same planting on the leeward will provide a long zone of protection for the same structure. This is so because of the combined effect of the vegetation and the land form.

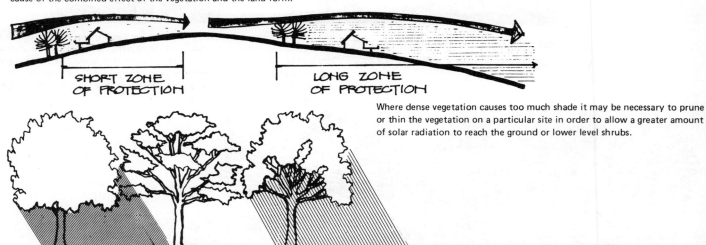

SHORT ZONE OF PROTECTION   LONG ZONE OF PROTECTION

Where dense vegetation causes too much shade it may be necessary to prune or thin the vegetation on a particular site in order to allow a greater amount of solar radiation to reach the ground or lower level shrubs.

As mentioned earlier in the discussion of the climatic impact of natural elements, all heat is exchanged by radiation, by conduction or by convection. The degree and efficiency to which this is done depends upon the medium through which this takes place. Radiant heat transfer depends upon the radiation or absorption capability of the surface. The reflectivity or "albedo" of a surface may vary from 1.0 which is perfect reflectivity of everything which shines on it without receiving or retaining any of the heat or light to 0 which is a perfect matte surface absorbing all of the heat or light and reflecting none of it. The following are illustrative of the albedo rates for common surfaces.

fresh snow	0.80—0.85
cloud surfaces	0.60—0.90
old snow	0.42—0.70
fields, meadow tillage	0.15—0.30
sand and heath	0.10—0.25
forest	0.05—0.18
surface of the sea	0.08—0.10

The albedo rate for rough textured, wet or dark colored surfaces tend to be lower than those of lighter, smoother or drier surfaces. Therefore to insure greater absorption and retention of solar radiation by various paving and construction surfaces darker colors and rougher textures should be used where possible. On the other hand, in order to reflect all incoming solar radiation smooth, light colored paving, wall and construction surfaces should be utilized in landscape design and construction. For instance on the cool north side of a building a light colored fence or wall and a light colored, smooth paving may help to reflect sunlight into the area and thus to make it lighter and warmer than would otherwise be the case.

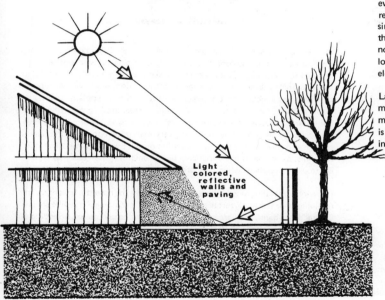

Light colored, reflective walls and paving

The conductivity of a surface refers to the speed with which heat passes through the surface. Materials with high conductivity transmit heat through themselves very quickly. The conductivity of natural material decreases as these materials are drier, less dense and more porous. Therefore a loose, light mulch of peat moss on a planting bed will conduct less heat than will a similar mulch of moist wood chip or bark mulch.

Heat is also transmitted by convection through the air and thus is carried by wind movement. Therefore the speed of the wind itself has a significant effect on cooling or heating. For instance, a 30 mile per hour wind with an air temperature of 30°F. A 12 mile per hour wind which is slowed to 3 miles per hour saves one half the fuel consumption in a residential structure. Therefore by controlling the convective transfer of heat or cold through the use of walls, fences or vegetation it is possible to alter or adjust the microclimate and thus to conserve energy and to provide more usable and pleasant areas outside of buildings.

All natural and man-made objects outdoors store or retain a certain amount of heat. An object with a larger mass is able to store a greater amount of heat and releases it more slowly. Therefore a massive, thick concrete or masonry wall helps to make a building cooler in warm weather and warmer in cold weather. Large water bodies serve the same function by moderating abrupt changes of weather, especially in late fall and early spring. Therefore, outdoor surfaces with a high albedo rate and with low conductivity lead to microclimates of extreme or abrupt temperature changes. If, on the other hand materials with a low albedo rate are combined with those with a high conductivity the surrounding microclimate has a better chance to be moderated. This is so since the excess heat is absorbed then released when the air temperature drops. This is the reason why lakes, oceans, parks, golf courses or even wetlands help in evening out or moderating climatic extremes. At the same time paved parking lots, snow covered earth or cities with large masonry buildings have temperatures which are more extreme. The range of mid-summer temperatures, in one specific area was from 106° on asphalt paving to 52° on nearby grass. The air temperature only varied by 45° in the same area. In an area when the air temperature was 77°, the surface temperature on concrete was 95° and on a dark slate roof on a house was 110°. Since water usually has a lower albedo rate and greater conductivity than does dry soil it follows that the drainage of wetlands increases the albedo and decreases conductivity. Thus, such drainage makes the local mesoclimates or even microclimates less stable. When this is done the levels of atmospheric humidity will fall and the cooling effect of evaporation from the moist soil will be lessened. Since water is usually highly absorptive of solar rays it may have an albedo rate as low as .10. However when early morning or late afternoon sun strikes the same water body at a low angle it is much more reflective of both heat and light. Thus, the water acts as a mirror and may have an albedo as high as .80. This may be acceptable or even pleasant for a structure on the west side of a water body which will receive a great amount of early morning sunshine. On the other hand, a similar structure on the east side of a rather large water body will receive the excessive heat and light reflected off the water's surface in late afternoon. This may be very uncomfortable and could raise the air conditioning load unless relieved or screened by some form of planting or architectural element.

Large masses of unrelieved paving or man-made structural building surfaces may increase the albedo and thus the temperature in the summer urban micro microclimate. Planting when used in parks and on streets throughout a city is able to decrease the albedo and thus serves to not only provide shade during the day but helps to moderate diurnal temperature extremes.

Selection and Use of Site Elements

This, obviously is a logical follow through and extension of some of the site design decisions. During the previous phase the options, alternatives and consequences were outlined in the design selection and use of site elements. The specific naming and dimensioning of the elements takes place in this particular stage or phase. Questions, such as the following, are answered by the site designer at this time: "Exactly how high a fence?" "What does the detail of this part of it look like?" "What are the connecting materials?"

During the selection and use of the site element phase, the contract documents are prepared by the site designer. These would include the plans, specifications, details, estimates and contract drawings. Even after the basic site design decisions are made, more precise, specific and detailed evaluation of the options, the alternatives and the possibilities must be made in regard to solar energy utilization and energy conservation. The increased cost of energy permeates the entire site design process. The detailed design decisions, the selection, placement and joining of materials all reinforce the design concept and are based to a larger degree than in the past on the solar energy utilization and energy conservation. Basically, the selection and use of site elements is prelude to construction. It is derived from the detailed site analysis, client analysis, circulation analysis, use analysis and a review of the constraints and the form determinants. It followed closely on the detailed site design and is, in essence, the detailed material selection phase. Each of the traditional natural and man-made elements typically used in site design and development must be evaluated as to their solar energy utilization and energy conservation factors, both in usage and in procurement.

The following are some considerations in the selection and use of various natural and man-made site elements for solar radiation utilization and for energy conservation.

Natural Elements

1. Vegetation

Oftentimes in the site design process the environmental designer makes a basic decision that he wishes to use a tree, a shrub or a ground cover in the specific location for a precise purpose. During the actual selection of this tree, the environmental designer has to decide precisely which tree and why he wants to use it from the standpoint of better solar radiation utilization and energy conservation on the site itself, in relationship to the architecture or to the solar collecting device. Obviously, each region has specific vegetational restrictions and limitations. The following, however, are general guidelines, realizing that not all of the forms, the plants or suggestions can be used in each area. It also must be realized that vegetation will have to be used in conjunction with either other vegetation, with architectural extensions, with paving surfaces, or altered or restructured ground forms.

The most commonly mentioned use of vegetation is the deciduous tree placed in the south side of the building which loses its leaves in the winter time. This makes what is generally considered the optimum solar control device. The plant has leaves in the summertime when the solar radiation must be blocked or in some way have its impact lessened, and loses its leaves in the wintertime when the sun's rays are desired or are more beneficial, at least in the more temperate or cooler regions of the nation. The following diagram illustrates that capacity of the deciduous tree.

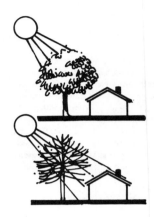

Most commonly suggested for wind control purposes is the conifer planting. There are certain limitations, however, in using a single row of conifers for wind control purposes. Since these evergreen plants are largely pyramidal in shape, a single row protects very well in cross section, but in an elevational view provides a picket fence appearance which must be supplemented by either another row of conifers or by supplementing these forms with various other deciduous or ericaceous plants. In some cases, older conifers have a tendency to become leggy, providing very little wind control at the base and are extremely tall plants with sharply pointed pyramidal foliage at the top. This, of course, is not envisioned by the environmental designer wishing to use these plants when they are young and small.

For wind control purposes, the ericaceous tree forms have the advantage of retaining their leaves all through the year and of providing a tree-like shape. On the other hand, they are extremely slow growing and dense, providing a great amount of shade at all seasons of the year. They have the disadvantage of usually only growing in the south, particularly in the hot-humid regions.

It must be kept in mind that vegetation of all types is a growing, changing natural element and, as such, it is not possible to predict a precise or exact form for every single plant used on a particular project. In the use of any vegetation, the following questions must be asked in regard to the ability of the plant to assist in solar radiation utilization and in energy conservation:

— the height and the spread of the single plant;

— the spacing of multiple plants;

— the shape or form of the plant;

— the density of the plant;

— whether the plant is a conifer, deciduous or an ericaceous plant; and

— the ultimate size and shape of the plant.

Basically, plant forms may be divided into the following categories:

— *Trees:* major and minor; these may also be categorized as canopy trees or understory trees. The division is generally made between those trees under 20 feet high and trees which are over 20 feet high.

118.

— *Shrubs:* may be arbitrarily divided into low shrubs, which mean those below four feet high; medium shrubs, those from four to eight feet; and tall shrubs, those eight feet and over.

— *Ground cover:* may be categorized as to their height, though most of them are 18 inches and under, their color, their density and whether they are coniferous or deciduous plants.

— *Vines:* have a wider variety, whether they are clinging, twining, or rambling vines. Quite often vines may be used to shade the wall of a building or may be used as part of an architectural canopy to provide additional climate control.

The following are series of diagrams illustrating some of the considerations concerning the precise selection of various forms of vegetation for solar energy utilization and for energy conservation.

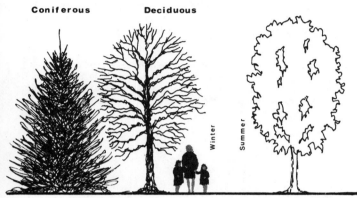

TYPE OF TREES

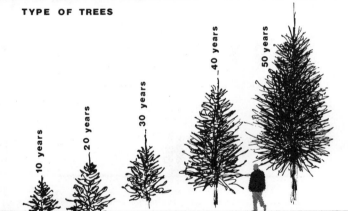

AGE OF TREES

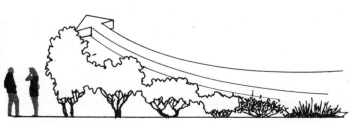

CONFIGURATION OF PLANTING

2. Earth Forms

In nearly every situation, when a building is to be built, the earth will have to be formed, shaped and altered to some degree. The cost to move earth for climate control so solar utilization purposes is relatively minimal, if anticipated early enough in the design and construction process. The major concern in the utilization of earth forms is the basic type of earth or soil to be utilized, whether sand, soil of any types, gravel or stone, and the basic angle of repose, it is necessary to have some sort of surfacing or means of stabilization. Also, in regrading or reforming the earth, it is necessary to take into account the drainage problems in all regions. Basically, earth forms take a great deal of horizontal room for the vertical advantages which they may provide. It is necessary to evaluate very carefully the height and the width of the desired earth form and to very carefully develop a grading plan in order that that precise earth form may be realized in the location and at the angle for which it is desired, either to utilize more fully the available solar radiation or to conserve more completely the available energy.

It is essential to be aware of the limitations of soil in specific areas before specifying certain land configuration for solar radiation utilization or energy conservation. Since it may not be possible to obtain the landform configuration desired, the site design may have to be modified to accommodate the limitations of the soil or extensive work may have to be undertaken to insure soil stabilization and erosion control on steeper slopes. This may entail excessive costs in irrigation and maintenance and may not be, in fact, justified by the ultimate savings in energy. These are the tradeoffs which the site designer must make in specific instances in such situations. Oftentimes the earth forms, as mentioned previously, can be very carefully ipulated·

manipulated to increase the height or the location and hence the effectiveness of either man-made architectural forms such as fences or walls, or vegetation. A precise grading plan should be developed by the site designer at this stage and the necessary specifications should be developed for soil stabilization and erosion control. The following illustrations indicate some of the detailed considerations in the use of earth forms for these purposes.

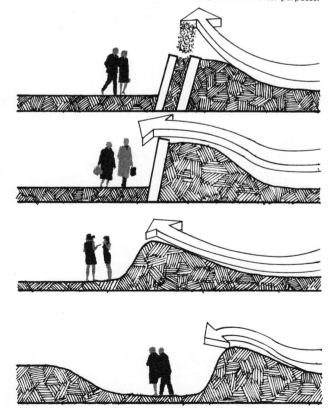

Man Made Elements

1. Vertical Elements

Usually the building or the architecture itself is the largest man-made element on a specific site. Quite often, for reasons of economy and aesthetics, the material and forms of the architecture will be that which is indigenous to the specific region and will dictate to a large degree the material used in the vertical landscape elements.

a. Walls

Precise design decisions must be made by the environmental designer at this stage in regard to the fenestration, the height, the material, and the precise location for walls, fences and screens. The most common construction elements for walls are obviously brick, stone, and concrete block. Such walls, depending on their height, may be able to block solar radiation and have certain degrees of ability to block the wind. The designer should study the cost effectiveness of various heights, widths, and materials for the man-made vertical elements. These walls may be either free-standing or they may be retaining walls, holding back a section of the earth's surface. Certain walls may be utilized to increase either the height or to provide a more optimum location for vegetation or architectural materials. In such cases, it is necessary to determine that the wall materials are obviously strong enough to hold back the earth or to sustain the vegetation or architectural forms. The following diagrams indicate some of the possible configurations in the use of walls as man-made vertical elements for solar radiation utilization and for energy conservation.

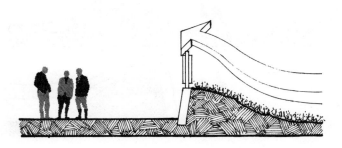

b. Fences

Fences obviously are free-standing architectural elements, usually made out of wood, metal, glass or plastic. Obviously, once again, certain basic decisions are able to be made concerning the material, the height, and the width of the fence on the basis of the surrounding architectural materials in use, the availability of various possible fencing materials, and the effect desired from the fence, either in solar radiation control, wind control or in energy conservation. Oftentimes fences may have openings or fenestration or they may be made of a relatively solid material.

Fences may, of course, by the same token, be fully closed all the way to the ground or they may have openings at the lower levels to allow for the passage of wind under the fence. There are, of course, certain obvious height limitations with fences, both as to how high they may be constructed or how low they can be and still have any effectiveness at all. Generally, fences below three feet are largely for the purposes of traffic control only. Over that height they do have some ability to control the microclimate of the outdoor environment. The following illustration shows a number of possible fence types.

There are, of course, nearly as many types of fences as there are materials and designers. The precise fence must be carefully designed, detailed and constructed to be appropriate for the specific site and the specific climate control problems for which they are used.

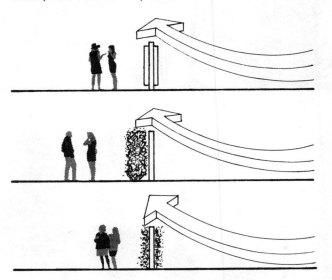

Fences can, of course, be used in conjunction with architecture, landforms, and overhead canopies. Heat and glare from the sun can be controlled by fences in several ways. Glare can be reduced through the use of plastic panels or glare-reducing glass. These materials can also turn dark corners into softly lighted areas. For dappled shade, sunlight can be filtered through a louvered slat or basketweave fence or trellis. Vertical sun screens around a patio or terrace will keep out the late afternoon sun which slants in under trees or a roof overhang. This type of screen can also be used to cool the western wall of a house by stopping or filtering the sun before it reaches the walls and windows.

The wind is not as easy to control as the sun because its behavior is less predictable. Before a screen or fence can be designed for wind control purposes, the environmental designer should have a complete understanding of the wind's behavior in a specific site or situation. The prevailing direction of the wind in a certain area is not necessarily the direction it will blow through a specific site. Houses and other large objects such as fences, walls and planting can act as baffles for the action of the wind. A particular small site can be affected by the fence's relationship to the house. Before planning and specifying a precise windscreen, it is advisable for the environmental designer to study and chart the wind currents by possibly hanging small flags about the site and noting their movement during windy periods. There are normally a number of choices in selecting materials for wind control structures. A transparent material such as glass might seem a logical choice for preserving a view as well as for controlling the wind. But, surprisingly, it is seldom the most efficient material for wind control. A windscreen made of closely woven slats or one of the slanting baffle on top will break up a strong wind more efficiently. A fence for wind control and a tree for solar radiation control when used in conjunction may alter or adjust the comfort levels significantly and reduce the need for extra energy usage.

A solid fence may not be the most effective means of controlling or reducing windflow. Due to the fact that a complete blockage of wind may not be essential to raise the temperature to a comfort level it is not vital to completely block the wind. A solid fence which reaches all the way to the ground creates an eddy which has a tendency to pull the flow of the wind down to ground level. A wind speed of 10 miles per hour, an air temperature of 68° and a shaded situation gives a feeling of 62°. Whereas an air temperature of 68° with full sun and the wind blocked gives a feeling of 75°.

The Sunset book *How to Build Fences and Gates* makes reference to studies regarding the effectiveness of various fence configurations to slow the wind and effect the temperature on the leeward side of a fence. Rather than using standard fence details, the environmental designer should be encouraged to design a specific fence, in elevation and sectional view, for specific wind control problems on the site and the activities which will take place on the leeward side of the fence. An outdoor play area, a dining area, an area for sunning all require different types and degrees of wind control and thus comfort levels. It may be possible, and in some cases even advisable to design and develop a fence with moveable louvers which are able to be moved in order to direct winds and breezes at different seasons of the year and times of the day to provide the desired comfort levels.

In areas of consistently heavy prevailing winds a solid fence is not a good wind screen to the reverse flow or eddy formed leeward of it.

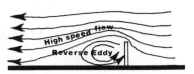

A 6 inch space left at the bottom of a solid fence relieves some of the pressure and reduces the leeward eddy.

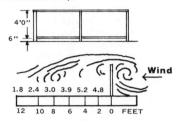

By adding a 45° baffle to the top of a six foot fence the amount of wind protection behind the fence is increased considerably depending upon the direction which the baffeling device faces.

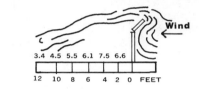

238.

Various types of louvered fence configurations have differing effects upon wind flow patterns by both lowering the wind velocity and the temperature leeward of the perforated fences.

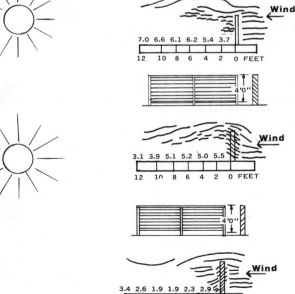

239.

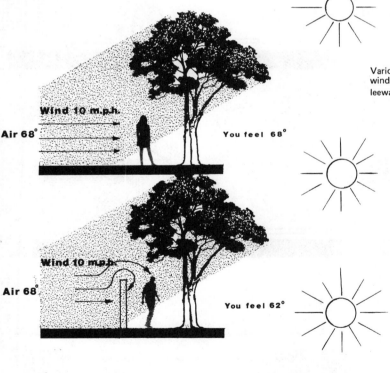

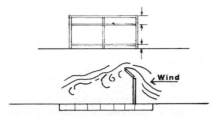

240.

In addition to the form and materials of a fence the color or lightness or darkness of a fencing material may contribute to its reflectivity or ability to absorb or deflect light or heat. This obviously has an impact on the energy conservation or solar radiation utilization of site elements since a fence is thus able to reflect or radiate heat or light into colder areas and to retain daytime heat for release during the cooler evening hours.

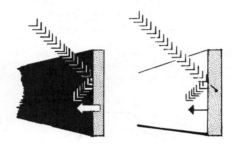

A lighter colored fence will reflect a greater amount of lighted heat. As mentioned earlier it is possible to use a light colored fence or wall facing south into a dark north side of a building as well as other places where the heat and light of the sun does not reach directly.

2. Horizontal Surfaces

Basically, man-made horizontal surfaces are of at least four types. These are: pavings, decks, artificial turf and canopies.

a. Pavings

The precise specification and detailing of the paving material used in various areas of the site should be very carefully studied. It is not possible to use the same material throughout an entire site without having a great number of climatic effects in different areas. The basic paving materials which are generally able to be specified are *concrete* (of all types and all colors); *asphalt* (of all varieties and with many different surfacing colors and material); *brick* (of various sizes, shapes and types); *stone* (such as slate, flagstone and limestone, bluestone and granite paver); *wood* (either wood rounds or wood blocks used as paving rather than decking material); and *cast stone materials*. Basically, in the precise design detailing of paving materials, the reflectivity, the ability to absorb and the ability to retain heat are all significant factors which need to be studied and reviewed.
The following illustrations show some of the various considerations in specifying, in detail design and dimensioning of horizontal surfacing or paving materials.

b. Decks

Various sorts of elevated deck structures are usually constructed of a variety of colors and types of wood. Climatically some of the advantages of an elevated deck are that it provides for ventilation from underneath the deck structure and the wood itself provides a lesser amount of reflectivity and hence a greater amount of heat absorption on warmer days. Since the wood deck is able to be removed from the horizontal surface in which its footings are found, it is able to place the human users of the deck at a higher

elevation where they are able to obtain a greater degree of climatic comfort from possible cooling breezes available at these higher elevations. In the detailed microclimatic analysis of a specific site, it may be found that there is either an exceptionally warm or exceptionally cool location on the site which is not at ground level. In such cases it may be that the designer may wish to place the deck at a higher elevation, either as an extension of the house or at some other area of the site. In such a location the users of the deck may take advantage of the specific detailed climatic situation. The following are some suggestions as to ways in which decks might be used in a variety of situations to provide access to specific areas on a site which is either more comfortable or more desirable at certain times of the year or time of the day. These illustrations also show some of the possible climatic effects and impacts of decks in regard to solar energy utilization and energy conservation.

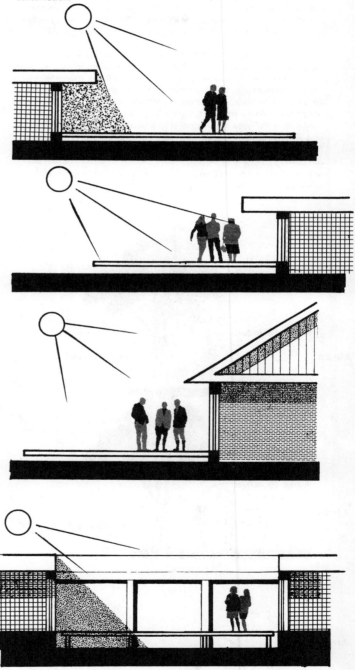

Usable Space Between Buildings

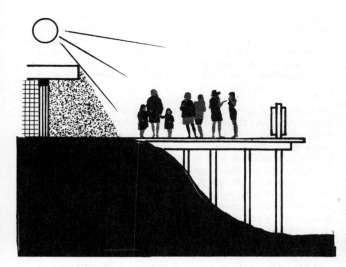

Access to Solar Exposure

A fence, if it is in the right location, of the optimum height and design is able to control wind flow and increase comfort and thus are useful architectural extensions to help in energy conservation and solar energy utilization.

Paving Over the Water

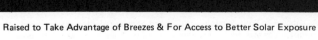

Paving Without Grading

Raised to Take Advantage of Breezes & For Access to Better Solar Exposure

excessive solar radiation and thus heat on a structure or on a paved area the design and development of a canopy structure extending from the architecture is a logical landscape architectural solution. For energy conservation and optimum solar energy utilization there are myriad options in canopy design, location and covering. Before any decisions are made, however, there are a number of basic principles which would govern the usage, design and placement of such elements.

The south side of any structure receives the majority of the solar radiation all during the day at all seasons of the year. The east side receives strong early morning sun which is usually not as warm as the afternoon sun which strikes the west side of a building. The area directly on the north side of any building does not receive any mid-day sun in any section of the United States. The early morning and some of the late afternoon sun may strike the north side but it will be for a very short period of time when the solar radiation has less strength and intensity. Therefore for all practical purposes the north side is shade most of the day by any structure. A multi-story structure will, of course, cast a longer shadow and a longer building will cast an appropriate shadow. Seldom is a canopy of any type, thus, needed on the north side of any building.

Architectural extensions are actually those structural elements which extend into the landscape to perform a design, climate control, or space articulating function. A canopy may consist of a structural framework of wood or metal with a covering of any number of materials. The framework will depend upon the surrounding architecture, the weight of the covering and the concept of the designer. The covering material will depend on the degree of solar radiation which is needed and precisely where it is located.

The degree or amount of solar radiation, the season of the year and the time of the day it is needed and whether the shade is needed on a building, a paved surface, or turf or earth all dictate the form covering and location of the canopy. To conserve energy in outdoor areas it may be necessary to design a canopy to shade a paved area but not the structure.

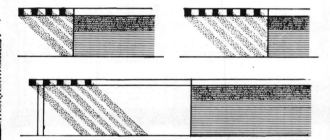

If the structure is either an "L" or a "U" shaped with a canopy to be fitted into the opening it will change the character and the canopy needed for energy conservation.

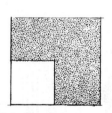

West facing terrace

NORTH ⇧

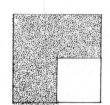

East facing terrace

South facing enclosed terrace

Air Circulation Under Deck

c. Artificial Turf

Earlier in this document research was cited in regard to the abilities of artificial turf materials to alter the microclimate of the air immediately above the surface on which they are placed. Generally, artificial turf, which it has many of the engineering characteristics of actual turf as far as being able to carry foot traffic does behave far differently than natural turf in regard to climate control. Generally the base in which artificial turf is found is normally asphalt or asphaltic byproducts. Because of this, artificial turfs of all types generally have some of the same general climate control characteristics as the asphalt itself. It does retain a good deal of the heat and warmth of the sun and provides a pleasant surface in cool weather in the evening in a protected location. However, it is oppressively warm on a summer day. Designers should keep this in mind before using the material extensively. Further studies will have to be conducted in years to come on artificial turf in a number of situations beyond its use on baseball and football fields which have been traditionally the major areas of its usage.

d. Overhead Canopies

As a device to limit or control the impact of solar radiation the overhead canopy, as an architectural extension, is most effective in energy conservation. Where there is no way with either land forms or vegetation to curtail

The materials which can cover a canopy for energy conservation or more effective solar radiation utilization may consist of:

wood—solid wood, wood lath, slats, louvers

metal screen—solid, perforated, metal lath, screening, mesh or expanded metal

glass

fabrics—canvas, burlap

planting—on the canopy framework

The choice of materials will depend on the degree of sun control needed, the cost, rain control factors, and permanence. Shading of paving will cut down on its reflectivity and degree of heat retention during the cooler evening hours.

Many environmental designers believe it is possible to control the overly warm rays of the afternoon sun by building an overhead canopy. It it possible that an overhead canopy of various materials can delay the penetration of the sun into a particular building for limited periods of time. But even such a canopy will not keep the sun away from people sitting under it on a paved terrace or patio when the low slanting rays come in under the roof and make this area somewhat unpleasant late in the afternoon. In such situations, the best possible protection is a row of closely spaced tall shrubs or trees. However, it is not possible or advisable to wait the three to five years that the plants require to grow to a screening height. The cost of purchasing fully grown plants or extremely large plants for such use is often prohibitive. Therefore, one of the better solutions is to provide vertical screening. This might be done through walling in the western side of an overhead structure or by hanging sun screens from the roof line which might be rolled up when not needed. It may be possible to use this sort of temporary screening and also to plant vegetation which can later grow to an adequate size at maturity. Generally, the orientation of any outdoor paved area is the chief determinant of the precise microclimate, either to be enjoyed or evaded in that particular area. This is also the primary reason for using canopy roofs over specific paved areas. The path of the sun, the amount of sun, and the intensity of the sun through the seasons of the year and the hours of the day affect the location, kind and amount of outdoor living and, at the same time, the site and usage of any outdoor paved area. As mentioned previously, the sun's angle on any day depends on how far north of the Equator any specific site is located. The effect of latitude makes only a relatively minor change in the sun's angles throughout the United States.

If a particular structure is located in a hot-arid or a hot-humid climate and the only location for outdoor areas is on the south, it may be necessary to use canopies with a vine covering or some sort of mist-spray device in order to make the area usable in that situation. Generally, any site development on the southern side of a building requires some sort of a roof that can be left up all year and one that provides maximum sun protection. Because of its exposure to the sun, it can be safely covered with perishable material such as canvas, reed or bamboo because they will be dried out quickly after a heavy dew or rain. Louvers and east-west running lath provide effective protection. Solid coverings, likewise, do well.

A west-oriented outdoor area is liable to be extremely warm in the late afternoon when it receives the full force of the sun's rays. Without overhead protection, the sun will radiate against the west wall of the building, with six times as much heat in the summer as it does in winter, and much of the heat will radiate back into an outdoor area. A canopy will make such a paved area usable from 11 o'clock in the morning until sometime after four o'clock which it may not have been before it was roofed. This overhead canopy, however, will not block out the low, hot rays of the afternoon sun. They will, as mentioned previously, only be caught by some sort of vertical sunscreen.

In a hot-humid or hot-arid region, the design of an outdoor overhead roof structure requires careful attention. It is necessary to provide for air circulation and avoid light colored building materials which would assist in building up a heat trap. Because the west-oriented outdoor area does not receive much morning sun, it should not be roofed with perishable materials unless they are taken down for winter. Damp canvas, bamboo or reed might not

dry out enough during the winter to prevent mildew growth. In cool winter areas, environmental designers should work with homeowners and developers to provide all the sunlight they can get from the west during the cold months. In these localities, removable coverings or deciduous vines would seem advisable because they do not block out the afternoon sun which is desired during these colder indoor months.

When an outdoor area faces the east, it benefits from the morning sun and begins to cool off in the afternoon. It is desirable orientation for a hot climate. In many climates overhead protection is not essential unless it is required to shed rain or fog or is desired as a means for holding the heat of the day into the evening. Fiberglass and glass are popular solutions in drizzly areas. Louvers or laths set to block the hot morning sun make good sense in hot-arid locations. Perishable covering should not be used except in arid areas, because once damp they may not receive sufficient sun to dry out properly. Removable coverings, panels, screens and rolls are favored in cool winter areas when the sun is not intense from sunrise until noon through the cold months.

Obviously, the coldest site for an outdoor living area is a northern exposure. Part of the north-facing terrace may never receive any direct sun, even in the summer when the sun in on its northward leg. If such a paved area is in back of a multi-story building, the entire area may pass through four seasons in complete shade. Because of its sheltered position, it does not need an overhead canopy, except as a protection from rain or as a visual necessity.

In a damp climate, glass, plastic or fiberglass roofs will shed rain and let in the light. In hot climate, a north-south lath or an egg crate will block glare and yet provide a pleasant feeling of protection. In cool summer areas, if there is a vertical screen to block the western sun, this may be all the protection you need. Perishable coverings of various types may be used in overhead canopies in dry climates. To leave them up through the winter, however, may be to invite damage through rotting or mildew.

Roofing over an outdoor area which is located in the pocket of an "L" or a "U" formed by walls of a building presents some special factors to be considered by the environmental designer. When an outdoor living area is enclosed by two or more walls of a structure, it may present the designer with additional problems if one wall of the building is a window wall. For example, an "L" with west and north-facing walls will need secure sun protection from the west-facing wall. But if the north-facing wall is a window wall, the designer will have to consider what affect the roof will have on the windowed room, particularly in winter. In such instances, the environmental designer has a number of possible options or choices. One of these is to build an adjustable or movable roof so that light can be admitted through the window in the wall as needed. The covering, on the other hand, could be of a removable material such as bamboo, reed, canvas, etc., which can be taken down in the winter. It is also possible for the environmental designer to specify a deciduous vine cover for the canopy. It is also possible to leave an open space between the roof and the north-facing wall. Ventilation in an "L" location may also be troublesome if the designer knows from experience that a particular outdoor living area is liable to be inadequately ventilated during the summer. It is advisable to make provision for letting air through and out of the area. This might be done by opening up the eaves or providing open slots or by placing various canopy sections at different levels so that ventilation can take place between the individual canopy. If an outdoor area is in a "U" or a fully enclosed court, the biggest problem for the environmental designer will be to keep from building an extremely hot box-like environment which will, in effect, not be usable for much of the year. In this case, ventilation in any outdoor living area is essential. A canopy that blocks the sun but admits air through closely spaced lath, netting, reed, bamboo, etc., should be the best solution in this case. Heat transmitting covering, such as glass, fiberglass, or plastic, should be installed with a great deal of caution. They should be provided with means for ventilation and should, if possible, have a layer of lath, reed, bamboo, louver or other sun-filtering cover fitted in the place above the glass. The unfortunate consequences of an incorrect specification or installation by the environmental designer are such that a great deal of study should be given to such areas before site decisions are made.

In using a lath or louvered overhead canopy for energy conservation or solar radiation control rather precise calculations are possible in regard to the amount of sun which may be curtailed or admitted at different times of the day and seasons of the year. The angle of the summer sun at midday in relation to the surface of the earth varies from 69° at 45° north latitude which includes the northern tier of states in the cold region to 84° in the south tier of states along the 30° north latitude including the hot-arid and the hot-humid regions. The sun angle at 40° north latitude is 74° and at 35° north latitude it is 79°. On the west coast 40° north latitude is in the temperate region, while 35°north latitude is in the hot-arid region. In all of the rest of the United States both the 40° and the 35° lines are in the temperate regions. The altitude of the sun at 8 a.m. and at 4 p.m. is nearly 37° in all four latitudes. It is possible to gain sun control with louvers in a variety of ways by calculating carefully the sun's angle and direction on a specific site at different times of the day and seasons of the year. For instance all three of the following louver angles will control the sun's rays, but certain ones will require more materials or will provide a more complete all day shade if the sun's angle is at 75°.

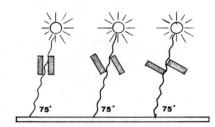

241.

The angle of the louvers and the direction which they are oriented will determine their effectiveness in sun control at different times of the day. For instance east-west louver slanted towards north block the mid-day sun.

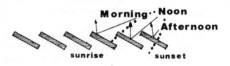

242.

North-south louvers may be slanted toward the east to pick up the morning sun and to block the afternoon sun. On the other hand, they may be slanted towards the west to pick up the afternoon sun and to block the rays of the morning sun.

3. Other Architectural Extensions

It may be advisable for the landscape architect working on the site itself to work very closely with the designer of the architecture to suggest ways in which architectural extensions might be utilized more fully to modify the climate, or the climatic effectiveness of the site. It also may be possible for the site designer to develop architectural extensions during the site development stage which would assist materially in the effectiveness of the site elements. The height, width, and direction of these architectural extensions should be very carefully selected by the environmental designer as to their long-range impact at all seasons of the year and at all times of the day. These should be studied in conjunction with the architecture, with vegetational forms and with land forms of the specific site. Such architectural extensions may consist of:

- walls or fences extending from the architecture;
- overhead canopies designed and developed in conjunction with and extending from the architecture;
- the development of courtyards inside of the individual piece of architecture; as well as
- balconies at various locations on the building and, in some instances,
- windscreens on the roof of the building itself.

Some of these architectural extensions may be stationary or movable to provide climatic amelioration at different times of the day or seasons of the year. These extensions should be very carefully designed in conjunction between the architect of the building and the site designer to provide the maximum potential impact on the climate surrounding the building itself. The following illustrations show some of the ways in which an architectural extension may assist in climate control on the site itself.

The utilization of structural wind scoops on architectural or landscape architectural elements is nearly as old as building in many parts of the world. Indigenous architecture often responds to hot, calm conditions by utilizing devices which catch and direct what wind there is into an area where it is needed. It is possible to use this technique in conjunction with fences, with terraces and porches to pull the maximum natural ventilation through an area, in order to cool it without the use of expensive and energy consuming air conditioning systems.

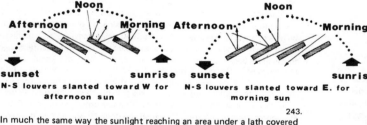

243.

In much the same way the sunlight reaching an area under a lath covered canopy is lessened depending upon the width, the spacing and the location of the lath in relation to the supporting structure. Therefore it is not enough to simply specify a lath covered canopy, but for it to be fully effective in solar radiation control in order to conserve energy, it must be designed in detail for a specific site and for particular climate conditions. For instance, the normal placement of lathing, one lath apart on top of a supporting structure will provide a certain amount of solar radiation control. The alternate placement of lath, with the same spacing as before on the top and the bottom of the structural member will significantly alter the amount of sunlight which reaches the area under the shade control structure.

244.

245.

Architectural extensions for wind control for natural ventilation

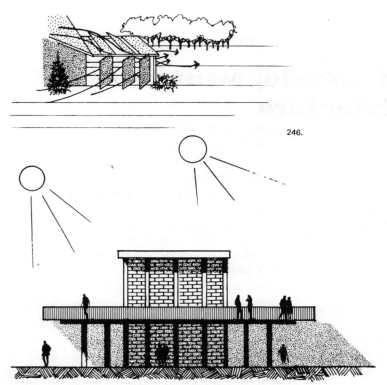

246.

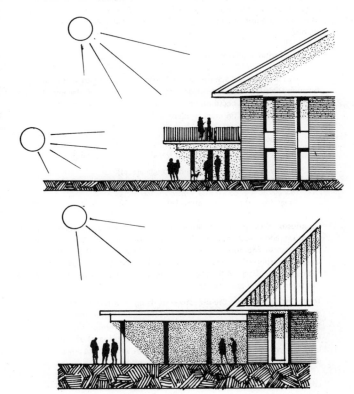

Architectural extensions for solar radiation control

127.

Site Planning and Development for Solar Energy Architecture

All buildings are placed or oriented in a precise location on a specific site by someone involved in the design process. There are certain site implications affecting the accommodation of solar collection systems. The site on which a solar heated or cooled structure is placed must be selected with care, modified as necessary, and utilized fully to maximize the solar collection impact and minimize the loss of energy generated.

Since each building site has unique characteristics and conditions, the same solar structure placed on various sites may require completely different site planning and design decisions. Yet while the site and the design of the structure may vary significantly from one project to another, the process of site modification or development is replicable and may be altered to include design factors associated with solar heating and cooling.

The Site Planning Process

Site planning is concerned with applying the same objective analysis and design process to specific site-related problems at increasingly smaller scales. The higher cost of energy and the utilization of solar collection devices call for a rethinking of the cost of site design so as to avoid the energy intensive architectural and mechanical modifications necessary to correct errors in site selection, in building orientation or placement or in modification of development of the site. Proper site planning can lessen the need for extensive and expensive architectural or mechanical corrections. The following steps are essential to economic design:

   . . . proper selection of a site within a region prevents problems in location on a structure on the site;

   . . . proper location of a structure on a site prevents problems in site planning;

   . . . proper site planning decisions prevent problems in site design, materials selection and detailing;

   . . . proper site design decisions prevent problems in architectural design; and

   . . . proper architectural design prevents problems in the engineering systems utilized in solar collection and in heating, ventilating and air conditioning.

The total planning and design process begins in the evaluation of a region and ends with the individual and his immediate environment within the structure itself. Site planning is a part of that total process.

Site Selection

At times, a builder, developer, or designer of dwelling units utilizing solar collectors may have the option of actually selecting a site or of determining the precise location on a larger site for the unit or units. In such instances, the best site for utilizing solar energy should be chosen by very carefully analyzing and evaluating all of the options and alternatives. The problems created by choosing an unsatisfactory site or location will have to be overcome by architectural or engineering modification of the structure, or the solar collection system. In terms of life cycle costs, both of these options are more expensive than proper site evaluation and selection.

Site selection decisions may deal with the choice of the optimum site within a region or an area or with the location of a specific activity or a particular building within a given site. Analytical data for the site(s) must be gathered in order to insure optimum solar utilization and should include information on the following factors:

   . . . geography of the area surrounding the site;

   . . . angles of slope on the site;

   . . . orientation of slopes on the site;

   . . . geology underlying the site;

   . . . existing soil potential and constraints (engineering and agricultural);

   . . . existing vegetation (size, variety and location);

   . . . climatically protected areas on the site;

   . . . climatically exposed locations on the site;

   . . . natural access routes to and through the site;

   . . . solar radiation patterns on the site;

   . . . wind patterns across the site;

   . . . precipitation patterns on the site (fog, dew, frost, snow drifting);

   . . . temperature patterns on the site (daily and seasonal); and

   . . . water drainage patterns on the site.

This analysis may be accomplished through the utilization of air photos or other remote sensing data such as maps, charts or by direct observations on the site. Site selection at whatever scale must take into account the distinctive characteristics of the major climatic regions of the United States. The analysis data having been collected and organized, it is then used to evaluate, rate and eventually select a specific location as the optimum site for a dwelling unit with a solar collection device and its related site activities. Obviously, the importance of each of the factors will vary in each climatic region.

Building Siting and Orientation

The next step toward achievement of optimum solar energy utilization through the site planning process is the determination of placement of the activities which will occur on various areas of the site. The most common activity areas found on most residential sites include:

   . . . the housing structure;

   . . . means of access (entrances to the site and the building);

   . . . means of service (service and storage areas);

   . . . areas for outdoor living (patios, terraces, et cetera); and

   . . . areas for outdoor recreation (play areas, pools, courts, et cetera).

On housing sites where solar energy is used for heating and cooling of the structure, additional areas for the accommodation of the collection device are necessary. The importance of interception of maximum solar radiation makes the orientation of the structure and the collector of great importance in this process. Each of the four climatic regions has its own distinctive site characteristics and constraints.

In the hot, arid region the objectives of siting orientation and site planning are to maximize the period of time of solar radiation on the collection while providing shade for outdoor areas used in late mornings or in the afternoon. Outdoor living areas should be oriented to the south-east of the dwelling units in order to utilize the early morning sun and to take advantage of the shade provided by the structure in the afternoon.

Glass wall areas on the structure should face south but should be shaded either by structural roof overhangs or by deciduous trees. Placement of win dows on the east and west sides of the structure should be avoided in order to minimize radiation into the house in early mornings and later afternoon. Where possible, multiple buildings should be arranged in clusters for heat absorption and protection of the east and west exposures. The buildings and activity areas in the hot-arid climate should take optimum advantage of cooling breezes by raising the local humidity level and lowering the temperatures. This may be done by channelling prevailing breezes across water bodies and by locating those areas to be cooled on the leeward side of the water. Also, lower hillsides provide sites which benefit from cooler natural air movement in early morning. The spaces beneath detached solar collection panels should be utilized as shaded areas. Excessive glare and radiation in the outdoor environment in hot arid regions may be reduced by providing:

... small shaded parking areas or carports;

... turf adjacent to the dwelling unit;

... tree shaded roadways and parking areas;

... parking areas removed from the dwelling units;

... east-west orientation of narrow roadways.

In hot-humid regions the solar collector should be given maximum exposure. Provision for shade involves south-oriented primary outdoor living zones. Placement of windows on the west side of the house should be avoided to minimize radiation from the low angle of the late afternoon sun.

Roofs, other than those incorporating solar collection devices should be light colored, reflective and well insulated. Glass walls on the south side of the structure should be shaded by trellises, roof overhangs or plantings. However, no vegetation on the south side of a dwelling unit should be allowed to shade a roof-mounted solar collection device.

Building siting and orientation in the hot-humid region requires:

... individual unit orientation for maximum solar impact;

... unit dispersal and setback variations for maximum ventilation and air flow;

... street orientation to facilitate the flow of the prevailing winds;

... high canopy planting with little understory growth to curtail desired wind flow;

... planting on property lines to provide for dual use of the same vegetation to provide both morning and afternoon shade on different houses;

... open cluster planning on multi-family housing to insure maximum air flow.

In the temperate region it is absolutely imperative to assure maximum exposure of the solar collection panel. The primary outdoor living areas should be located on the southwest side of structures and protected from north or northwest winds. Only deciduous vegetation should be used on the south side of the dwelling—this provides summer shade and allows for the penetration of winter sun. The impact of winter winds should be lessened. The structure itself should have steeply pitched roofs on the windward side, thus deflecting the wind and reducing the roof area affected by the winds. Blank walls, garages or storage areas should be protected with earth mounds, evergreen vegetation, walls or fences. Outdoor areas used during warm weather should be designed and oriented to take advantage of the prevailing southwest summer breezes.

In cool regions the structure and the solar collector should, whenever possible, be placed on slopes facing south to southeast. Outdoor living areas should also be located on the south side of buildings, and oriented toward the south. Exterior walls and fences should be used to create sun pockets providing protection from chilling winds. Buildings should be located on the lee side of hills, in the "wind shadow" of prevailing north-west winter winds. Evergreen vegetation, earth mounds (berms) or heavy blank exterior walls should be used to protect the north and northwest exterior walls of buildings. If the roof of the building does not contain the solar collector, it should be flat, or shallow pitched to collect and hold snow for added insulation. Where possible, structures should be built into hillsides or partially covered with earth and planting for natural insulation.

Site Planning and Design

After the specific site is selected and the locations of the building and the functions determined, it is then necessary to develop the site for optimum solar utilization and energy conservation. This entails the use of all types of vegetation, of paving, of fences, walls and overhead canopies and other architectural extensions to improve the comfort and energy efficiency of the site and the building. The materials used in site design have the ability to store, radiate, absorb and deflect the warmth of solar radiation as well as to channel warm or cool air flow. Trees of all sizes and types block incoming and outgoing solar radiation, deflect and direct the wind, and moderate precipitation, humidity and temperature in the area around buildings. Shrubs deflect wind and moderate temperature and glare. Ground cover controls absorption and radiation on the earth's surface. Turf controls and day-night temperature differential is less reflective than most paving materials. Certain paving surfaces, fences, walls, canopies, trellises and other site elements may be located on the site to absorb or reflect solar radiation, channel or block winds, and expose or cover the dwelling or solar collector.

The architectural configuration to incorporate the solar collection device may have to be modified in order to either preserve or plant vegetation on the south side of the building.

Multi-layered vegetation usually has canopy trees as well as understory trees or shrubs. Such a configuration provides a multiple braking effect on wind patterns or air flow. The site designer may use this planting pattern to substantially decrease the wind velocity moving over a site.

A mass planting of trees provides a dead air space in the air under and around itself. It also provides decreased air flow velocity in an area windward of itself up to five times its height. The area twenty-five times its height is protected from excessive air flow leeward of the planting.

Planting on the windward site of a hill or a change in grade provides a shorter zone of protection than does planting on the leeward side of a rise in elevation. The ability of planting to provide wind protection can be materially increased by using it in conjunction with topographical relief.

Careful selective thinning of existing vegetation may have to be done in order to place the building and the solar collection device on the site. Full solar collection device may require further vegetation clearing and removal. Analysis of the possible activities to take place on the site and an approximation of their spatial requirements should be made in order to analyze the options in vegetation clearance and other site disruption early in the site planning process. Where possible, minimum existing vegetation should be disturbed and any replacement planting should use similar species and varieties as much as possible. Deciduous trees should be used for summer sun shading effects and for winter sun penetration. Coniferous (needled evergreen) trees provide shade but generally do not perform as well as canopy plants or shade trees. Ericaceous (broad-leaved evergreen) plants form an extremely dense all-season canopy in areas where they are hardy. Vegetation may be placed, in the course of the site design and the planting design process, in such a way as to channel or block typical air flow patterns which cause unexpected pleasant or unpleasant climatic conditions in the areas around single or multi-family dwelling units. Fences, walls, or vegetation may all serve to block natural daily or seasonal air flow patterns. Great care must be taken in the site design process to provide the necessary visual control and yet not create adverse climatic situations or conditions. As cooler air flows downhill in the evening, fences, walls or plantings should not serve to block or dam this flow and thus create a cold air pocket.

Composite site planning considerations for multi-family town house development to utilize solar radiation and to conserve energy in the cool or temperate regions include:

... the orientation of the units;

... the amount of window areas on the south side of the units;

... the use of open cultivated ground to absorb reflected heat; and

129.

. . . the use of solid walls in the direction of the poorest exposure.

Site planning for solar energy utilization and energy conservation based on the forms suggested in the vegetation and topographic analysis suggests southern exposure, northern protection and unimpaired air movement in the cool or temperate regions.

The site planning principles for solar energy utilization and energy conservation for the cool and temperate regions indicate:

   . . . the use of windbreak planting;

   . . . the orientation of road alignment with planting on either side to channel summer breezes;

   . . . the location of units in a conformation suggested by the topography;

   . . . the use of the garage to buffer the dwelling from northwest winter winds;

   . . . the use of berms to shelter outdoor living terraces; and

   . . . the use and location of deciduous trees to block or filter afternoon summer sun.

Site Accommodation of the Solar Collection Device

Solar energy collection often requires the utilization of extensive mechanical and engineering equipment. This structural equipment may be placed either on the building to utilize the energy, on a related or nearby structure or at some point on or near the site itself. The incorporation of solar collection devices into the architecture, the use of common collectors or the use of individual solar collection devices on the site all have site implications.

If the collector is placed on the structure itself, there are minimal site implications. Most of these have to do with proper siting and orientation of the building itself, selective trimming of existing vegetation, protection of remaining planting and screening or masking of any visually disharmonious areas or elements created by the collection structure or device on the building. The following illustrations show methods for determining the location and types of site development which may be able to be utilized without disrupting or interfering with the roof mounted solar collector on an individual dwelling unit.

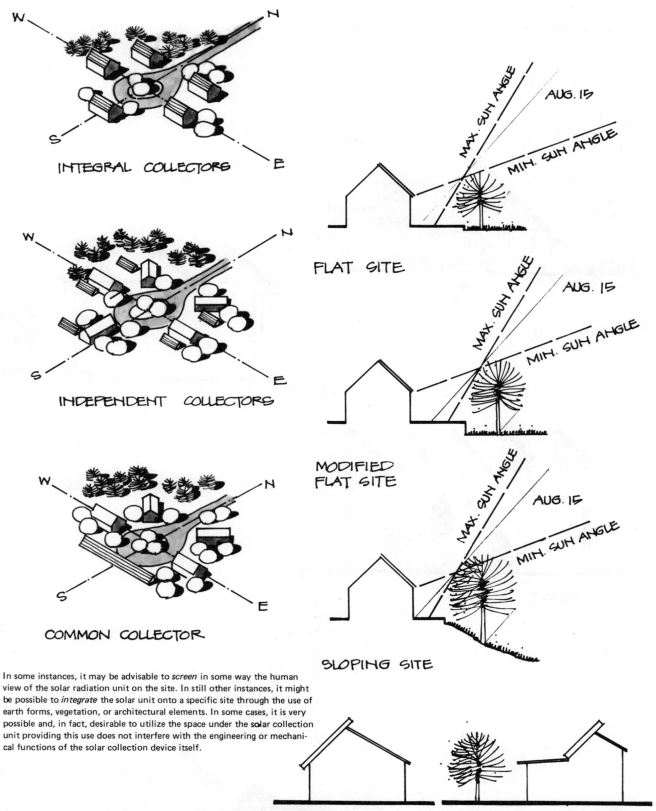

INTEGRAL COLLECTORS

INDEPENDENT COLLECTORS

COMMON COLLECTOR

FLAT SITE

MODIFIED FLAT SITE

SLOPING SITE

MODIFIED TO ALLOW PLANTING ON SOUTHERN EXPOSURE FACE.

In some instances, it may be advisable to *screen* in some way the human view of the solar radiation unit on the site. In still other instances, it might be possible to *integrate* the solar unit onto a specific site through the use of earth forms, vegetation, or architectural elements. In some cases, it is very possible and, in fact, desirable to utilize the space under the solar collection unit providing this use does not interfere with the engineering or mechanical functions of the solar collection device itself.

In certain situations, it is advisable and possible to *emphasize* the quality and character of the solar collecting device. In such cases the environmental designer should exercise careful control while seeking to effect the most positive design solution.

131.

At times, it may be appropriate to consider the placement of the solar collection device on structures other than the dwelling unit itself. Such locations might include the use of the roofs of parking garages, carports, overhead trellises, or even the embedding of collection devices in paved areas.

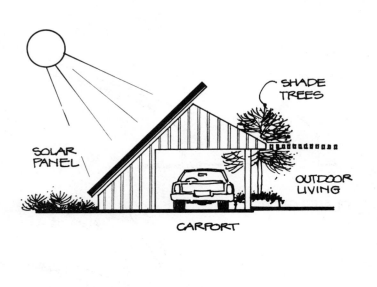

SHADE TREES

SOLAR PANEL

OUTDOOR LIVING

CARPORT

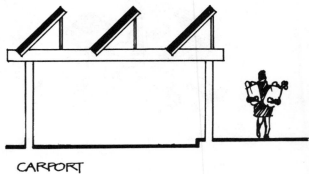

CARPORT

TRELLIS

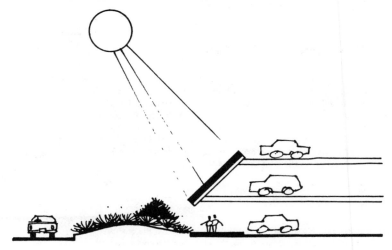

The following illustrations show some of the ways in which the on-site solar units can be hidden, screened, integrated or emphasized, or the space under the unit utilized in a variety of ways through the manipulation of earthforms, planting and architectural elements.

In still other instances, it is possible to consider the placement of the solar unit at some place or point on the site itself. This may be done to improve the location or orientation of the collector, or to avoid some screening element which would interfere with collection in the optimum location for the dwelling unit. In such cases there are at lease five various approaches which the site designer may take to this solar collection unit.

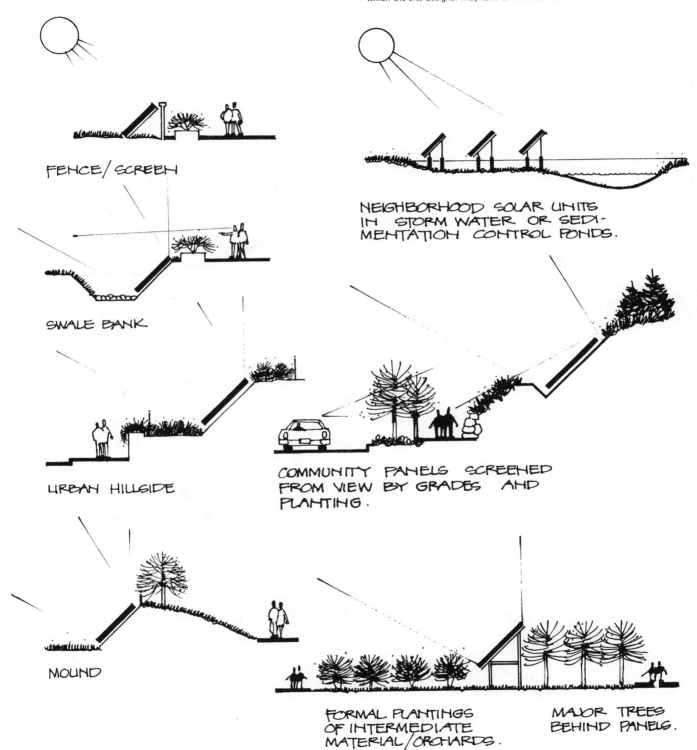

FENCE/SCREEN

SWALE BANK

URBAN HILLSIDE

MOUND

NEIGHBORHOOD SOLAR UNITS IN STORM WATER OR SEDIMENTATION CONTROL PONDS.

COMMUNITY PANELS SCREENED FROM VIEW BY GRADES AND PLANTING.

FORMAL PLANTINGS OF INTERMEDIATE MATERIAL/ORCHARDS.

MAJOR TREES BEHIND PANELS.

The solar collection devices may be screened or hidden by fences or screens, behind the planting on an urban hillside, behind a planted mound or within a swale bank and screened.

Solar collection devices could be integrated into the areas around transportation facilities, on the south facing slopes of depressed expressways, or within buffer zones surrounding parking lots.

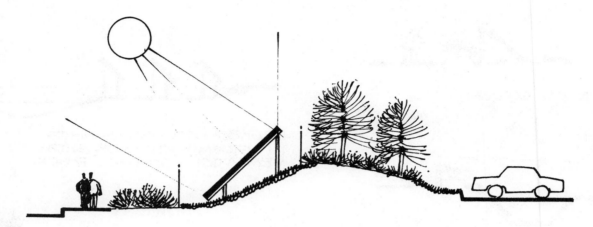

SOLAR PANELS LOCATED WITHIN BUFFER ZONES ADJACENT TO LARGE PARKING AREAS.

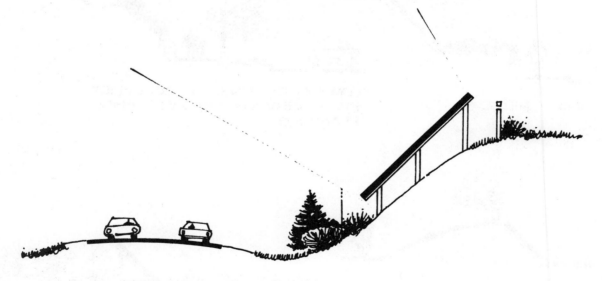

SOUTH SLOPES OF DEPRESSED URBAN EXPRESSWAYS.

134.

At times, these units are somewhat unsightly because of their engineering apparatus. Also, they may present safety hazards on the site or be subject to vandalism. Therefore, when they are placed on the site it may be advisable to find a way either through the use of earth forms, planting, or architectural elements to completely hide the solar radiation unit from view from most, if not all, directions.

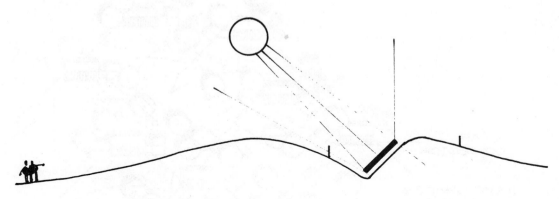

'HA·HA' TREATMENT
IN OPEN COUNTRY, WHERE
SCREENING WITH PLANTING
IS EITHER IMPRACTICAL OR
UNDESIRABLE.

Solar collection devices may be incorporated into the site in a wide variety of ways. This may be done through the use of the slopes alongside of elevated promenades or the use of the area under hillside decks.

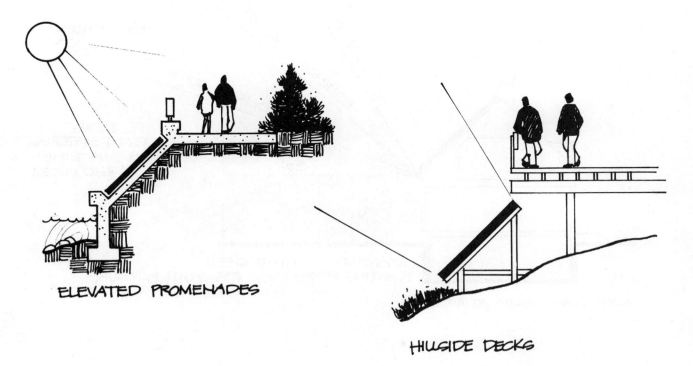

ELEVATED PROMENADES

HILLSIDE DECKS

135.

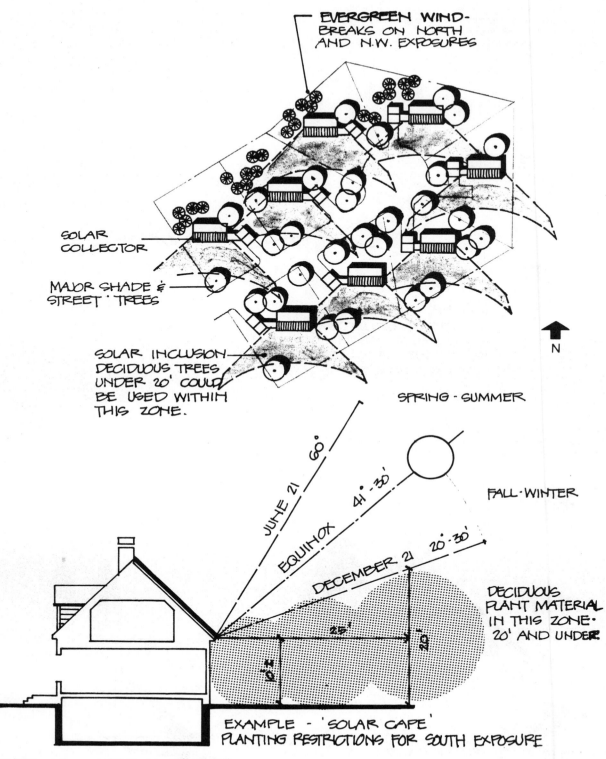

EVERGREEN WIND-BREAKS ON NORTH AND N.W. EXPOSURES

SOLAR COLLECTOR

MAJOR SHADE & STREET TREES

SOLAR INCLUSION DECIDUOUS TREES UNDER 20' COULD BE USED WITHIN THIS ZONE.

N

SPRING · SUMMER

JUNE 21 60°

EQUINOX 41° · 30'

FALL · WINTER

DECEMBER 21 20° · 30'

DECIDUOUS PLANT MATERIAL IN THIS ZONE · 20' AND UNDER

25'

20'

10'

EXAMPLE - 'SOLAR CAPE' PLANTING RESTRICTIONS FOR SOUTH EXPOSURE

NOTE - SUN ANGLES AT 40° N. LAT.

136.

Retrofitting for Energy Conservation

The term "retrofitting" is one which comes out of engineering and space technology. It refers to the modification redesign, remodeling or renovating of something which is already existing in order to enable it to accommodate new materials and elements within an existing structure. It is, in essence, the fitting in of something new after a basic structure has been completed.

In order to redesign or remodel an architectural structure to accommodate a desired solar collector, extensive and somewhat expensive changes may be required. However, the redesign or retrofitting of a site to accomodate a solar collection device or to enable it to function more efficiently in utilizing solar energy or to conserve a greater amount of the existing energy is not nearly as substantial or costly.

It is entirely possible and feasible to regrade, replant or in other ways to redesign the site and its elements to function more efficiently in solar collection and energy conservation. Earthforms may be created or modified to provide proper orientation for solar panels or to give greater protection from winter winds. Existing plantings may be able to be removed and replaced where they excessively block incoming solar radiation. They may be replaced with windbreak plantings or with smaller or less dense shade trees.

Therefore, when a site is "retrofitted" for solar energy utilization it does not necessarily imply that the structures on the site are retrofitted. It only indicates that an existing site is able to be redesigned and redevloped in order to make it more capable of utilizing natural energy sources and indigenous elements to assist in conserving the existing energy. Certainly a great many existing sites throughout the United States could be made much more efficient and effective by careful analysis of existing climatic conditions and implementation of the necessary modification.

Obviously, site planning and design for energy conservation and solar energy utilization is much easier to plan and carry out if conceived before a site is selected and a building is placed. However, there are a great many things which can be done in the way of site modification or "retrofitting" on a site which has been selected and modified and on which a building has been placed. The first step in such a process must involve analysis of the local microclimate and the site itself in some detail. This involves the:

- amount of solar radiation striking the site
- time of day the sun strikes each area of the site
- existing impediments which restrict the sun from reaching some areas of the site on which it is needed
- prevailing daily and seasonal wind patterns moving across the site
- existing impediments to desirable wind movement through the site
- existing areas on the site where snow piles up and is retained until late in the spring
- existing areas on the site which are unusually moist or damp
- existing places on the site which are continually dry, or hot, or cool

These of course are only a few of the specific factors to be analyzed on any particular site in order to see if it needs redesign or retrofitting in order to make is more effective in the use of existing energy and requiring less artificial energy in order to heat or cool the building or to make the site more comfortable and usable for a greater part of the year. The site may be modified by adding, subtracting, thinning or moving elements on the site. The site may be modified to:

- increase or decrease solar collection or impact
- increase or decrease wind impact
- increase or decrease humidity or moisture

Since in site retrofitting situations the dwelling units are already oriented and fitted on the site, there are limitations as to what can be changed or modified easily or relatively inexpensively. The following are some general guidelines as to the ways sites could be retrofitted apart from the incorporation of solar collection devices either on the structure or on the site:

. . .To insure a greater amount of solar energy reaching the site:

— Remove, thin or replace large existing deciduous shade trees on the south side of the dwelling unit.
— Remove any coniferous (needled evergreen) or ericaceous (broadleaved evergreen) trees from the south or south east of the dwelling unit. Those on the south west may be removed, thinned or replaced depending on the precise climatic situation on the site.
— Use only deciduous trees which cast a light shade on the south and south east of the dwelling unit.
— Create outdoor "sun-pockets" by orienting certain areas toward the south or southwest, with no overhead shade and providing a solid wind screen from the north, northeast and northwest.
— Remove and replace turf or paving on the south sides of dwelling units to give paving which is not highly reflective but which retains heat into the evening hours.

. . . To conserve energy inside the dwelling unit:

— Use evergreen plantings on the north or northwest sides of the architectural structures.
— Regrade to create earth mounds (or berns) on the north or north west sides of structures.
— Reorient outdoor activities which require warmth and sun to the south side of dwelling units.
— Reorient or relocate outdoor activities which require shade to the north side of structures.
— Use louvered fence panels to allow for ventilation while blocking unwanted cooler winds and providing privacy.

The following illustrations show the ways in which mobile home parks, single family, multi-family or high-rise sites might be retrofitted to make them more effective in solar radiation collection and in energy conservation.

The following illustration shows a prototypical single family site as modified and retrofitted to increase efficienty and ability in energy collection and modification. This is done by providing solar collection devices incorporated into a fence which would also give privacy and winter wind protection. The use of coniferous trees on the north as windbreaks and deciduous trees on the south use as summer sun control devices are utilized more fully in this redèsign of the site. In one instance, the solar collection device has been incorporated into the patio shade cover or a trellis.

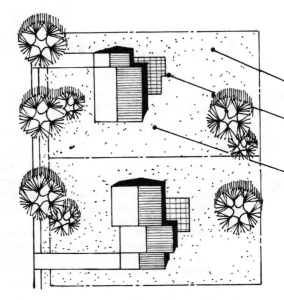

Cold winter winds are generally from the north and should be protected against

In the summer the sun's radiation is too hot for human comfort and generally requires shading

During the winter the sun's radiation is desirable for warmth and should not be shaded

TYPICAL EXISTING SITES

NORTH

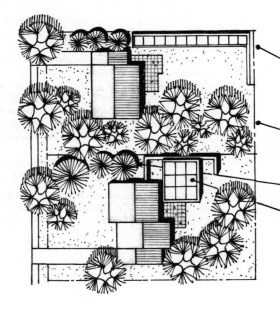

Solar collectors designed into a fence for privacy and protection from winter winds as well as collection of solar radiation

Deciduous trees planted on the south to provide summer shade yet allow winter sun

Evergreen trees planted on the north to provide winter wind protection

Solar collectors designed as a patio shade cover or trellis

HOUSE SITES 'RETRO-FITTED'
FOR CLIMATIC ADVANTGE

NORTH

The following drawing shows a typical townhouse site as retrofitted for solar energy utilization and for energy conservation. Windbreak plantings and the addition of outside fencing provide wind control in the cooler seasons of the year. Solar panels are added to the roofs of dwelling units or carports. Louvered fences are used in some instances for desirable air flow control and deciduous trees provide summer shade and winter solar radiation passage.

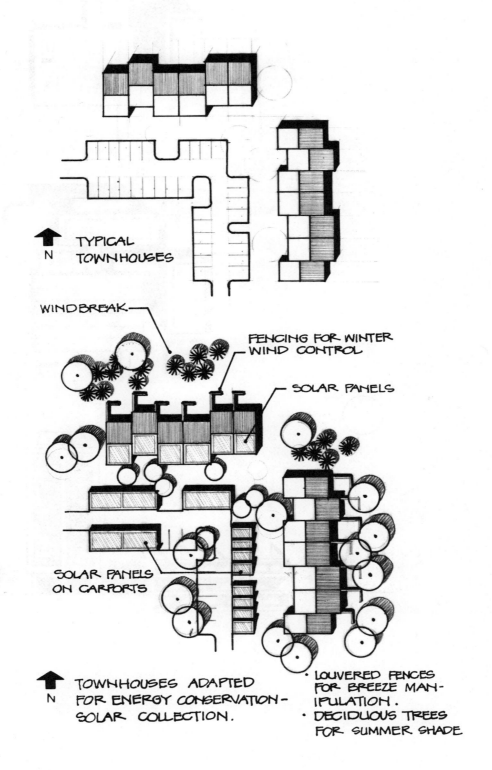

N TYPICAL TOWNHOUSES

WINDBREAK

FENCING FOR WINTER WIND CONTROL

SOLAR PANELS

SOLAR PANELS ON CARPORTS

N TOWNHOUSES ADAPTED FOR ENERGY CONSERVATION - SOLAR COLLECTION.

· LOUVERED FENCES FOR BREEZE MANIPULATION.
· DECIDUOUS TREES FOR SUMMER SHADE

The following illustration shows the way an existing high rise site may be retrofitted in order to make it usable for solar energy utilization and for energy conservation. Solar collection panels are placed on the roof of the parking structure, planting is placed on the north-west side of the building, planting screens the parking garage and covered canopies provide access from the parking garage to the high rise building.

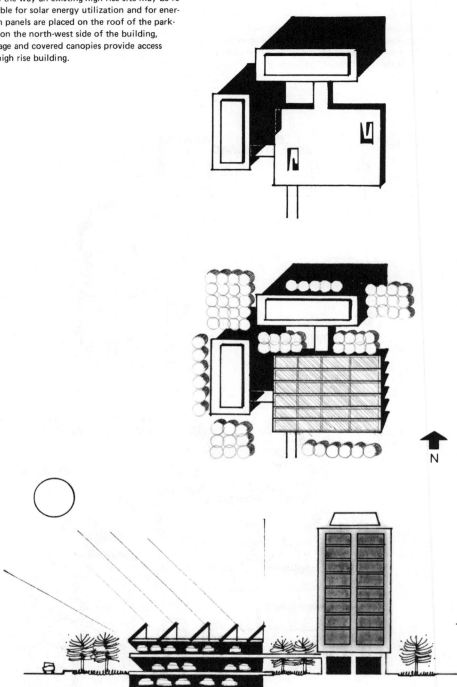

The following sketch illustrates the modification of the site in a mobile home park to accommodate solar collection devices and to provide for vegetational sun and wind screening and control.

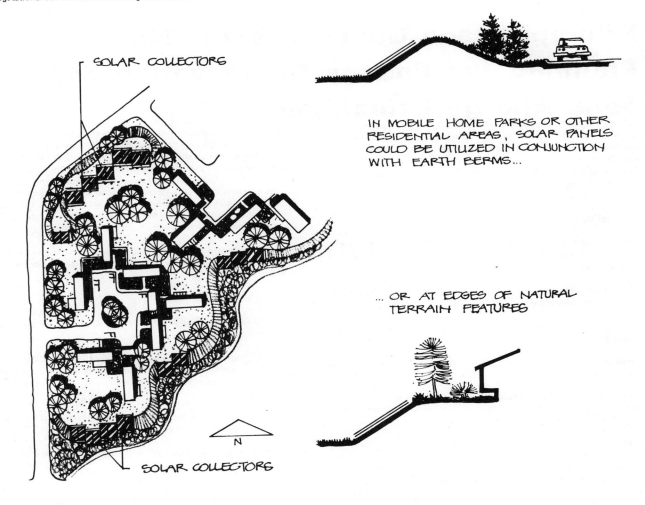

SOLAR COLLECTORS

N

SOLAR COLLECTORS

IN MOBILE HOME PARKS OR OTHER RESIDENTIAL AREAS, SOLAR PANELS COULD BE UTILIZED IN CONJUNCTION WITH EARTH BERMS...

... OR AT EDGES OF NATURAL TERRAIN FEATURES

141.

Principles and Guidelines for Site Planning for Energy Conservation and Solar Energy Utilization

The following are generalized principles and guidelines for site selection, building siting and orientation and site planning and design, primarily for solar energy collection and utilization and incidentally for energy conservation. These are given for use at a variety of scales and for all of the regions. These are not given as complete or definitive, but merely as capsulizations of some of the basic considerations in the entire site selection, planning and design process.

Site Selection

It is not possible to control or modify, in any significant way, major land masses. It is therefore necessary to recognize them, to accept them, and to work with them.

For purposes of solar radiation utilization, buildings should be on the south or west slopes and not in the north.

Given the option, all buildings should be done as high as possible on sloping landmasses.

If possible, no building should take place in water gaps, fog gaps, or openings in landforms.

The climatic effects of large land masses may be modified to a certain extent by microclimate control through manipulation of smaller land forms, vegetation and architectural elements used in the landscape.

It is necessary for everyone making decisions concerning site selection, building siting and orientation, site planning and design to understand fully the site analysis, planning and design process. This process should be made visible and clear to insure its usage to fully understand the problem, the site, the client and the limits, barriers, constraints and potential interest in the project. The available decision making options should be communicated at each stage in the process and these should be used to assist in elucidating the costs of various alternatives at each stage in the process.

The environmental design professional responsible for site selection should obtain the best available data concerning a specific site or a series of sites (within the time and money constraints) from the widest possible sources.

All gathered or accumulated data should be oriented toward or evaluated in light of solar energy utilization.

Sites should be as near to large water bodies, with their modifying influences as possible.

Sites for optimum solar radiation collection should be on moderate slopes oriented southeast to south to southwest, depending on the specific region.

Sites should be selected where the occurrence of coastal fog is limited or where fog is now known to be retained for long periods of time during the diurnal cycle.

Sites should be selected in areas where the north and west winter winds are at a minimum or where protection is easily able to be provided from them.

In the hot arid region the buildings should not be sited in narrow valleys but should be at lower hillside locations in order to gain maximum benefit from cool air flow. Locations above the valley floor tend to be cooler in the afternoon and warmer at night than lower sites.

In the hot arid region, sites should afford the opportunity to orient buildings and activate in an east-south-east direction. The wide daily temperature ranges makes easterly exposure desirable for daily heat balance. During a large portion of the year, afternoon shade is required. Sites should provide access to the cooling influences of prevailing east-west breezes. East-west winds are important cooling factors during the day, from March to November. Buildings should not be located in concave landforms on the site. In the hot arid region, concave landforms are usually "cold islands" at night, often too cold to insure proper durmoil radiation balance.

Sites should be selected where it is not necessary to remove much of the existing vegetation in order to place the structure and the solar collection device.

Building Siting and Orientation

Buildings should be placed so as to minimally disturb the existing natural landforms on the site.

Building siting should be planned so as to cause the least possible amount of existing vegetation to be removed to make provision for the building and the solar collection device and its necessary access to the sun at all seasons of the year.

Select site and orient and locate structures to reduce specular heat and glare from adjacent water bodies.

Locate buildings to be cooled by natural means on the leeward side of bodies of water.

The solar collection device should be oriented so as to obtain optimum exposure to the sun's rays and yet not give unnecessary reflection and radiation on to neighboring areas.

Dwelling units should be sited in the hot arid region so as to shade one another as well as to shade outdoor living areas and circulation paths. The concentration and clustering of dwelling units should be designed to maximize the attachment or close spacing of one dwelling to another so as to eliminate as much as possible direct solar impact on east and west wall of any dwelling.

Integration of Building and Site

In the hot arid region, the construction and utilization of subterranean and quasi-subterranean living areas should be encouraged where structural and soil properties of the site permit.

In the hot humid region, the building should be elevated and opened in order to take advantage of cooling breezes as much as possible.

Site development should reflect as much as possible indigenous forms and materials in paving, walls, fences, canopies, fountains, pools and accessories. The solar radiation effect of each of these materials should be carefully assessed.

Site Planning

Roads on the site should be oriented to avoid winter winds and to channel summer breezes.

Buildings, fences, walls, hedges and other plantings should be located so as to control unwanted winds in cool or temperate areas.

In the hot arid region, principal streets and pedestrian ways should orient as closely as possible along an east-west axis to capture prevailing cooling breezes. Large non-permeable paved areas such as parking areas and paved court game areas, should be located leeward of, and as far as possible from dwelling units and related outdoor living areas.

Careful selective thinning of existing vegetation may have to be done in order to place the building and the solar collection device on the site. Full solar access to the collection device may require further vegetation clearing and removal.

Analysis of the possible activities to take place on the site and an approximation of their spatial requirements should be made in order to analyze the options in vegetation clearance and other site disruption early in the site.

The vegetation should be studied as to the impact of reflection from the solar collection device on its growth cycle.

The screening, hiding, masking or integration of the solar collection device may be an important part of the site planning process. This can be done with vegetation, earth forms or architectural elements.

Functions which take place in the late evening should never be placed at the bottom of a slope unless adequate air drainage is provided.

Prevailing winds are from the water on to the land during the day, and from the land back on to the water in the evening. Functions taking place on the waterfront either in the early morning or late evening should be located so as to utilize these natural wind patterns rather than in opposition to them.

Streets, roads and walkways should be located where possible on the north sides of structures rather than being placed on the south side which is the prime solar collection area of the site.

Activities should be located on the site in such a way that screening is only required on the north side of the function or activity. Screening devices also act as shading devices and such shading on the south side reduces solar radiation utilization.

Site Design and Selection and Use of Site Materials

Deciduous trees should be used for summer sun shading effects and for winter sun penetration.

Coniferous (needled evergreen) trees provide shade but generally do not perform as well as canopy plants or shade trees.

Ericaceous (broad-leaved evergreen) plants form an extremely dense all season canopy in areas where they are hardy.

Consciously, not accidentally, trap or channel cold air flowing down hill out of a valley.

Paving with hard reflective surfaces should be either reduced as much as possible on the south side of the structures or the paving should be shaded to reduce excess heat and reflectivity.

Lighter colored paved areas reflect the sun's rays and are most effective for buildings up to two stories high.

Vines growing on walls, trellises or canopies are effective in shading the south and west walls of buildings, keeping them cooler.

Windbreak plantings of coniferous trees should be of at least two and preferably three rows and should be planted in staggered arrangement because of their traditional pyramidal form.

Decks provide a less reflective and less hot paving surface and also provide for air circulation from underneath the decking.

Decks can be used as extensions of the architecture to place people in a more favorable climate on the site.

Use earth forms to shade or screen the exposed walls of buildings.

Use earth forms as windbreaks.

Use earthforms to channel winds or breezes.

Use sloping sites to practically bury buildings or use earth berms to reduce heat transmission and solar radiation.

Do not use vegetation, fences or walls to dam cold air flow. Provide for cold air drainage.

Provide shade for dwellings and outdoor living areas through the use of high deciduous trees.

Tall evergreen trees on the south and southwest should be avoided in the cool or temperate regions.

Punctuate large lawns and other major usable open spaces on the site with groupings of shade trees.

Windbreaks should be used against winter winds and should be placed close to the structure or area being protected.

Hard surfaced materials should be used for terraces and other outdoor sitting areas, since solar heat on hard surfaced terraces, patios and courtyards will increase the length of evening use.

Lawns and grassy materials should be used in the immediate area of the dwelling structure since grass is a material capable of keeping a relatively even temperature throughout the day.

Employ medium colors on sun exposed surfaces, use dark colors only in recessed places protected from summer sun. Light colors will generally be too reflective in the hot arid region. Dark colors, except in special places, absorb more radiation than desired in the hot humid or hot arid regions.

Avoid overplanting of coniferous or ericaceous canopy trees since they create a cool and somber effect during protected cloudy winter days.

In the hot arid region, all east and west walls should be shaded.

In the hot arid region, the total non-permeable paving areas, such as parking lots, sidewalks, and streets should be kept to a minimum. Those paved areas which must be used should be shaded through the use of vegetation, landforms, walls, screens, canopies and overhangs. Grass ground cover, gravel or other permeable surfacing material should be used in lieu of non-permeable paving around all walls of a building since the lower light reflectivity may be employed as an element of cooling. Place trees so that they overhang and shade roof areas.

In the hot humid and the hot arid region, trees and medium-to-high hedges within ten feet of a building should be placed relative to openings in the building so that they assist in channeling and directing beneficial breezes through the openings.

Lighter colored materials should be used on the north side of structural elements, with darker colors on the south side.

Solar collection devices either on the building or independent of the building on the site offer highly reflective surfaces which may reflect light or glare off the site with unpleasant consequences for surrounding neighbors. Screening of this reflectivity should be done on the site with either architectural or vegetational materials.

Planting or other obstructions in the valley floors or on the side slopes act as dams, creating cold air pockets uphill from the obstruction to make it warmer.

Locate the area on the windward side of water bodies.

Use the south and west sides of structures for outdoor activities.

Maximize the exposure to solar radiation by placing activities out from under shading devices, up against the south and west walls of buildings and on hard paved surfaces.

Expose large paved areas to solar radiation.

Create and enhance sun pockets where possible, by exposing south-facing areas and protecting them from the north and west winds.

Utilize windbreaks and cold air flow diverters to block cooling winds.
Use structural canopy devices which impede outgoing radiation at night.

Expose walls and roofs of buildings to maximum solar radiation.

Decrease the use of water in pools, fountains, and irrigation.

Use hard masonry materials for paving, walls and screens.

Use very light, lacy, thin deciduous trees for shade; use fences or screens on the north and west to block prevailing winds or breezes.

To Make It Cooler

Locate activities on the leeward side of water bodies.

Orient all activities to the north and east of structures.

Use extensive coarse textured deciduous, coniferous shade trees and vines.

Provide shade on the south side of all activities and areas.

Do not block or curtail down-hill airflow.

Use moveable overhangs, awnings and canopies during the day which may be moved aside at night to allow for release of the trapped warm air.

Use extensive turf and ground covers throughout the site.

Use a minimum of hard, paved surfaces. Shade all paved surfaces with structural or vegetational canopies.

Use raised decks for paving where possible.

Use vines, shade trees or canopies over all exposed wall surfaces.

Prune lower growth on all trees to allow for increased air circulation.

Plant to divert winds or breezes throughout the site.

Provide pools, fountains, spray devices and irrigation as extensively as possible throughout the site.

To Make It Less Windy

Screen the north and west side of all activity areas.

Use windbreaks, baffles and diverters (either vegetational, structural or geological) where possible to block winds, breezes and air currents.

Do not build in valleys or on hillside slopes.

Provide enclosed or semi-enclosed outdoor living areas.

Depress outdoor areas below ground level, if possible.

Use heavy coniferous screens or windbreaks in multiple rows for optimum vegetational windbreak.

To Make It Breezier

Orient activities to the north and west quadrants of the site.

Minimize or remove all windbreak elements or devices which block wind, breezes or airflow.

Ust structural, vegetational or geological elements to direct and focus desirable winds or breezes to areas desired.

Prune lower branches of all trees in order to allow for easier air circulation.

Remove understory or low plant growth which blocks wind.

Locate activities either on a hilltop, or a narrow valley floor or on a sloping hillside.

Orient openings in mature vegetation to accommodate roadways and walkways so as to channel or direct wind flow.

To Make It More Humid

Use extensive overhead planting to slow evaporation and increase transportation.

Provide windbreak devices 6' and lower.

Use extensive irrigated turf and ground cover.

Reduce paving and hard surfacing on the site.

Use extensive pools, cascades, fountains and sprinklers on the site.

Locate activities on the north and east sides of structures in order to minimize solar exposure.

To Make It Dryer

Locate activities on the south and west sides of any structure or building to maximize solar exposure and evaporation.

Encourage and direct as much airflow across the site as possible.

Provide the most efficient drainage system possible by locating high on a hill in a site well drained gravelly soil and by providing an extensive under-drainage system.

Reduce ground covers, turf and planting on the site and pave all surfaces.

Eliminate fountains, pools, sprinklers.

Reduce all vegetation on the site as much as possible, including canopy trees.

It is not possible to control all climatic factors in a specific area no matter how careful the site planning, design and element selection. Most weather in the United States is characterized by extremes of discomfort. In most areas temperatures range from sub-zero to unpleasantly hot weather. The periods of winter are usually too cold, the summers too hot and the spring[and fall may bring rains or storms. The summer periods are usually hotter, in most areas, because the higher angle of the sun concentrates the solar radiation. The longer daytime periods allow for greater heat accumulation and the short nights of summer lessen the period of heat loss. The microclimatic areas of a particular site, yard, or small area are the only ones which can be controlled by individual efforts or activities. It is possible through site selection, building orientation, site planning and design to increase or lessen, to a certain extent, wind impact, local humidity, and thus temperatures to a certain extent. Even this activity does have certain limitations since they must work within the general weather pattern of a macroclimate. The modification of microclimates does have the capacity to vary seasonal climates enough to improve outdoor living and to conserve energy in both winter and summer. It may not be possible to make a climate ideal but it can be improved so as to consume less energy.

The mesoclimate of an area is able to be changed, once again, to a certain extent by combined and cooperative human activity. A city, because of its heat production and the heat absorptive, storage and re-radiation characteristics and natural radiation properties of the building materials is usually much warmer both during the day and at night than surrounding suburban or rural areas. Cropland which is irrigated, lakes, reservoirs and transpiration from vegetation may often raise the humidity of valleys which may have been dry and arid.

Obviously the best climatic controls are a good site, proper orientation of structure and activities and appropriate architecture. The angle of slope and orientation of a lot determines the amount of solar radiation it may receive. Nearby land forms regulated daily and seasonal wind flow patterns. In certain situations the problems may not have been properly solved, or may be out of the control of the individual at a particular time or place. In such situations or instances it may be necessary to correct any errors or deficiencies at the site planning or design stage. In site planning it is possible to modify in a variety of ways. The following will give some suggestions as to how this might be accomplished.

In order to make a site or an area warmer site planning should:

- emphasize and utilize all of the maximum solar exposures on a particular site
- provide for paved areas and rock or masonry surfaces on south facing slopes or surfaces
- provide for structural or vegetational outdoor canopies to reflect outgoing radiation at night
- provide, allow for or retain "sun pockets" which are potentially on a site or already in existence
- provide for wind breaks and cold airflow diverters either with vegetation, walls, fences or the architecture itself
- locate paved areas and terraces on the south side of structures and remove shading devices during the day
- use heat retaining structural materials such as concrete, masonry and wood
- remove shade or shading devices from the south and west sides of a structure
- locate outdoor terraces, used in the afternoon, on the south or southwest sides of structures
- remove water features or water bodies
- prune and thin existing shade trees to allow for maximum solar penetration.

In order to make a site or an area cooler the site planner (whether amateur or professional) should:

- make extensive use of shade trees as an overhead canopy
- use vines, either on an overhead trellis or canopy or on south and west facing walls

- use overhangs, trellises, arbors or canopies where possible (this makes an area cooler in the daytime and yet warmer at night since it limits the release of colder air into the colder night air)
- use ground covers or turf on earth surfaces rather than paving
- prune lower branches of tall trees and remove or thin lower trees and shrubs to improve and increase air circulation
- provide for evaporative cooling from sprinklers and pools
- use areas on the north and east sides of structures for outdoor activities
- remove windbreaks, either natural or man-made which would limit or hinder airflow especially during the warmer months.

In order to make a site or an area less windy or breezy the site planner should:

- use extensive vegetational, man-made or earth formed windbreaks, baffles, diverters or wind direction or guidance devices
- use, where possible, outdoor living areas which are semi-enclosed either by the architecture or landscape elements
- stop pruning or thinning lower branches on tall trees or under-story vegetation in heavily wooded areas
- locate outdoor activities in areas protected by natural windbreaks
- excavate and place activities partly below ground level in order to use the earth to block winds and to require lower additional windbreaks.

In order to increase the wind flow and the cooling effect of wind on a site or in an area the site planner should:

- remove all restrictions to natural air flow patterns on a site
- prune all lower branches of taller trees
- curtail and limit all low plant growth between 1 foot and 10 feet high which would inhibit or limit wind flow
- locate outdoor activities in areas which have the maximum access to wind on a particular site
- build decks or platforms on the windiest areas on the site in order to take advantage of natural breezes
- create natural wind tunnels or breezeways using either plant materials, earth forms, architecture, fences or walls
- locate activities or areas on the sides of a valley wall to take advantage of the day and night wind flow patterns.

In order to make a site or an area more humid the site planner should:

- allow standing water to remain on the site and limit drainage to the minimum
- encourage or increase overhead planting which slows evaporation and adds moisture through transpiration from the plants
- add, to the extent possible, pools, fountains, sprinklers and waterfalls which increase the moisture in the air. The sound of water also increases the sense of coolness and humidity
- use turf or ground cover on all surfaces where possible rather than paving
- use relatively low windbreaks, below four feet in height to preserve moisture transpired by turf or ground cover
- use natural wood chip or peat mulch under all plantings.

In order to make a site drier a planner should:

- maximize solar radiation exposure on the site and reduce shading devices
- maximize air flow and ventilation across the site
- provide an efficient water drainage system for ground water and for storm drainage
- pave all horizontal ground surfaces

- reduce planting, especially ground covers and turf
- eliminate all water bodies, pools and fountains.

From all of the above number of factors are obvious concerning the ability of landscape site planning to moderate microclimates and thus to reduce energy usage and needs. Windbreak plantings are able to reduce the operating expenses of heating and air conditioning up to one-third when the plants are mature and when they are properly selected and placed. If a house is properly shaded it has less need for a mechanical air conditioner and even if it is used, it will only need to operate half as much as it would to provide the same cooling effect if the building were unshaded.

A tree located to shade the wall and roof in the afternoon will keep house temperatures more comfortable and may reduce the wall and roof temperatures by as much as 2o to 40°Farenheit. This helps to eliminate the well-known "attic furnace" where rooftop temperatures of 140 degrees may be recorded. By having a tree shade the west wall and roof of the house it will be protected from the hot sun when and where such protection is most needed.

In the spring, fall and winter, the tree shading a house will not interfere with the sun, which at that time sets in the southwest. Moreover, by use of a tree which sheds its leaves the sun will be certain to shine on the property during the cold season when all the natural warmth which can be obtained is needed.

Large paved areas near a structure store and radiate heat for many hours after sundown and may cause stifling conditions in the house at night, making sleep difficult. Plants near a building, on the other hand, transpire and evaporation of the moisture rising from them makes the air cooler.

A solidly paved walk or driveway absorbs as well as reflects heat, and it also causes glare. By making a walk of small squares so that grass grows between them the heat is lowered to considerable extent and glare is reduced.

Patios and swimming pool decks absorb and reradiate a great deal of solar energy. Tree and plants in containers can be placed so as to provide areas of shade. Shade houses—gazebos, garden shelters, teahouses or belvederes—provide an oasis for relaxing or entertaining.

At night the insolation is shut off, but the ground continues to radiate its heat. Net outgoing radiation is greater than incoming, and the ground cools.

It is possible to use the absorptive qualities of different surfaces to increase their heat storage in chillier climates, or to lessen it in warmer ones.

By absorbing more insolation during the day, a surface slows down its heat loss at night. This is illustrated on some chilly mornings by frost patches on the grass but not on adjacent soil.

Ground temperature can be altered through the use of materials with different absorptive properties.

It is possible to modify ground conditions with mulches. Marble chips, rocks, or bricks used as decorative mulches around the bases of plants carry more heat to the soil. They are useful for warming effects where higher soil temperatures are desirable. On the other hand, loose mulches such as grass clipping, leaves, or straw act as an insulation, reducing the solar radiation which reaches soil underneath.

Large surface areas of different substances radiate and reflect solar energy in varying degrees, producing changes in the temperature of the air. Rocky areas, pavements and masonry can create extremely hot mesoclimates in the surrounding environment. At times this can be advantageous as a little sun can be radiated into nearby cold areas, making enjoyable warm sunpockets.

As light surfaces reflect more than dark ones, the color value of a material is another microclimatic control. A wall painted in a high key can send light into an otherwise dark or cool area especially one on the north side of a building or on a shaded situation.

When light becomes glare it can be a nuisance, however. A window properly shaded from direct sun can admit reflected rays from light-colored concrete below it. One cure is the substitution of dark paving, flagstone, or brick, but grass would be a better light regulator. Though its reflective qualities are high, it is a cool surface and its rough surface diffuses the glare.

Obviously, shade trees should be planted on the south and west sides of a building to do the best job of cooling. In colder climates species with compound leaves are especially effective because they have fewer and coarser twigs than those with simply foliage. The temperature of plants is many degrees cooler than that of pavement in the hot sun. By use of shrubbery and grass the rays of the sun are not reflected against the house from the pavement to make the house doubly hot in summer.

Differences of 8 degrees have been measured on the outer surfaces of shaded and unshaded buildings. Contrasts within are even greater with differences up to 20 degrees recorded. Heat within an unprotected structure, in fact, can build up to a greater temperature than that outside. A single tree on the afternoon side of the house, compared to the unshaded building, can reduce the temperature 10 degrees or more.

Trees with high, arching contours and without low growth to impede air circulation about a structure should be selected for shade plantings.

To a lesser degree, climbing and clinging vines also afford summer shade and heat control. They are effective sun screens, supported by trellises, over outdoor areas. Grown on masonry walls, they act at insulating blankets for both hot summer sun and cold winter winds.

High canopied deciduous shade trees or vines which act as "plant awnings" in the summer time, when in full leaf, shade windows. In winter they allow the transmission of the full sun. A trellis or a wire netting close to the house above a window may be used to support such deciduous vines. Use dedicuous trees and vines when cold seasonal temperatures are a consideration. Their foliage screens hot summer sun, bare branches admit the winter insolation, and screen it during the transitional spring and autumn months. In the summer the sun sets in the northwest in the temperate zone and nearer to due west as one goes south. The hottest part of the day is in the early afternoon, when the more direct rays of the sun strike the roof of the house.

Often we see solid walls of a wood, brick or stucco house out in the sun where they absorb the full blast of the sun's heat. The heat is stored all through the house to cause many sleepless nights. Where this situation exists, vines, shrubs, or espaliered plants provide cooler house walls in the summer, and, as with the "plant awnings," if they are the kind that shed their leaves, give the house the full benefit of the sun's warmth in winter.

The relative effectiveness of vegetation in providing shade and cooling for a single family house is shown in the following illustration from the Ortho Chemical Company's book, *Weather Wise Gardening*.

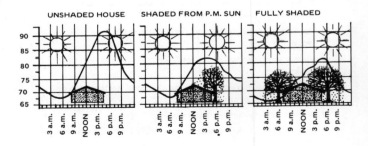

Case Studies

Hot-Humid Region
Edward D. Stone, Jr. and Associates

Stretching across the lower portions of the Gulf States, then encompassing Florida and the coastal areas of Georgia and South Carolina, the Hot-Humid climate region presents an environment unique to the continental United States. Beyond the borders of the contiguous states, however, the State of Hawaii, U.S. possessions in the Carribean, a number of Latin American nations, and many other countries around the world share the advantages and problems of the Hot-Humid zone.

Within the state, a pure Hot-Humid climate is perhaps best typified by Miami, Florida and vicinity. In general, data presented henceforth will be derived from an analysis of the Miami area, and may not necessarily generalize to other parts of the region. What follows is a guide to those site and solar conditions which should be considered in the *local context* in order to achieve maximum site efficiency for energy use and conservation.

Included examples and illustrations are simplified in order to show the principles of analysis, site planning, and design which can identify and optimize positive natural influences, reduce or neutralize the negative impacts of climate, and provide opportunities for collection of the sun's energy. By no means should these be interpreted as design *standards*—for all land parcels are in themselves unique, and require careful individual analysis before appropriate planning and design goals may be realized.

The climate of coastal South Florida presents an environment quite unlike those commonly found in more northern states. Sunshine averages 66% of the possible daylight hours, and helps produce a typical warm climate with small yearly temperature variations. Three-fourths of the year's temperatures fall within the 65° to 85° range, with higher temperatures occurring only 11% of the time, to a record high of 96°. (Olgyay)

The effects of temperature are magnified by humidity, however, the average annual rainfall of 60 inches combines with ocean and undrained lowland evaporation to produce an unpleasantly high humidity during the majority of the year. Lynch reports that the "limit of human tolerance"—the maximum air temperature at which extended work may be performed without raising body temperature—has a threshold of 150° in dry air, but is reduced to only 90° in full humidity. This limit is completely within reach in the Hot-Humid region.

Due to the adverse effects of solar radiation, it is necessary to utilize the positive natural moderating influences of shade and wind to produce comfortable living environments during overheated periods. Olgyay suggests that shading is required 88% of the year, while wind is advantageous 62% of the time in order to reduce temperature and humidity within the comfort range.

An advantage of the sun's almost continuous warmth, of course, is manifested in the area's extremely mild and pleasant winters. Temperatures fall within the 45°-65° range only 14% of the time, and this is usually during winter nights. Only rarely are freezing conditions encountered, and they have never continued throughout the day.

The actual perception of climate conditions is a variable, however, depending on the "normal" average conditions and ranges encountered in an area. Winter fronts often create uncomfortably cool temperatures as they pass through the region and indicate a need for wind protection and solar warmth even in Miami during 12% of the year. For this reason, the northern parts of the region, which encounter more frequent cold conditions, will wisely incorporate some principles from the Temperate zone into climatically-oriented planning efforts.

The climate conditions found in Hot-Humid areas dictate the need for climatic improvements to improve human comfort. When viewed from an energy conservation perspective, the broad goal is clearly to minimize the effects of heat and corresponding use of energy-consuming climate control systems. The subject of solar energy in the Hot-Humid region presents an interesting paradox. The sun's rays are an intense and abundant resource when properly utilized as an energy source, but solar radiation also is a primary cause of human discomfort. Solar energy technology is now evolving beyond the water heating stage (available in this area for years) into the realm of solar powered air conditioning, which will be of major importance to the region worldwide. Making appropriate use of the prolific solar resource is a long range goal of engineering technology; reducing the *overall need* for energy-consuming climate control systems by improving microclimatic conditions requires solutions in architecture, site planning and landscape design which will reduce the impact of solar radiation while also lessening the effects of high humidity.

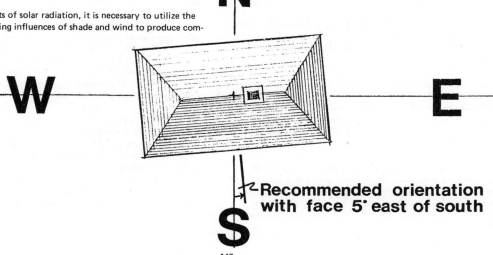

Recommended orientation with face 5° east of south

ORIENTATION. The physical placement of buildings in relation to the sun's path is a highly important factor in the Hot-Humid region. The majority of adverse solar loading occurs when the low angle sun in the East and West penetrates windows and other wall openings and heats a building's interior. When the sun is high during midday only minor wall exposure is encountered and major radiation falls on horizontal surfaces. Since building roofs are usually of a monolithic character (without openings) and often constructed of positively insulating materials, roof surfaces are fairly efficient in the reflection (especially if light in color) and/or insulation of radiation.

The most appropriate building shape for the region is rectangular and optimum orientation places the short sides toward East and West, in order to limit the area which is exposed to a low-angle sun. Because of the easterly prevailing winds in the Miami area, Olgyay recommends a minor compromise by shifting the building's face to 5° East of South. From this recommended orientation, buildings may be shifted up to 10° in either direction in response to individual site requirements without seriously impeding functional efficiency.

It must be stressed that while the East-West general orientation goal will be constant throughout the Hot-Humid region, local prevailing winds must be determined and used as a basis for minor shifts in orientation in each specific locality. For purposes of illustration, the recommended Miami orientation has been used in examples which follow.

SHADING. Obviously, if the surfaces of structures can be shaded from the sun, the impact of radiation will be significantly reduced. Shading solutions are possible through architectural treatment of the building, but may also be greatly influenced by the use of vegetation—locating buildings beneath mature tree canopies and the strategic placement of trees which will provide required shade at various times of the day. Careful analysis of the dynamics of solar movement will make such functional planting possible.

WINDS. In the Miami area, a fairly predictable and frequent ocean breeze is a highly important climate moderating factor. Breezes can greatly reduce air temperature and the perceived degree of humidity, and significantly increase human comfort. A prime goal, then, is to maximize the flow of cooling breezes by careful placement of plant and building masses. In harmony with solar orientation goals, intelligent planning of building relationships and circulation paths can channel winds through a community and greatly improve microclimatic conditions.

SOLAR ENERGY UTILIZATION. As the viability of solar technology continues to improve, it will become more important to combine the previously stated climate improvement goals with solar collecting objectives. The general abundance of the solar resource in the region will facilitate high utilization of solar systems as they become available. Proper site planning and architecture will need to provide sufficient unobstructed space for the mounting of solar collecting hardware. For the most part, it is assumed that rooftop collectors provide the greatest functional efficiency, and most examples are founded on this assumption.

SITE ANALYSIS. In order to achieve the desired climate improvement and energy conservation goals, it is imperative to carefully analyze and understand the solar, climatic and physical characteristics of individual sites which are influential to the planning and design process.

Proper analysis techniques will record existing physical site conditions such as topography, vegetation, water bodies and hydrology, and uses of adjacent land, couples with data on the effects of wind, temperature, precipitation and solar orientation information characteristic to the locality. Only from the foundation of accurately collected data may viable climate-oriented planning evolve.

SITE SELECTION. Despite the diversity of sub-regional and local conditions which make generalized site selection criteria unworkable, several key goals which are common to the region may be identified. Based on the knowledge that all sites are in themselves unique, however, it remains imperative to perform careful analysis of specific sites prior to making site selection decisions.

One of the greatest climate moderators in the Hot-Humid region is wind, and certain areas naturally provide more cooling breezes. Coastal zones generally exhibit fewer temperature extremes than inland areas; they are cooler in summer and warmer in winter. Great care must be exercised in development of beach zones, however, in order to avoid environmental disruption of sensitive and important dune ecosystems.

The depth of wind-moderated coastal zones is often highly variable, due to local conditions. In lowland areas, breezes may be dampened by the windbreak effect of dense plantings or overly dense building masses in contiguous areas. In mountainous or hilly coastal areas, however, sites will be more open to breezes and the windbreak effect may be greatly negated for some distance inland. Flat areas usually require greater building separation and hence lower density in order to maximize the wind's usefulness; greater topography may allow a higher density of development.

Similar considerations apply as well to inland areas, where higher lands and slopes present opportunities for capturing available breezes. Proper drainage of interior sites is of importance also, for low, poorly drained sites increase humidity and associated discomfort. In areas of extreme humidity, the limit of human tolerance to temperature may be as much as 60° lower than in an arid environment, therefore, a lowering of humidity is highly desirable. While the presence of water may cause problems in areas of little wind, inland water bodies may provide substantial cooling effect where breezes are prevalent. Again, an understanding of the interrelationships of climate factors is important.

The final desirable condition which raw land may offer is relief from the extremes of solar radiation in the form of shade from existing trees. Wooded sites may be highly desirable as long as air movement can be maintained. Overly dense and low tree canopies may prevent cooling breezes and trap humidity in a pocket of "dead air." Higher canopy trees, however, may provide needed shade, while allowing breezes to pass beneath them. In view of solar energy goals, the advantages of shade canopies must be weighed together with the need for unobstructed collector space in formulating an overall development strategy for particular sites.

BUILDING SITING AND ORIENTATION. As previously outlined, the best orientation of solar homes should be along an East-West axis for efficient presentation of collectors toward the sun, and to reduce radiation loading of eastern and western wall surfaces, which receive the majority of adverse solar exposure in lower latitudes. Because of the need for cooling breezes, however, it is recommended that this axis be slightly rotated toward the direction of prevailing winds in the local area.

In areas of moderate topography, the siting of structures in relation to the land form can serve to improve microclimatic conditions. Buildings located toward the windward edge of plateaus, on ridge crests, and on facing slopes receive the benefit of increased air flows and higher wind velocities, although Eastern and Western clopes will also receive more solar radiation. In areas which encounter winter cold fronts, South facing slopes can greatly moderate the chilling North winds.

Where possible, the siting of individual structures should attempt to use existing vegetation to maximum functional advantage. Large trees to the East and West of buildings are generally desirable in reducing the wall radiation loads produced by low sun angles, provided that they do not impede the flow of cooling air. On the contrary, dense windbreaks of existing vegetation to the North of a structure will reduce exposure to winter winds, and further reduce the already low heating needs of the region.

ARCHITECTURAL ELEMENTS. Many design elements of architecture and site are of use in moderating microclimate and reducing energy consumption. Utilizing key characteristics of the Hispano-Moorish tradition, a functional architectural form has evolved in past years. Typically, wall and roof surfaces are painted either white or in light pastel shades. White roof surfaces may be up to 95% reflective, significantly reducing radiation gain. When constructed of positive insulating materials such as tile and masonry, residential structures may compensate for much of the adverse heat effect.

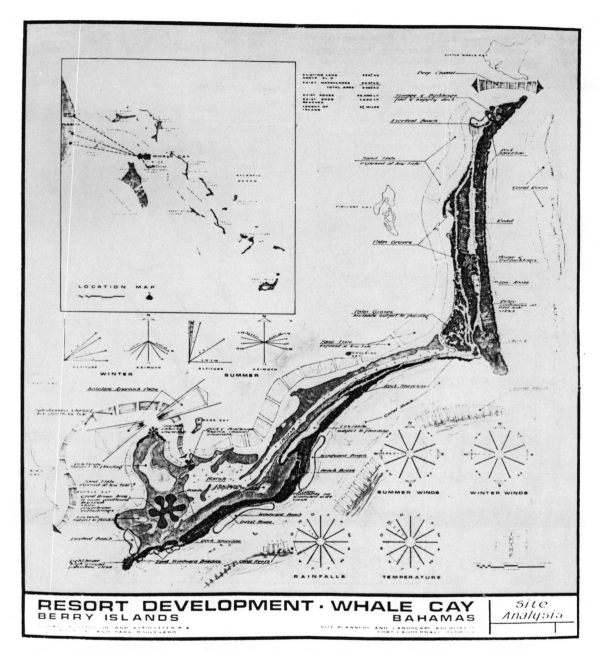

SITE ANALYSIS: Detailed inventory and analysis of natural site characteristics.

Window openings are typically designed to admit maximum amounts of fresh air and flow-through ventilation is encouraged by the layout of interior spaces. Because of the impact of radiation on windows, they are wisely shielded from the sun's rays by the use of architectural or natural shading devices. Roof overhangs, recessed fenestrations and other design elements can add considerably to the comfort level within the structure.

ARCHITECTURAL TYPES. As the conditions of topography and wind vary within the region, the basic architectural and site planning determines change as well. In flatland areas, development is idealistically an open plan of relatively low density and scattered individual structures, allowing for maximum circulation of air. Elements such as large roof overhangs, often incorporating exterior circulation corridors or verandas, are widely used. Beyond the building itself, covered walkways which provide both shade and free air circulation are often seen on intensely developed sites.

As the top slope of an area increases, changes in basic architectural concept become evident. Buildings are sited in harmony with the landform, and the vertical offset of structures allows tighter clusters and a generally higher density without interruption of cool air movement. Thick walls, substantial overhangs and deep set openings add to the character is the clusters and intriguing tunnels, alleys or other "openings" admit air into the cluster's core.

INTEGRATION OF ARCHITECTURE AND SITE ELEMENTS. An example of the combined use of architectural and site design elements in order to achieve energy efficiency is demonstrated by early studies for the Life of Virginia Corporation headquarters in Richmond, Virginia. The building, designed by Hardwicke Associates, Inc., was programmed to minimize energy use by lessening the impact of solar radiation on wall surfaces, and thereby increasing systems efficiency. While highly successful in reducing direct radiation, a large plaza, which covered the integrated parking facilities, presented the problem of indirect glare and heat radiation. Using the basic principles of plant function, it was simply demonstrated that the incorporation of a deck-top planting of deciduous trees could substantially solve the problem—reducing summer heat while capturing winter warmth.

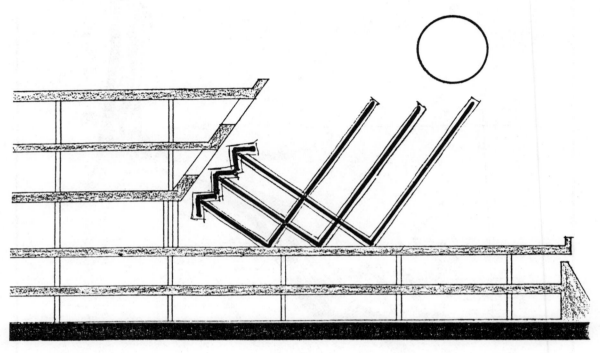

1. REFLECTED HEAT PROBLEM

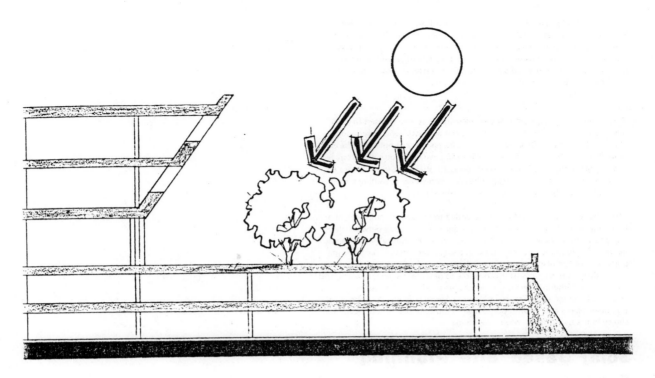

2. SUMMER : TREES BLOCK SUN KEEP FACADE COOL

3. WINTER : TREES ALLOW SUN THRU HELP WARM BUILDINGS

SITE ELEMENTS. The desirability of proper utilization of natural elements on a building site has already been stressed. Trees provide shade and can selectively channel desirable breezes or block adverse winds. In the proper situation, water in the landscape may be a useful element, vines trained on walls or roofs will reduce radiation, and the character of slope and land form may also be used to improve microclimate.

Certain other landscape architectural elements may also be used to great advantage in improving the solar loading characteristics and general microclimate of structures. Overhead trellis structures, for example, may be extended beyond building walls to provide effective radiation relief from even lower angle sunlight, and also provide pleasant, shaded outdoor spaces for residents. The effect of shading overheads is immediate as well; one must not wait for newly planted shade trees to reach maturity.

When the scope of development is expanded to include an entire neighborhood or designed community, the importance of the spaces between dwellings and building groups is magnified. Because a community's open spaces define its basic fabric of circulation and social interaction, it becomes important to utilize planning principles which will reduce climatic discomfort in these public spaces. If it is possible to provide shaded walkways, and meeting and gathering places, it is reasonable to assume that use of such outdoor areas will increase. The combination of protected places, along with true open spaces and warmth pockets will tend to maximise the freedom of choice in the use of the outdoor environment.

Solar Based Site Planning Process

The following series is intended to illustrate a simplified process by which solar data may be used to identify problems of adverse radiation and graphically show functional solutions. Each hour of the day, the sun's position varies in both direction (bearing) and height (altitude), causing differing radiation effects. In order to solve solar problems, it is necessary to fully understnad the dynamics of the sun's movement at a given latitude. The illustrations are based on data for Miami, Florida (approximately 26°N. Lat) on the longest and highest radiation day of the year, June 22.

For purposes of simplifying the example, the following factors are assumed as given:

- Residence is of rectangular shape, with reasonable roof overhang.
- Unit orientation is 5° East of South, as recommended.
- Lot size is approximately ¼ acre.
- Prevailing winds from East should not be blocked at ground level.
- Roof surface should be left unobstructed for solar collection.

Each hourly study will demonstrate the bearing and altitude of the sun, state problems encountered, and show a functional solution which will either eliminate or greatly reduce radiation impact. As the day progresses, the cumulative study will built a strictly functional design plan for the hypothetical residence. The final study in the series is a "Functional Composite" which adds minor elements to improve wind control, but is nonetheless a final reflection of the additive solar planning process.

In any locality, a similar process may be followed by substituting a specific site, planned structure(s), and solar and wind data which apply to the particular area.

Solar-Based Site Planning Process

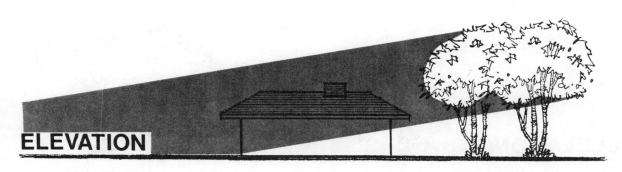

ELEVATION

problem-
SOLAR LOAD ON EAST
& NORTH WALLS

solution-
CANOPY TREES AT N.E.
PARTIALLY SHADE WALLS,
BUT ADMIT BREEZES.

sun data-

bearing	111° 18' E
altitude	10° 03'

radiation (btu/sq ft/hr)

east wall:	113	north wall:	44
south wall:	0	roof:	22
west wall:	0		

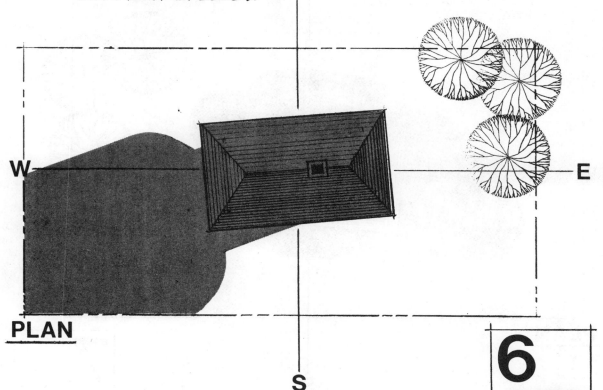

PLAN

Miami, Florida - June 22

6 am

Solar - Based Site Planning Process

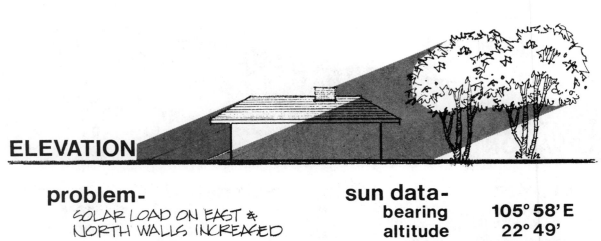

ELEVATION

problem-
SOLAR LOAD ON EAST &
NORTH WALLS INCREASED

solution-
CANOPY TREES TOTALLY
SHADE WALLS – NO
ADDITIONAL ELEMENTS
REQUIRED

sun data-

bearing	105° 58' E
altitude	22° 49'

radiation (btu/sq ft/hr)

east wall:	185	north wall:	53
south wall:	0	roof:	81
west wall:	0		

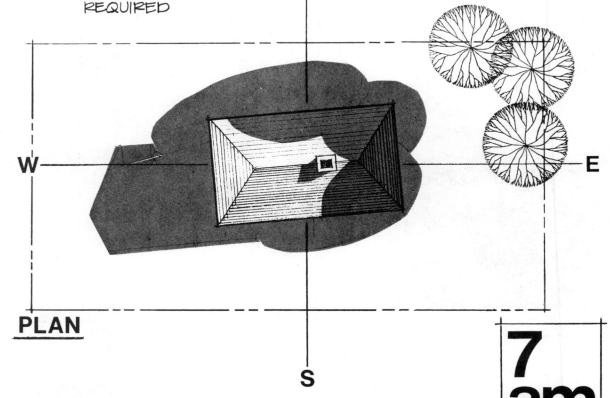

PLAN

Miami, Florida - June 22

7 am

Solar - Based Site Planning Process

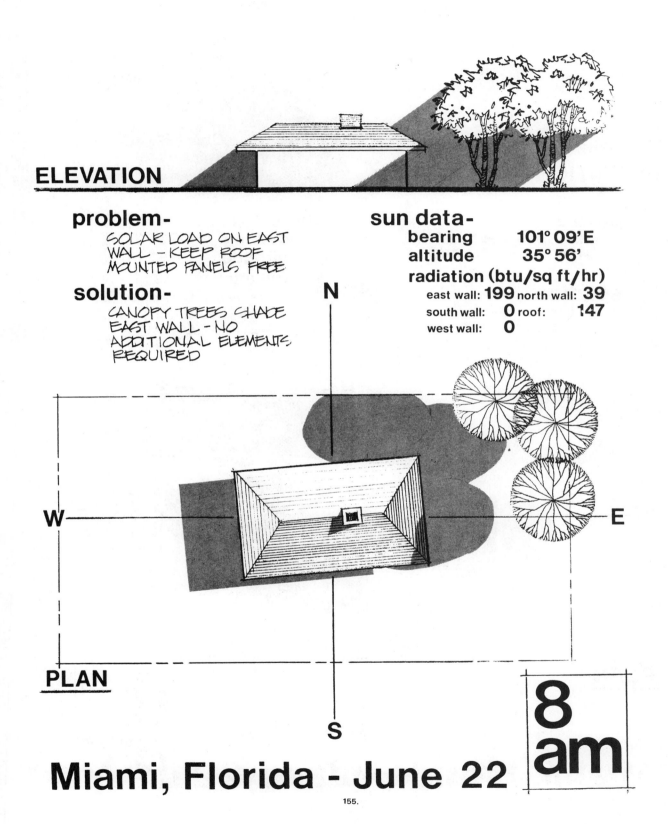

ELEVATION

problem-
SOLAR LOAD ON EAST
WALL - KEEP ROOF
MOUNTED PANELS FREE

solution-
CANOPY TREES SHADE
EAST WALL - NO
ADDITIONAL ELEMENTS
REQUIRED

sun data-
bearing   101° 09'E
altitude   35° 56'
radiation (btu/sq ft/hr)
east wall: **199** north wall: **39**
south wall: **0** roof: **147**
west wall: **0**

N

W   E

S

PLAN

8 am

Miami, Florida - June 22

155.

Solar-Based Site Planning Process

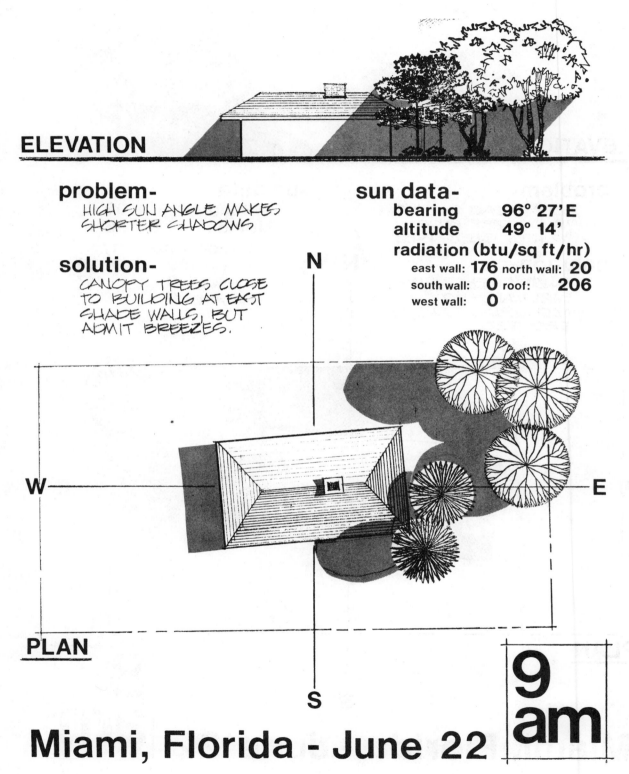

ELEVATION

problem-
HIGH SUN ANGLE MAKES
SHORTER SHADOWS

solution-
CANOPY TREES CLOSE
TO BUILDING AT EAST
SHADE WALLS, BUT
ADMIT BREEZES.

sun data-
bearing   96° 27'E
altitude   49° 14'
radiation (btu/sq ft/hr)
east wall: **176**   north wall: **20**
south wall: **0**   roof: **206**
west wall: **0**

N

W   E

PLAN

S

Miami, Florida - June 22

9 am

Solar - Based Site Planning Process

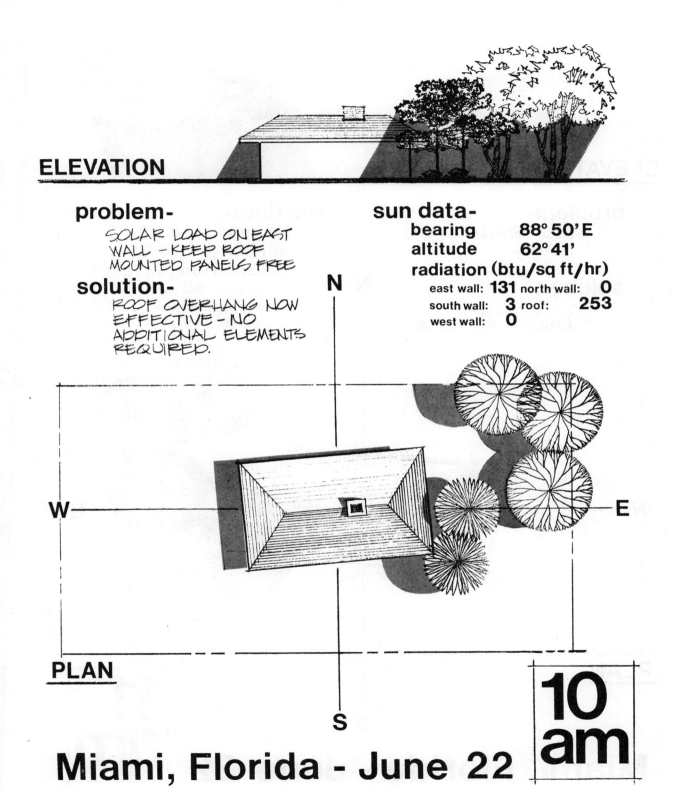

ELEVATION

problem-
SOLAR LOAD ON EAST
WALL - KEEP ROOF
MOUNTED PANELS FREE

solution-
ROOF OVERHANG NOW
EFFECTIVE - NO
ADDITIONAL ELEMENTS
REQUIRED.

sun data-
bearing   **88° 50' E**
altitude   **62° 41'**
radiation (btu/sq ft/hr)
east wall: **131** north wall: **0**
south wall: **3** roof: **253**
west wall: **0**

N

W   E

PLAN

S

10 am

Miami, Florida - June 22

Solar - Based Site Planning Process

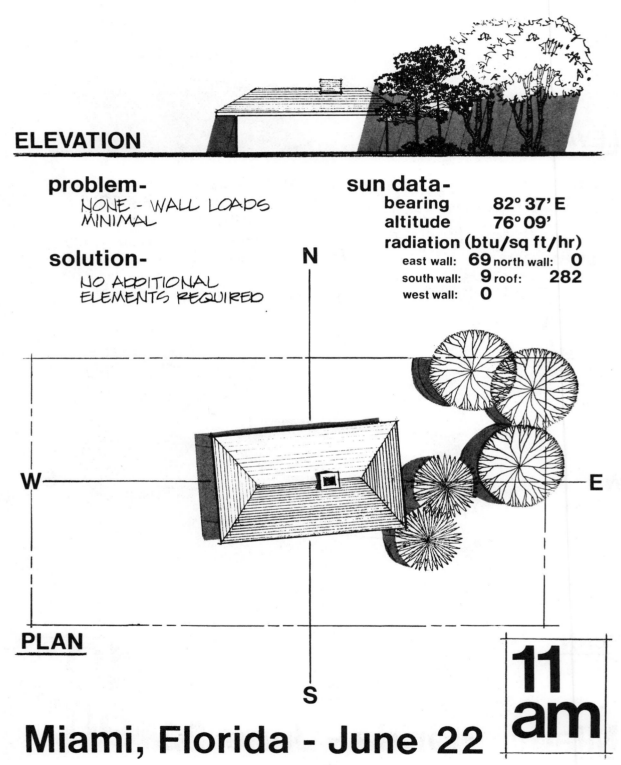

ELEVATION

problem-
NONE - WALL LOADS MINIMAL

solution-
NO ADDITIONAL ELEMENTS REQUIRED

sun data-
bearing   82° 37' E
altitude   76° 09'
radiation (btu/sq ft/hr)

east wall: **69** north wall: **0**
south wall: **9** roof: **282**
west wall: **0**

N

W ——— E

PLAN

S

Miami, Florida - June 22

11 am

Solar - Based Site Planning Process

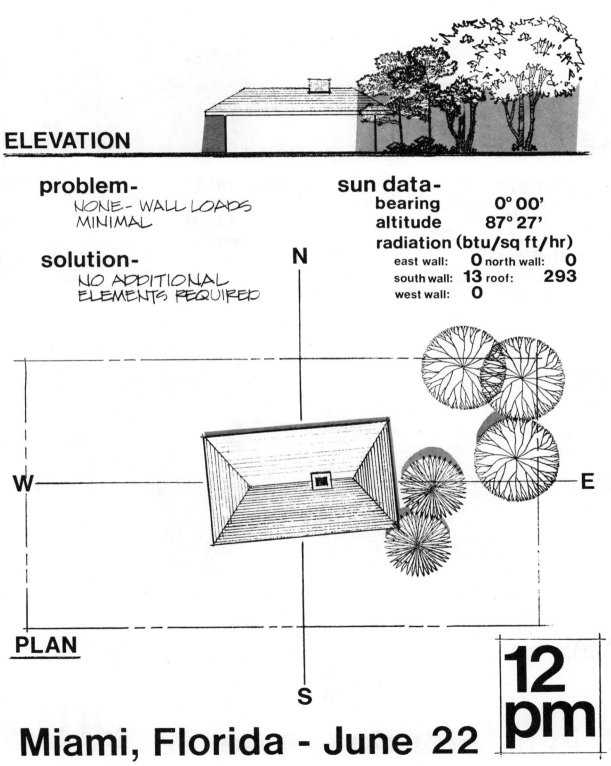

ELEVATION

problem-
NONE - WALL LOADS MINIMAL

solution-
NO ADDITIONAL ELEMENTS REQUIRED

sun data-

bearing	0° 00'
altitude	87° 27'

radiation (btu/sq ft/hr)

east wall:	0	north wall:	0
south wall:	13	roof:	293
west wall:	0		

N

W   E

S

PLAN

12 pm

Miami, Florida - June 22

Solar - Based Site Planning Process

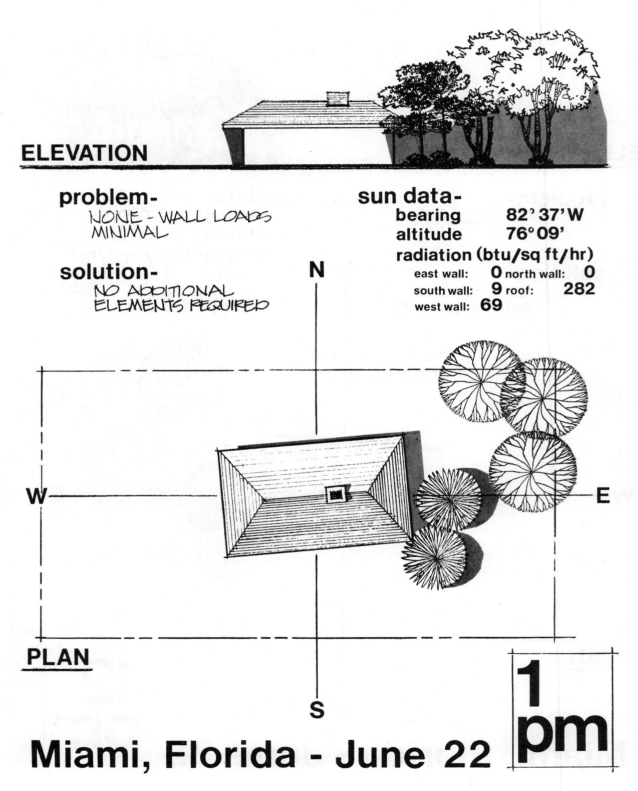

ELEVATION

problem-
NONE - WALL LOADS
MINIMAL

solution-
NO ADDITIONAL
ELEMENTS REQUIRED

sun data-
bearing   **82' 37'W**
altitude   **76° 09'**
radiation (btu/sq ft/hr)
east wall: **0** north wall: **0**
south wall: **9** roof: **282**
west wall: **69**

N

W   **E**

PLAN

S

Miami, Florida - June 22

1 pm

Solar - Based Site Planning Process

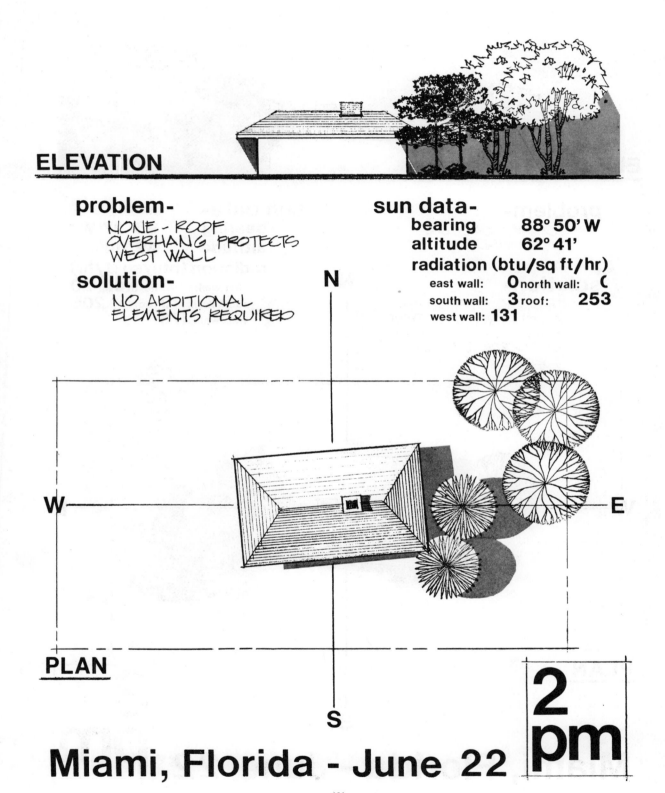

ELEVATION

problem-
NONE - ROOF
OVERHANG PROTECTS
WEST WALL

solution-
NO ADDITIONAL
ELEMENTS REQUIRED

sun data-
bearing   88° 50' W
altitude   62° 41'
radiation (btu/sq ft/hr)
east wall: **0** north wall: **0**
south wall: **3** roof: **253**
west wall: **131**

N

W   E

S

PLAN

2 pm

Miami, Florida - June 22

Solar-Based Site Planning Process

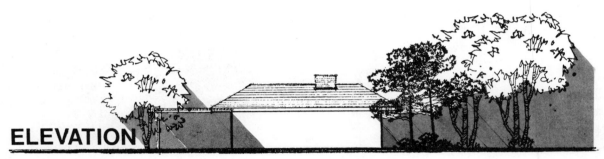

ELEVATION

problem-
SOLAR LOAD ON
WEST WALL

solution-
TRELLIS STRUCTURE
& ORNAMENTAL TREE
SHADE WALL, DEFINE
OUTDOOR LIVING
AREA

sun data-
bearing   **96° 27'W**
altitude   **49° 14'**
radiation (btu/sq ft/hr)
east wall: **0** north wall: **20**
south wall: **0** roof: **206**
west wall: **176**

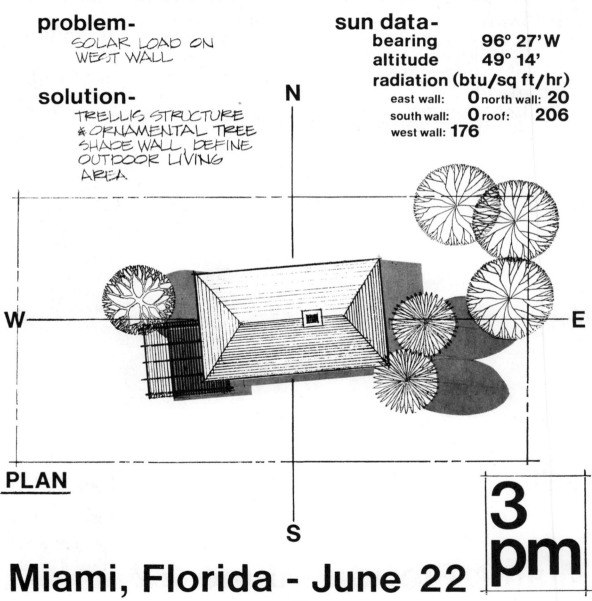

PLAN

Miami, Florida - June 22

3 pm

Solar-Based Site Planning Process

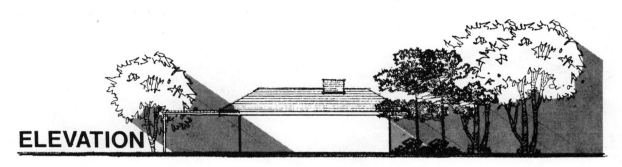

ELEVATION

problem-
SOLAR LOAD ON
WEST WALL

solution-
TRELLIS & TREE SHADE
WALL - NO ADDITIONAL
ELEMENTS REQUIRED

sun data-
bearing   101° 09' W
altitude   35° 56'
radiation (btu/sq ft/hr)
east wall: **0** north wall: **39**
south wall: **0** roof: **147**
west wall: **199**

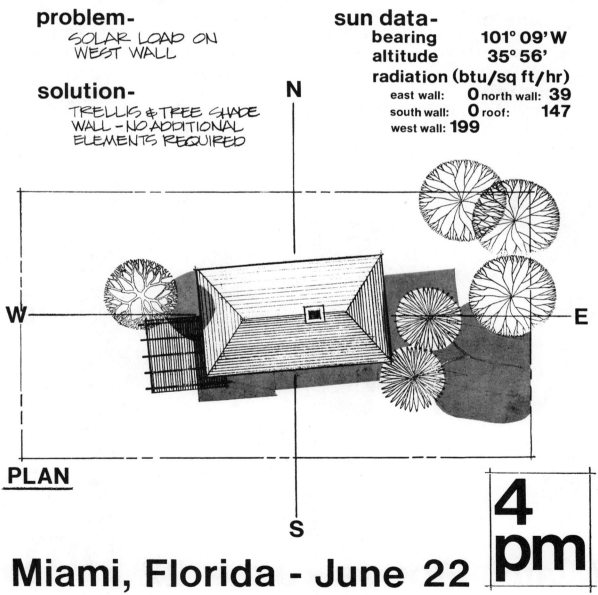

PLAN

Miami, Florida - June 22

4 pm

Solar-Based Site Planning Process

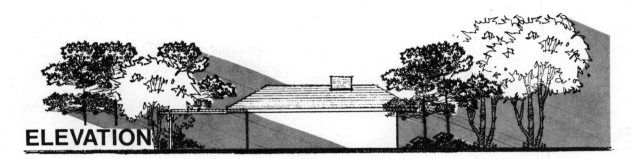

ELEVATION

problem-
SOLAR LOAD ON WEST
& NORTH WALL - LOW
SUN ANGLE

solution-
CANOPY TREES AT N.W.
PARTIALLY SHADE
WALLS, CUT GLARE

sun data-
bearing   105° 58' W
altitude   22° 49'
radiation (btu/sq ft/hr)
east wall: **0** north wall: **53**
south wall: **0** roof: **81**
west wall: **185**

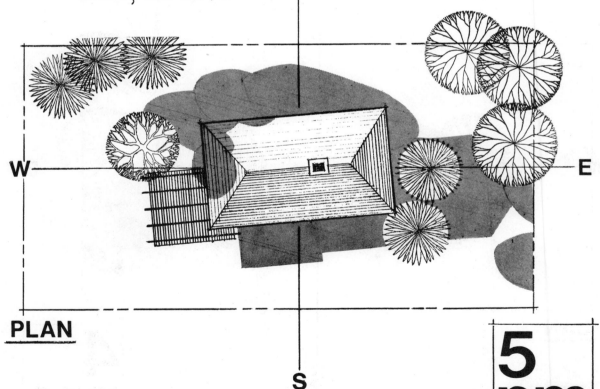

PLAN

Miami, Florida - June 22

5 pm

Solar-Based Site Planning Process

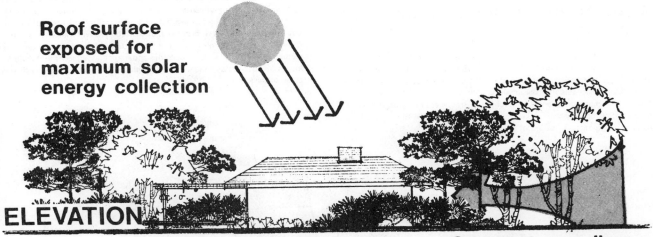

Roof surface exposed for maximum solar energy collection

ELEVATION

Canopy trees allow breezes to pass

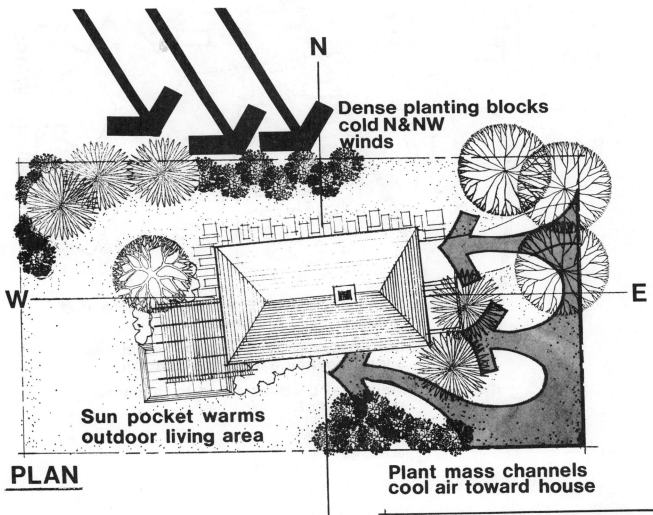

Dense planting blocks cold N&NW winds

N

W

E

S

Sun pocket warms outdoor living area

PLAN

Plant mass channels cool air toward house

Miami, Florida -

FUNCTIONAL COMPOSITE

Site Planning Principles

The individual energy conservation goals for the Hot-Humid region can and should be expanded beyond individual sites to larger scale development. The goals of orientation, shading and wind movement may be wisely incorporated into a site design program along with the complex of more traditional planning determinants. When climatic effects are recognized, certain patterns develop which may exert substantial influence on the eventual form of the layout and design of a development.

SINGLE FAMILY HOUSING. Many of the functional solutions found in the individual residence example may be utilized in single family groupings as well. The planting of canopy shade trees along traditional lot property lines can provide shade to neighboring houses at different times of the day, but still allow breezes to pass beneath. Windbreaks will shelter structures from cold winds, but also serve to channel desirable easterly breezes. When studied at the larger scale, it becomes apparent that adjacent units will allow more air circulation when the individual structures are sited with staggered setback lines.

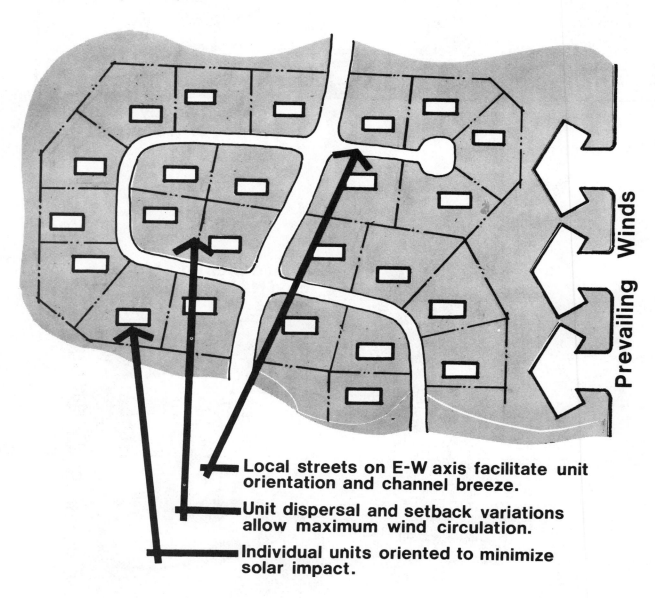

Prevailing Winds

Local streets on E-W axis facilitate unit orientation and channel breeze.

Unit dispersal and setback variations allow maximum wind circulation.

Individual units oriented to minimize solar impact.

SUN

Shared property line trees shade one house in the morning . . .

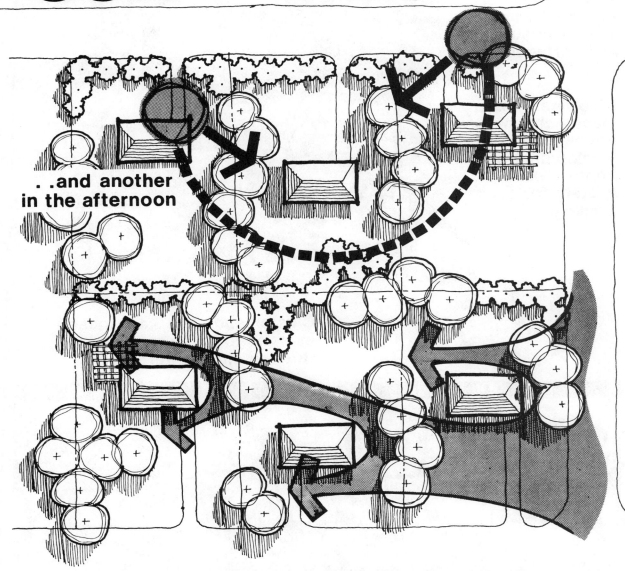

. .and another in the afternoon

Breezes flow beneath canopy trees & between staggered structures

& WIND

MULTI-FAMILY HOUSING. The same planning goals apply to multi-family dwellings in either townhouse or apartment form. When the linear axis of building complexes is East and West, radiation exposure of wall surfaces is greatly reduced. Additionally, this orientation is highly compatible with the prevailing winds in the Miami area. Townhouse or apartment cluster planning should recognize the need to allow breezes into the center of the cluster and also utilize open spaces as functional wind channels. The broad scale planning of multi-family communities should additionally provide centrally located community facilities which may be reached via shaded pedestrian walkways. The convenient and comfortable pedestrian accessibility of such facilities can reduce unnecessary automobile use and corresponding fuel consumption.

167.

1 Units form linear clusters, oriented on an East-West axis for minimum solar exposure of wall surfaces.

2 Wide, linear open spaces channel breeze through development and create a scenic amenity for each dwelling.

Ocean Breezes

THE BREAKERS   PLANNED UNIT DEVELOPMENT
FLAGLER SYSTEM INC.   CORPORATE OPEN SPACE   5A

3 Cluster ends are kept loose in order to guide winds into & through the center.

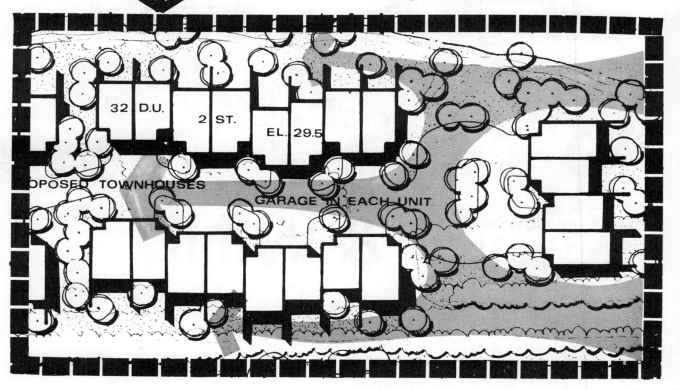

32 D.U.   2 ST.   EL. 29.5

ROPOSED TOWNHOUSES   GARAGE IN EACH UNIT

168.

MOBILE/MODULAR SINGLE FAMILY HOUSING. Mobile and modular home planning determinants are similar in many respects to those encountered in traditional single family developments. A primary difference, however, is that by virtue of the unit's mobility requirements, its typical size and shape are automatically compatible to solar goals in the Hot-Humid region.

The orientation possibilities provided by mobile home lot planning is the key determinant to successful energy-conserving solutions. As long as individual units may be properly oriented and wind movement may be encouraged, the efficiency of the solar-sensitive planning process may be maintained.

Solar Collectors

During the previous discussion of solar site planning for the Hot-Humid climate region, it has been assumed that solar collection hardware would be mounted in conjunction with architectural structures. While this seems a generally desirable and logical assumption, there may arise instances where the remote location of solar panels is deemed necessary.

The amount of physical space required for solar collectors is considerable, especially in high density situations. Since open space is a valued amenity in such developments, whenever possible the placement of solar equipment in open space areas should be avoided. It would seem appropriate to investigate possibilities for the use of "utility spaces" as collector locations. In much the same manner as rooftops may be utilized, a doubling of function may occur in parking lots and similar areas without jeopardizing the character and usefulness of open space. Remote site structures such as utility buildings, garden trellises, or carport roofs are obvious possibilities for solar panel mounting In carports, for example, not only are automobiles shaded and p protected from the elements, but collected solar energy may be stored in some central location from which it may be tapped by all units in the housing cluster.

An extension of this idea would cover entire lots with a trellis-like structure which could be of more lightweight and simple construction. Such an approach would provide similar benefits of centralized energy collection and shaded parking, and could also greatly improve the general aesthetics of traditional parking lots.

The typical surface material used in parking areas is asphalt concrete. When asphalt is new or recently seal-coated it is essentially a "black body" with high heat absorbing characteristics. With further development research into its workability, it could be possible to incorporate collector piping within parking lot and road pavements. Usual residential areas tend to find parking lots virtually empty during daylight working hours when radiation is most intense, so the idea may prove viable. Again, central storage of solar energy is possible, and the visual aesthetic benefits of such an approach are obvious.

Conclusion

The Hot-Humid regions of the world are highly amenable to the use of solar energy systems, primarily due to the intensity and the amount of available radiation resources. As planned cooling systems and future electrical generating equipment are developed, the use of solar energy will increase greatly in the region.

Because of the adverse effects of radiation encountered in the climate zone, however, the proper application of site planning principles assumes a high priority as an energy conservation objective. Intelligent consideration of the factors of orientation, shading, and wind utilization will alter the effects of climate and increase human comfort levels. Our long-range goal of energy and resource conservation will be immeasurably promoted by the creation of comfortable living environments in which the need for artificial, resource-consuming systems is substantially reduced.

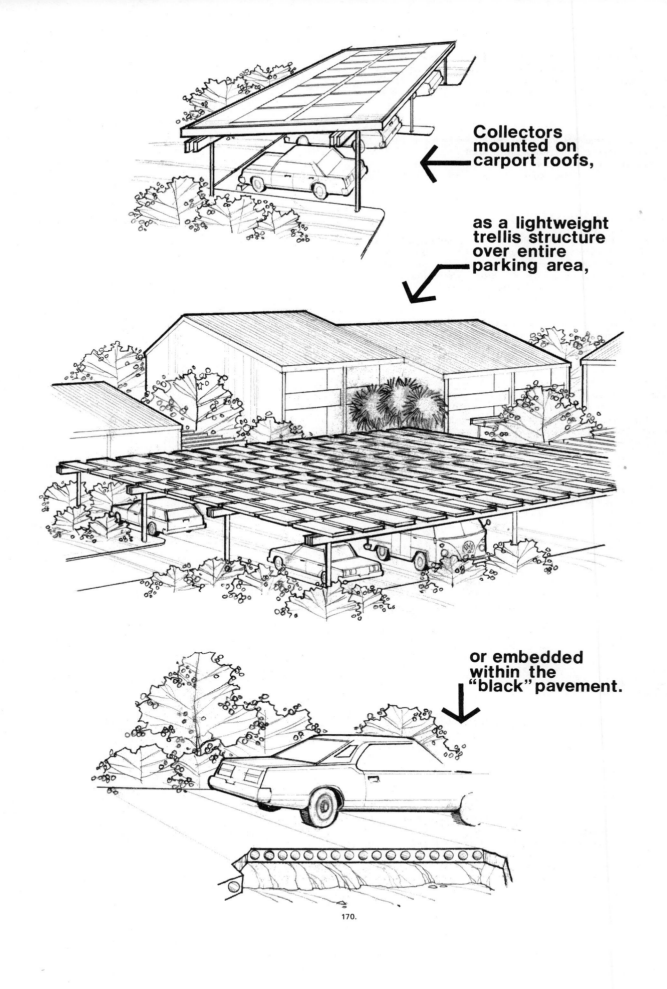

Collectors mounted on carport roofs,

as a lightweight trellis structure over entire parking area,

or embedded within the "black" pavement.

170.

Case Studies
Western Temperate and Hot-Arid Regions
Sasaki Walker and Associates

Western Temperate Region

1.
Site Selection

1.A REQUIREMENT. WITHIN A RANGE OF POSSIBLE SITES, SEEK THOSE WHICH POSSESS THE MOST FAVORABLE COMBINATION OF SLOPE, ORIENTATION, PROXIMITY TO LARGE WATER BODIES, AND AIR MOVEMENT CHARACTERISTICS TO MODERATE THE EXTREME EFFECTS OF SOLAR RADIATION.

1.A.1 CRITERION. Sites should be on upper or middle slope areas, which tend to receive larger amounts of radiation during underheated periods, and less at overheated times, than horizontal sites or those located at the foot or crest of slopes.

COMMENTARY. In the western temperate zone heat should be conserved but extreme conditions of heating or cooling should be avoided. On valley slopes intermediate temperature conditions are created by a mix of smaller circulations with the neighboring warm air. As a result, the higher sides of the slopes tend to remain warm while ridge crest and valley floor conditions are cold. In the temperate zone this "thermal belt" is most advantageous for site selection. However, if this location is exposed to unwanted crest winds that may offset higher temperatures, a more desirable place would be approximately halfway up the slope. On long slopes, locations over the crest or behind intermediate ridges would offer wind protection.

1.A.2 CRITERION. Sites should have slopes oriented south to southeast.

COMMENTARY. The total maximum insolation would come from an orientation facing directly south. But in the cold months maximum radiant gain is from the east of south. Generally speaking, exposures up to about 17 degrees East of South are permissible, but this would have to be verified for specific sites in the western temperate zone.

1.A.3 CRITERION. Sites should be as close to large water surfaces as possible.

COMMENTARY. Water, with a higher specific heat than land, is normally warmer in winter and cooler in summer, and usually cooler during the day and warmer at night, than land. Thus, proximity to bodies of water moderates extreme temperature variations, raising the winter minimums and lowering the summer heat peaks on adjacent land.

1.A.4 CRITERION. Sites should be selected where the occurence of coastal fog is limited, or where fog is not known to be retained for long periods of time during the diurnal cycle.

COMMENTARY. Fog normally represents extreme cooling conditions, and very foggy sites should be avoided where possible. Generally speaking, fog will linger less at higher elevations with good air movement.

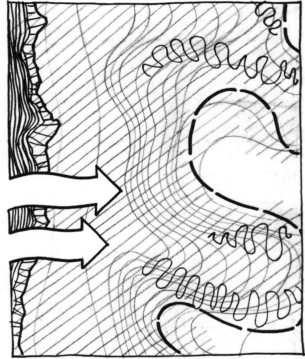

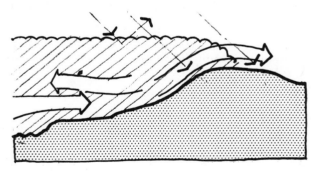

N COASTAL FOG/BREEZE FACTORS

Based upon studies prepared by SWR for Jenner, California.

- seek locations with good air movement
- but avoid strong air movements and evening breeze exposure
- seek higher elevation where fog retention is less
- site in areas protected from wind by native vegetation or topographic features
- cluster buildings in "villages" to afford protected enclaves

1.A.5 CRITERION. Sites should be selected in areas where north and east winter winds are of a minimum.

COMMENTARY. In general, January and February are the windiest months in the west temperate zone. It is desirable to find sites where those winds are from S & SE, which is the prevailing direction. NNE winds can be extremely cold in those localities where such conditions exist.

1.B REQUIREMENT. WITHIN A GIVEN SITE FOUND SUITABLE IN TERMS OF ITS GENERAL LOCATIONAL CHARACTERISTICS, SEEK SPECIFIC LOCATIONS WHERE FAVORABLE MICROCLIMATE CONDITIONS EXIST, OR CAN BE CREATED.

1.B.1 CRITERION. Specific locations should be able to provide protection from night breezes and strong air movements in general.

COMMENTARY. In most parts of the western temperate zone, orientation to wind is not a primary concern, but wind protection will be a factor. Evening breezes induce rapid cooling, often to extremes. Constant strong air movements, including constant on-shore movements, can bring with them a wind chill factor through extreme cooling effects.

1.B.2 CRITERION. Specific locations should have soil and moisture properties capable of supporting vegetation that can be used for wind protection.

COMMENTARY. In this zone it is probable that any suitable site will still require high trees for some degree of winter wind protection, even winds from the south and southeast. Satisfactory natural conditions to support protective vegetation should, therefore exist.

2.
Building Siting and Orientation

2.A REQUIREMENT. MAJOR BUILDINGS AND OUTDOOR LIVING AREAS SHOULD BE SITED WITH RESPECT TO ONE ANOTHER TO ENHANCE THE BENEFICIAL PROPERTIES OF AIR MOVEMENT AND RADIATION.

2.A.1 CRITERION. Buildings should be clustered with respect to another, with major terraces and outdoor living areas integrated within the building clusters.

COMMENTARY. The clustering of structures can reduce adverse wind effects by channeling air movements around a development, thus improving heat absorption within the cluster. The clustering need not be rigid; flexibility in building siting is still possible.

2.A.2 CRITERION. Sun nooks on south sides of buildings or clusters should be provided.

COMMENTARY. Solar nooks on south walls sheltered from wind where radiation can be concentrated will extend period of sedentary outdoor living in cool months.

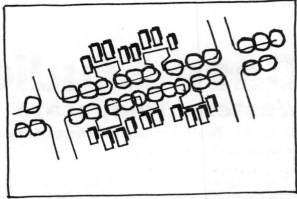

N SITING FOR HEAT ABSORPTION CONTROL

- cluster buildings for heat absorption, SSE orientation
  protect west & east exposure of buildings by "row
- protect west & east exposure of buildings by "row house" patterns
- protect street and parking areas from heat sink by E-W narrow streets and deciduous street trees

2.B REQUIREMENT. MAJOR BUILDINGS AND OUTDOOR LIVING AREAS SHOULD BE SITED TO TAKE MAXIMUM ADVANTAGE OF THE INTRINSIC ORIENTATION AND LANDFORM OPPORTUNITIES AFFORDED BY SUITABLE SITES.

2.B.1 CRITERION. On sites affording the opportunity, orient buildings in a south to southeast direction.

COMMENTARY. The total maximum isolation in this zone would come from an orientation facing directly south. But in the cold months maximum radiant gain is from the east or south.

2.B.2 CRITERION. Major entrances to buildings and exposed outdoor living areas should not be oriented in a south to southeast direction, if possible; if not possible, these entrances and areas should be protected from winter winds.

COMMENTARY. In most areas of the western temperate zone, strongest and chilliest winter winds prevail of the south and southeast. This may differ for different sites such as on-shore winds on coastal sites, so winter wind direction should be verified in all cases.

3.
Integration of Building and Site

3.A REQUIREMENT. BUILDINGS SHOULD BE SITED WITH PROPER CONSIDERATION OF THE ROLE THAT EXISTING NATURAL FEATURES OF THE SITE CAN PLAY IN MODERATING THE EXTREME EFFECTS OF SOLAR RADIATION.

3.A.1 CRITERION. Buildings should not be located immediately to the north of any significant stand of evergreens that may be existing on the site.

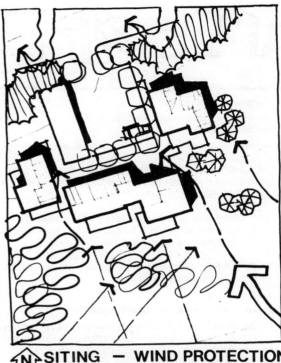

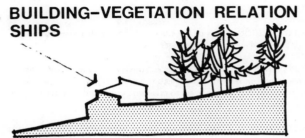

.locate south of evergreens

BUILDING-VEGETATION RELATION SHIPS

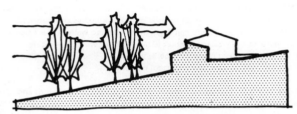

.plant in groupings for wind breaks

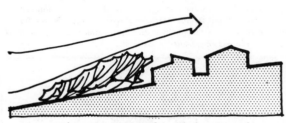

.native vegetation for wind protection

- locate south of evergreens
- plant in groupings for wind breaks
- native vegetation for wind protection

SITING — WIND PROTECTION
Based upon studies done by SWR for Jenner, California.

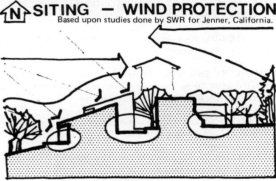

- cluster buildings for wind protection
- native vegetation protects W side of cluster
- plant groups of trees, especially in large open areas, E side of cluster
- north buffer of evergreens
- protect entrances and outdoor areas from winter winds

COMMENTARY. In the western temperate region as much as 75 percent of winter days are overcast, so unencumbered exposure to solar heat is necessary to capture as much of the limited radiation as possible. Deciduous, rather than evergreen trees are far more appropriate near sunstruck portions of a structure.

3.A.2 CRITERION. Buildings and outdoor living areas should be located to receive afternoon shade from any significant deciduous trees that may be existing on the site.

COMMENTARY. To make maximum use of the sun with the most comfort, southern exposure with provision for late afternoon shade is an appropriate combination. Existing trees should be evaluated for their usefulness in providing desired shade.

3.A.3 CRITERION. Buildings should be placed so as to minimally disturb the existing natural landforms on the site.

COMMENTARY. Undue disturbance of the natural topography may have consequences for the nocturnal release of ground-absorbed radiation.

4.
Site Planning and Design

4.A   REQUIREMENT. MAXIMIZE THE USEFULNESS OF THE SITE FOR VARIOUS DESIRED FUNCTIONS, BY MATCHING THE EXPOSURE REQUIREMENTS OF EACH FUNCTION WITH THE RADIATION CHARACTERISTICS EXISTING ON THE SITE.

4.A.1 CRITERION. Outdoor living areas should be located on the site where there are good opportunities for receiving solar heat.

COMMENTARY. The general mild temperatures in the western temperate region, in all seasons, encourages more outdoor activity than in cooler climates. The length of time available for outdoor living can be appreciably extended on a daily basis by taking advantage of solar radiation opportunities in those areas of the site where said opportunities exist.

4.A.2 CRITERION. In general, provide shade for dwellings and outdoor living areas through the use of high deciduous trees.

COMMENTARY. Shading devices for summertime use must not preclude summertime breezes or solar radiation in the wintertime. High deciduous trees are most desirable, because they permit maximum winter sun and cooling summer breezes to penetrate the site. It is especially important to avoid tall evergreens to the south and southwest.

4.A.3 CRITERION. Punctuate large lawns and other major usable spaces on the site with groupings of shade trees.

COMMENTARY. In large open areas a balance should be struck between sun and shade, for evenness in radiation. Vegetation planted in groupings is also more effective for wind protection.

4.B   REQUIREMENT. LOCATE ELEMENTS ON THE SITE FOR BENE-FICIAL UTILIZATION AND CONTROL OF AIR FLOWS ACROSS THE SITE.

4.B.1 CRITERION. Roads on the site should be oriented to avoid winter winds and to channel summer breezes.

COMMENTARY. As linear open spaces, streets and roadways can assist in channeling air movements. Since roads often have sidewalks for pedestrians, it is important that the roads are so oriented as to afford comfortable walking conditions.

4.B.2 CRITERION. Locate Buildings, fences, walls, hedges and other planting as shelters to control breezes; where necessary employ portable wall shelters.

COMMENTARY. It is important to maximize outdoor living opportunities and wind protection, in addition to solar orientation, plays an important role in this respect. Shrubbery can augment fences and walls in this respect in many parts of the western temperate region, since winds are not normally violent enough to cause damage to vegetation. Specific local wind conditions should be carefully analyzed, however.

4.B.3 CRITERION. Windbreaks should be used against winter winds and placed close to the structure or area being protected.

COMMENTARY. The chilling effects of winter winds should be mitigated. Variation in direction of summer and winter winds permits windbreaks close to a house without serious interference with summer breezes. Some protection may be needed from North and East winter winds by evergreen deflectors in localities where such winds exist.

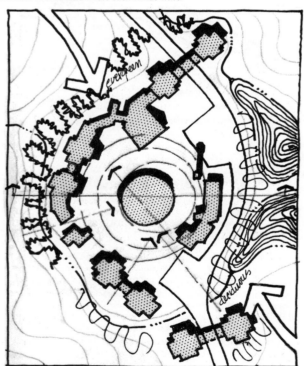

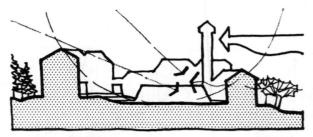

Based on studies prepared by SWR for Sun Valley, Idaho

- evergreen wind deflectors
- buildings and walls surround central outdoor space for wind protection and sun trap
- high deciduous trees to the south for summer shading

4.C   REQUIREMENT. SOLAR COLLECTORS NOT LOCATED ON BUILDINGS SHALL BE LOCATED ON THE SITE WITH CONSIDERATION OF THEIR AESTHETIC IMPACTS'

4.C.1 CRITERION. Solar collectors should be located in areas of the site where they least impair its visual quality, and they should be adequately screened from view from dwellings and outdoor living areas.

COMMENTARY. Solar collectors present potential aesthetic problems, even when they are not located on buildings.

Selection and Use of Materials

5.A   REQUIREMENT. MATERIALS USED IN THE DEVELOPMENT OF THE SITE SHALL BE SELECTED WITH CONSIDERATION OF THEIR CAPABILITIES TO ASSIST IN PRODUCING AN EVENNESS IN RADIATION CONDITIONS'

5.A.1 CRITERION. Hard-surfaced materials should be used for terraces and other outdoor sitting areas.

COMMENTARY. Solar heat on hard-surfaced terraces, patios, and courtyards will increase the length of evening use and contribute to a fuller enjoyment of the site in the western temperate zone.

5.A.2 CRITERION. Lawns and grassy materials should be used in the immediate area of and dwelling structure.

COMMENTARY. Grass is a material capable of keeping a relatively even temperature throughout the day.

5.A.3 CRITERION. Where possible, employ medium colors on sun exposed surfaces, and use dark colors only in recessed places protected from summer sun.

COMMENTARY. Light colors will generally be too reflective in the western temperate zone to achieve balanced radiation, and dark colors, except in special places, absorb more radiation than desired.

5.B   REQUIREMENT. MATERIALS USED IN THE DEVELOPMENT OF THE SITE SHOULD BE CAPABLE OF ADAPTING TO UNIQUE CLIMATIC EFFECTS IN WESTERN TEMPERATE REGIONS.

5.B.1 CRITERION. Materials should be selected with proper consideration given to their abilities to withstand dampness.

COMMENTARY. The wet winter season is more of a problem for site materials in the western temperate region than the relatively dry summers. Walkways, for example, should be impervious, well-drained, and have a textured (i.e., non-skid) surface. Where shrinking and swelling of carport roofs, wood decks, or trellises is a problem, hygroscopic materials should be sealed or avoided entirely.

N SITE DESIGN

SOME FACTORS INFLUENCING SITE SELECTION:
WESTERN TEMPERATE REGION*

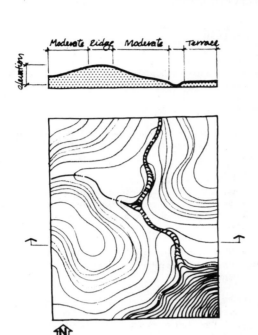

A. ALTITUDE AND SLOPE

- seek upper or middle slope areas

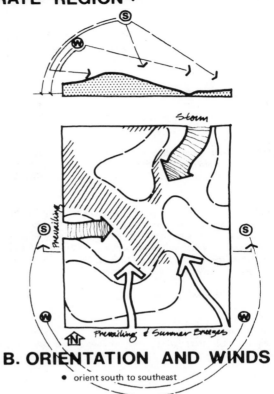

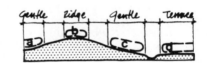

B. ORIENTATION AND WINDS

- orient south to southeast

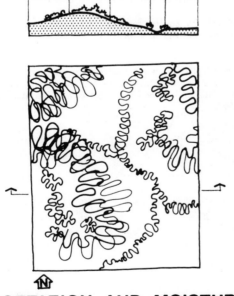

C. VEGETATION AND MOISTURE

- relate to wind protection possibilities
- relate to sun pockets

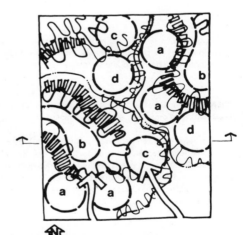

D. COMPOSITE SHOWING PREFERRED SITES

* *Based on studies prepared by SWR in Jenner, California and Monterey, California

5.B.2 CRITERION. In areas where sand is carried by the wind, such as beaches or large river valleys, choose trees which can withstand the effect and protect against it.

COMMENTARY. The desire to live near ocean and river shorelines in the western temperate zone brings with it some special consequences. Trees such as poplars, willows, cypress and pines are very effective in tolerating and mitigating the effects of blowing sand.

5.C REQUIREMENT. MATERIALS USED SHOULD BE INSTRUMENTAL IN CONVEYING A PSYCHOLOGICAL IMPRESSION THAT THE SITE IS A COMFORTABLE PLACE IN WHICH TO LIVE AND TO VISIT.

5.C.1 CRITERION. Materials, colors, and textures should, to the extent possible, be those which are natural or traditional to the area, and of demonstrated value in adapting human activity to western temperate climatic conditions.

COMMENTARY. Vegetation, materials, colors and textures recognized as indigenous to an area are likely to be associated with qualities adaptive to local climatic conditions. In fact, this is often the case. Continued wise use of the elements may be the best way to deal with solar radiation problems. Certainly, the feeling that the elements have been successful in the past demands active consideration of their use in the future.

5.2.C CRITERION. Avoid over-planting of coniferous and broadleaf evergreens.

COMMENTARY. Coniferous and broadleaf evergreens, if over-planted, can create a cool and somber effect during protracted cloudy winter days.

**Hot-Arid Region
1.
Site Selection**

1.A REQUIREMENT. WITHIN A RANGE OF POSSIBLE SITES, SEEK THOSE WHICH POSSESS THE MOST FAVORABLE COMBINATION OF ALTITUDE, SLOPE AND ORIENTATION TO REDUCE THE EFFECTS OF DAYTIME RADIATION.

1.A.1 CRITERION. Acceptable sites should not be in valleys, especially narrow valleys, but instead should be at lower hillside locations benefitting from cool air flows.

COMMENTARY. In general, in the hot-arid zone the desirability of heat loss overrules the demands of the cool periods. Locations above a valley floor tend to be cooler in the afternoons and warmer at nights than lower sites.

1.A.2 CRITERION. Sites should afford opportunity to oreint buildings and activities in an east-southeast direction, within acceptable limits.

COMMENTARY. The large daily temperature range makes easterly exposures desirable for daily heat balance. In a large portion of the year, afternoon shade is required. Accordingly, sites affording ESE exposure are preferred in the hot-arid zone. Generally speaking, exposures 25 degrees East of South secure balanced orientation, but this would have to be verified for specific sites in the zone.

1.A.3 CRITERION. Sites should be considered if they are not constrained from advantageous exposure to the cooling influence of prevailing east-west breezes.

COMMENTARY. East-west winds are important cooling factors during days from March to November. In certain areas of the hot-arid zone, however, uncomfortable desert breezes can occur, so care must be taken at the gross site selection level to understand nature and frequency of such occurances.

1.B REQUIREMENT. WITHIN A GIVEN SITE FOUND SUITABLE IN TERMS OF ITS GENERAL ALTITUDE, SLOPE, AND ORIENTATION CHARACTERISTICS, SEEK SPECIFIC LOCATIONS WHERE FAVORABLE MICROCLIMATE CONDITIONS EXIST, OR CAN BE CREATED.

1.B.1 CRITERION. Specific locations should have distinct evaporative possibilities.

COMMENTARY. On-site conditions in hot-arid areas can be such as to require as much protection as possible against thermal gradients of 90 degree F or more between outside and inside temperatures. This microclimate stress may be reduced by many measures including shading, air movement, insulation, and surface evaporative cooling.

1.B.2 CRITERION. Specific locations should, where possible, be leeward of large water bodies or irrigation areas, provided such locations maintain ESE orientation.

COMMENTARY. Moisture given off by large water bodies or irrigation areas will tend to reduce air temperatures on leeward lands, as it evaporates. However, the albedo of water increases for light striking it at a low angle of incidence which may be unpleasant for outdoor living areas facing westerly across a lake.

1.B.3 CRITERION. Specific site locations should have, or be related to landforms and vegetation that permit air movement to occur and be utilized as a cooling factor.

COMMENTARY. Winds are perhaps more influential at the microclimatic level than at the level of macroclimatology because there is more chance for variation and wind speed effect from the various land forms. The winds have a chance to be channeled and directed at this scale and therefore are important to site selection.

**2.
Building Siting and Orientatation**

2.A REQUIREMENT. MAJOR BUILDINGS, LIVING AREAS, AND CIRCULATION WAYS SHOULD BE SITED WITH RESPECT TO ONE ANOTHER TO REDUCE THE EFFECTS OF DAYTIME RADIATION.

2.A.1 CRITERION. Site unit dwellings in a concentrated way so that structures may shade one another as well as shade outdoor living areas and circulation ways.

COMMENTARY. A shaded and dense layout of buildings on a site assists greatly in the ability of the buildings to react against heat. The general liveability of the outdoor site environment in hot-arid areas is also thereby enhanced.

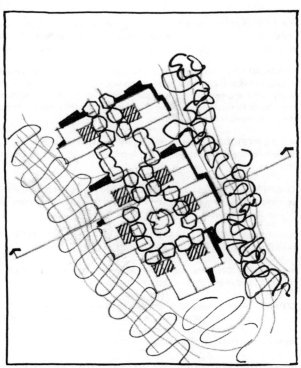

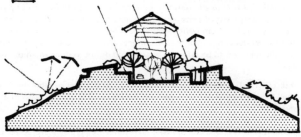

BUILDING RELATIONSHIPS

- shaded and dense layout desirable
- interior courtyards () provide shading opportunities
- small parking bays reduce sun pockets
- narrow streets

2.A.2 CRITERION. The concentration and clustering of unit dwellings should be designed to maximize the attachment or close spacing of one dwelling to another so as to eliminate as much as possible direct solar impacts on east and west walls of any dwelling.

COMMENTARY. In hot-arid zones east and west sides are most vulnerable to solar radiation, due to lower sun angle in morning and late afternoon. It should be noted that this is an especially challenging criterion, requiring an effective compromise with the desirability to maintain exposure to east-west winds.

2.A.3 CRITERION. Dwelling units or groups should create patio-like areas.

COMMENTARY. Small patio-like or courtyard areas provide good shading opportunities.

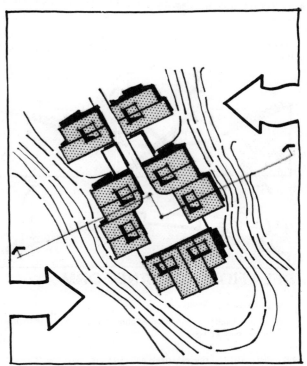

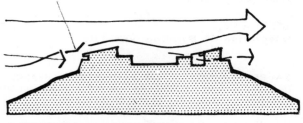

BUILDING ORIENTATION
(Based upon studies done by SWR for Cochiti Lake, New Mexico)

- ESE orientation to reduce solar impact
- E winds sought for cooling
- ridge location to avoid heat traps and to catch breezes

2.A.4 CRITERION. Streets and other circulation ways should be kept as narrow as possible consistent with other health safety and welfare requirements.

COMMENTARY. Subject also to the demands of carrying traffic safely, special efforts to be made to reduce generous right-of-ways to permit adjoining building shadows to have a significant effect on cooling these open areas.

2.B REQUIREMENT. MAJOR BUILDINGS, OUTDOOR LIVING AREAS AND CIRCULATION WAYS SHOULD BE SITED TO TAKE MAXIMUM ADVANTAGE OF THE INTRINSIC ORIENTATION AND LANDFORM OPPORTUNITIES AFFORDED BY SUITABLE SITES.

2.B.1 CRITERION. On those sites affording the opportunity, orient buildings and outdoor activity areas in an east-southeast direction, within acceptable limits.

COMMENTARY. The large daily temperature range in most hot-arid zones makes easterly exposures desirable for daily heat balance. In a large portion of the year (May to November) afternoon shade is required. Accordingly, buildings sited to give major living areas an ESE exposure are preferred. For simple structures exposures 25 degrees east of south will secure balanced orientation, but this should be verified for specific sites and structures within hot-arid zones.

2.B.2 CRITERION. Buildings should not be located on concave land forms on the site.

COMMENTARY. In hot-arid zones concave landforms are usually "cold islands" at night, often too cold to insure proper diurnal radiation balance.

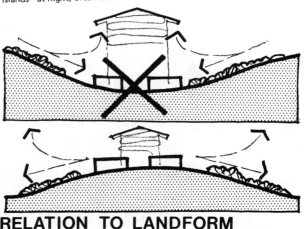

RELATION TO LANDFORM

- avoid concave landforms
- positive landforms minimize radiation trap during day, minimize cold air drainage at night

2.B.3 CRITERION. Principal streets and pedestrian ways should orient as closely as possible along an east-west axis.

COMMENTARY. East-west orientation of circulation ways best captures prevailing cooling breezes.

3.
Integration of Building and Site

3.A   REQUIREMENT. APPLICABLE LITHOSPHERE ARRANGEMENTS SHOULD BE A CONSIDERATION IN PLANNING BUILDING-TO-SITE RELATIONSHIPS FOR ANY SPECIFIC SITE.

3.A.1 CRITERION. The construction and utilization of subterranean and quasi-subterranean living areas should be encouraged where structural and soil properties of the site permit.

COMMENTARY. The coolness of below-grade structures is of benefit in hot-arid climates, potentially reducing energy needs. Not all sites possess soil bearing capabilities to permit subterranean development, however. The prospect of such capabilities should be examined and utilized where physically and economically appropriate.

3.B   REQUIREMENT. BUILDINGS SHOULD BE SITED TO TAKE MAXIMUM ADVANTAGE OF EXISTING VEGETATION.

3.B.1 CRITERION. Locate buildings to utilize appropriate existing trees for shade on east and west walls exposed to high levels of solar radiation.

COMMENTARY. Existing vegetation, especially mature trees, can be of immediate assistance in mitigating heat gain, so should be used to advantage if possible.

4.
Site Planning and Design

4.A   REQUIREMENT. MAXIMIZE THE USEFULNESS OF THE SITE FOR VARIOUS DESIRED FUNCTIONS, BY MATCHING THE EXPO- SURE REQUIREMENTS OF EACH FUNCTION WITH THE RADIATION CHARACTERISTICS EXISTING ON THE SITE.

4.A.1 CRITERION. Orient outdoor living and working areas according to the times during which they are most frequently used, seeking cooler locations for afternoon activities and warmer locations for evening activities.

COMMENTARY. Outdoor living and working activities demand as much consideration as comparable indoor functions, and the same general rules apply. Time and frequency of use of an outdoor area relative to its function should be carefully examined as the use is allocated a space in the outdoor environment of hot-arid zones.

4.A.2 CRITERION. Locate outdoor utility, storage and similar areas in those locations on the site not otherwise better used for living and working activities.

COMMENTARY. Areas not normally inhabited should not normally enjoy precious site locations better used for reducing the effects of daytime radiation on areas of human activity.

4.B   REQUIREMENT. LOCATE ELEMENTS ON THE SITE TO REDUCE SUN POCKETS AND RADIATION GLARE.

4.B.1 CRITERION. Total non-permeable paving area, such as in parking lots, sidewalks, and streets shall be kept to a minimum.

COMMENTARY. Areas absorbing a high degree of solar radiation in hot-arid zones can create hot spots, or "heat sinks", which can cause considerable heat gain in these and surrounding areas. Asphalt and concrete are leading examples, and such paving should be minimized.

4.B.2 CRITERION. Non-permeable paving areas should be shaded, through the use of vegetation, landforms, walls, screens, canopies, and overhangs.

COMMENTARY. In addition to minimizing the absolute amount of paving in the development of a site, shading devices ought to be employed. For example, shading from walls can exclude solar radiation and reflected radiated temperatures from hot ground surfaces around houses. Also, horizontal shading is known to be effective in areas and during seasons when the sun is high.

4.B.3 CRITERION. Grass, ground cover, gravel or other suitable material shall be used in lieu of non-permeable paving areas around all walls of a dwelling.

COMMENTARY. Because of lower light reflection, grassy areas, sod and the like can be effectively employed as an element for cooling.

4.B.4 CRITERION. Large non-permeable areas, such as group parking areas and paved court game areas, should be located leeward of, and as far as possible from dwelling units and related outdoor living areas.

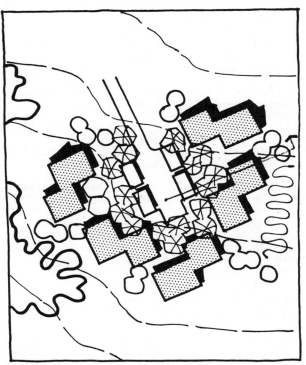

REDUCING RADIATION GLARE

(Based upon studies by SWR for Desert Ranch, Arizona and Cathedral Mountain, Texas)

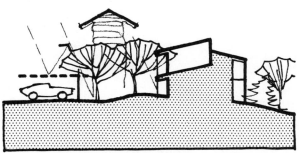

- small, shaded parking areas and carports
- grass adjacent to dwelling units
- tree-shaded roadways
- parking areas removed from dwelling units
- existing native vegetation shades West side of cluster
- planted native materials shade E side

COMMENTARY. Removal of potential "heat sinks" to locations remote from frequently inhabited areas can mean energy savings and increased livability in those areas.

4.B.5 CRITERION. Through grading or plowing, reduce the potential for heat storage in large areas of ground not subject to grass planting or other vegetation cover.

COMMENTARY. The heat capacity of air is quite low, so there is less heat stored in soils which have been plowed or loosely disturbed than those left hard or undisturbed. Decisions on grading or plowing, however, would have to take into consideration the dust problem in windy areas.

4.C REQUIREMENT. LOCATE ELEMENTS ON SITE TO REDUCE HEAT GAIN IN SURFACES PROTECTING OR

4.C.1 CRITERION. Where possible, place vegetation to overhang roof areas.

COMMENTARY. It may not always be possible to reduce heat loss in buildings through orientation or choice of materials. Roof areas are a prime example. In addition to efforts at using roof materials possessing low overall heat transmission coefficients, shading such areas would be helpful, assuming no conflict with rooftop solar collectors. Overhanging trees present the most economical and effective method of achieving this.

4.C.2 CRITERION. All east and west wall surfaces should be shaded.

COMMENTARY. In hot-arid zones, east and west sides of any structure are most vulnerable to solar radiation. Where those sides must be exposed, vegetation or other devices should be employed to shade or shield the surfaces from the effects of the sun.

4.D REQUIREMENT. LOCATE ELEMENTS ON SITE FOR BENEFICIAL UTILIZATION AND CONTROL OF AIR FLOWS ACROSS THE SITE.

4.D.1 CRITERION. For protection against unwanted anticyclonic winds provide shelterbelts of trees oriented generally east-west which follow local topographic changes such as spurs or ridges, and are moderately penetrable to the wind. The number, length, density and spacing of shelterbelts shall be determined by analysis of the microclimatic conditions on each specific site as they affect the location of buildings and activities.

COMMENTARY. In hot-arid regions some winter wind protection may be needed. The east-west orientation will best obstruct the continental winds blowing off the Gulf and the anti-cyclonic winds, usually from a southwesterly source. Each site has to be separately evaluated for the degree to which this protection is warranted.

4.D.2 CRITERION. Trees and medium-to-high hedges within ten feet of a building should be placed relative to openings in the building such that they assist in channeling and directing beneficial breezes through the openings.

COMMENTARY. It is important to carefully coordinate the design of buildings with plans for adjacent landscaping, so that integrated tree-hedge-building combinations are produced. From a solar radiation point of view the immediate environment of the dwelling unit may be the most important to control. Numerous combinations are possible, so flexibility in design is not necessarily constrained.

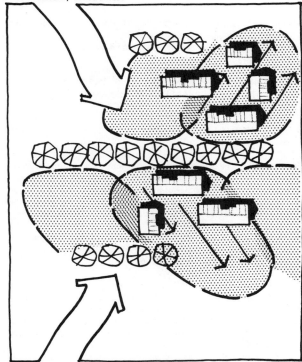

WIND PROTECTION

- tree rows (shelterbelts) oriented E-W
- spacing allows moderate wind penetration
- ridge line locations preferred

4.E REQUIREMENT. SOLAR COLLECTORS NOT LOCATED ON BUILDINGS SHALL BE LOCATED ON THE SITE WITH CONSIDERATION OF THEIR AESTHETIC IMPACTS.

4.E.1 CRITERION. Solar collectors should be located in areas of the site where they least impair its visual quality, and they should be adequately screened from view from dwellings and outdoor living areas.

COMMENTARY. Solar collectors present potential aesthetic problems, even when they are not located on buildings.

5.
Selection and Use of Materials

5.A REQUIREMENT. MATERIALS USED IN THE DEVELOPMENT OF THE SITE SHALL BE SELECTED WITH CONSIDERATION OF THEIR CAPABILITIES TO ASSIST IN REDUCING THE EFFECTS OF DAY-TIME SOLAR RADIATION.

5.A.1 CRITERION. All site materials, except those specifically intended as part of a solar collection system, should be of low conductivity and high reflectivity in areas of the site intended for frequent or occasional habitation, subject also to consideration of the thermal radiation characteristics of the materials.

COMMENTARY. Materials with low heat conductivity cool off rapidly and do not store heat at night, thus minimizing the possibility of day-to-day accumulation of heat. Except where specifically intended to trap winter radiation, reflective materials should be used. It is also important to evaluate the thermal properties of such materials. Those highly reflective of solar radiation may have to be shaded or used in shady areas.

5.A.2 CRITERION. Where possible, use light colors on sun exposed surfaces, and use dark colors on deep set surfaces and where reflections toward living spaces are expected.

COMMENTARY. Light colors have high reflection ratios. It is desirable to use light colors to help reflect heat as a result of solar radiation (but not thermal radiation). Deep set surfaces can be dark colored for winter radiation absorption.

5.B REQUIREMENT. MATERIALS USED IN THE DEVELOPMENT OF THE SITE SHOULD BE CAPABLE OF WITHSTANDING THE UNIQUE CLIMATIC EFFECTS OF HOT-ARID REGIONS.

5.B.1 CRITERION. All building materials must be able to withstand effects of excessive dryness.

COMMENTARY. The dryness of the climate limits the moisture available for the effective performance of many materials, such as certain woods. Care must be taken in the choice of materials in terms of their ability to endure arid conditions and still remain durable, pleasant in appearance, economically maintainable, and useful in solar radiation effectiveness.

5.B.2 CRITERION. Vegetation used in the control of solar radiation and related microclimatic conditions should be selected with consideration of their drought-and-dust resistant properties.

COMMENTARY. Trees and shrubs which conserve moisture and are unaffected by blowing dust and sand will be most suitable in hot-arid zones. Indeed, native vegetation is usually best. Where native species can be easily grouped or arranged to provide winter protection or reduce hot desert soil temperatures, they are preferable to imported species. Native species commonly used include artemisia, sagebrush, pinon, paloverde, acacia, and cottonwood; species should be used, however with a thorough understanding of their own growing requirements.

5.C REQUIREMENT. MATERIALS USED SHOULD BE INSTRUMENTAL IN CONVEYING A PSYCHOLOGICAL FEELING THAT THE SITE IS A COMFORTABLE PLACE TO LIVE AND TO VISIT.

5.C.1 CRITERION. Materials, colors and textures should to the extent possible be those which are natural or traditional to the area, and of demonstrated value in adapting human activity to hot-arid climatic conditions.

COMMENTARY. Vegetation, materials, colors and textures recognized as indigenous to an area are likely to be associated with many years of experimentation and experience in adapting to unique climatic conditions. In fact, this is often the case. Continued wise use of the elements may be the best way to deal with solar radiation problems in hot-arid areas. Certainly, the feeling that they have been successful in the past demands that they be actively considered for the future.

5.C.2 CRITERION. Water, in the form of pools and fountains, should be a key element in site design, where supply is available.

COMMENTARY. Water heats up rapidly, but also cools down rapidly. It has a cool appearance and pleasant sound, and can give psychological assurance that the site environment has a permanent and sustaining character, despite the harsh climatic condition. Recycling of water ought to be considered in locations with water supply problems: the energy required to recycle may be less than the energy required for water importation.

5.C.3 CRITERION. Vegetation (trees) should be placed and be of such height so as to be capable of movement under even the slightest of breezes.

COMMENTARY. The movement—and the sound—of trees rustling in the wind can be of great psychological benefit, even if the physiological effect is barely existent. Seeing and hearing the swaying of tall trees can assist the individual in feeling the presence of a comforting breeze which, depending on the individual and his activities, may cause the site environment to seem more comfortable in hot-arid areas.

180.

SOME FACTORS INFLUENCING SITE SELECTION:
HOT ARID REGION *

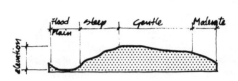

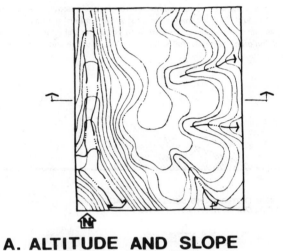

A. ALTITUDE AND SLOPE

- hillside locations

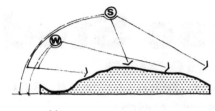

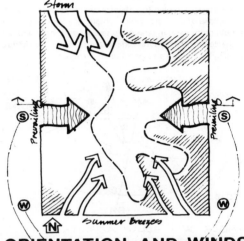

B. ORIENTATION AND WINDS

- ESE exposure
- E-W winds

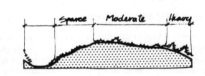

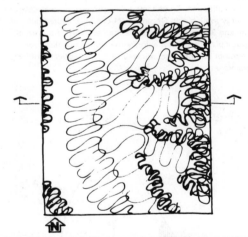

C. VEGETATION AND MOISTURE

- vegetation permitting air movement
- evaporative possibilities

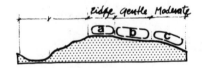

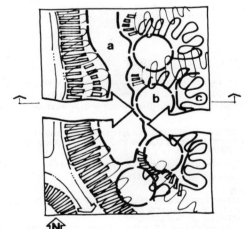

D. COMPOSITE SHOWING PREFERRED SITES

(Based upon a series of site studies by SWR for Desert Ranch, Arizona, Cathedral Ranch, Texas, Cochiti Lake, New Mexico)

Case Studies
Cool Region
Johnson, Johnson and Roy

Sun Angles and Shadows

In order for a designer to plan for effective use of sun and shade, he must first study the laws of solar mechanics. Obtaining a basic understanding of how the sun paths vary, in relation to season, date, hour and latitude, will allow the designer to moderate the amounts of sun and shadow through proper building orientation and use of plant materials.

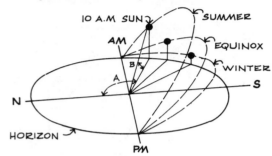

ANGLE 'A' = SOLAR AZIMUTH
ANGLE 'B' = SOLAR ALTITUDE

ILLUSTRATION OF SOLAR MECHANICS

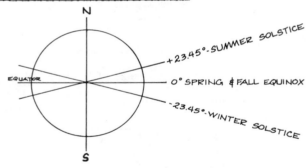

SOLAR DECLINATION ANGLES

By applying the principles and formulas illustrated, a designer can plot shadow contours on a plan, or in section, for a given day in a season to better understand the shadow pattern relationship between buildings and outdoor spaces.

FOR DETERMINING THE SOLAR ALTITUDE AND AZIMUTH ANGLE FOR ANY HOUR DURING THE YEAR, THE FOLLOWING FORMULA APPLIES:

$$\text{SIN B} = (\cos D)(\cos L)(\cos H) + (\sin D)(\sin L)$$
$$\text{SIN A} = (\cos D)(\sin H) + (\cos B)$$

WHERE:
A = AZIMUTH OF THE SUN, MEASURED EAST OR WEST OF SOUTH
B = ALTITUDE OF THE SUN ABOVE THE HORIZON
L = LATITUDE OF THE LOCATION
D = DECLINATION ANGLE OF THE SUN ABOVE OR BELOW THE CELESTIAL EQUATOR
H = THE LOCAL HOUR ANGLE OF THE SUN, EAST OR WEST OF THE NOON MERIDIAN. EACH HOUR EQUALS 15°.

FORMULAS FOR CALCULATING
SOLAR SUN ANGLES

To maximize the use of solar radiation through solar energy collection, the designer should first plot the maximum shadow length of the year. This will give maximum distances to work with in spacing architectural elements or locating plant materials that will not conflict with the ability to collect solar radiation.

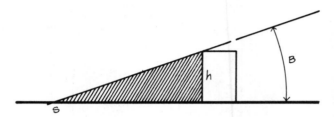

FOR DETERMINING LENGTH OF SHADOW THE FOLLOWING FORMULA APPLIES:

$$S = \frac{h}{\tan B}$$

WHERE:

S = LENGTH OF SHADOW
h = HEIGHT OF OBJECT
B = SOLAR ALTITUDE

SHADOW LENGTH CALCULATIONS

There are some practical limitations to the collection of solar radiation in regard to building spacing. Since shadows cast in early morning and late afternoon during the winter season are extremely long, it is not practical to prevent some shading at these times. In fact, calculations of solar energy during winter solstice show that very little energy in direct sunlight is available for collection. These same calculations show that 85% of the total solar energy available on December 21 can be collected between the hours of 9 a.m. and 3 p.m. Using this time span as a rule of thumb, the designer can lay out a plan that will maximize the ability to collect solar radiation, with some minor compromises allowed to solve practical problems and achieve spatial variety.

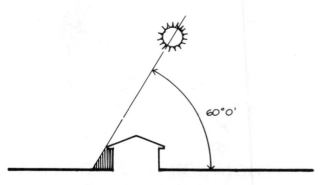

SUMMER SOLSTICE – JUNE 21

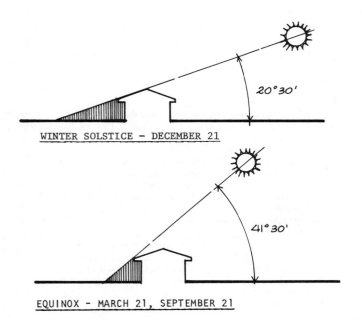

20°30'

WINTER SOLSTICE — DECEMBER 21

41°30'

EQUINOX — MARCH 21, SEPTEMBER 21

ILLUSTRATION OF VARYING SEASONAL SUN ANGLES AT 40 DEGREES NORTH LATITUDE

The greatest problem in preparing a site plan that will maximize solar energy collection arises when the tallest building element lies between the sun and the smaller building element. This can result in impractical distances between building elements in order to minimize shading. The ideal solution would be to place the highest elements downsun from the lowest elements and central in the group.

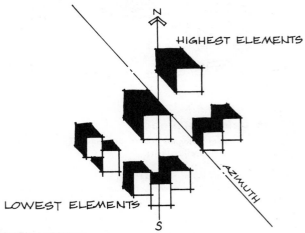

HIGHEST ELEMENTS

LOWEST ELEMENTS

AZIMUTH

ILLUSTRATION OF DESIRED BUILDING
ARRANGEMENT TO MINIMIZE SHADING

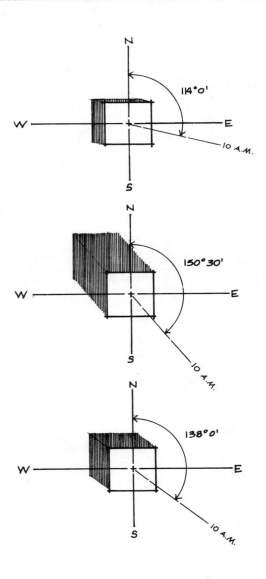

114°0'

10 A.M.

150°30'

10 A.M.

138°0'

10 A.M.

Site Planning and Solar Collectors

Since solar collectors in the northern hemisphere must fall south at a slope perpendicular to the sun's rays, there are site planning problems which develop when site dictates make it impractical to orient buildings on an east-west axis. In this situation it is most practical to consider having the solar collector detached from the building elements. This will allow the greatest flexibility in building orientation as well as solve some of the spacing problems discussed in the previous section.

Climatic Influences

Summer lake breezes prevalent along the west 3,100 mile long Great Lake shoreline is caused by the temperature difference between land and water resulting in pressure variances which generate air flow patterns.

During the day the solar energy warms both land and water with the land absorbing more heat than water. The air above the land mass becomes warmer expanding upward generating increased air pressure at higher levels. The air moves at higher levels out to a lower pressure area over the water. The air pressure builds up over the water causing the air to move downward to the cooler low pressure area at water level. The air is recycled, moving the water cooled air inland over the warm land mass. The air movement accelerates over the morning hours to its maximum during the afternoon when land temperatures are the warmest creating a natural air conditioning system. This land area near the shoreline will require little, if any, energy to power air conditioners.

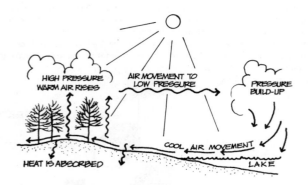

HIGH PRESSURE WARM AIR RISES

AIR MOVEMENT TO LOW PRESSURE

PRESSURE BUILD-UP

COOL AIR MOVEMENT

LAKE

HEAT IS ABSORBED

Great Lakes Influences

Michigan is in a unique situation in terms of weather. Three primary factors influence its unusual, erratic climate.

FIRST, a strong meandering band of strong wind referred to as the "jet stream" that controls storm movements moves from the south settling over Michigan during the spring. This "jet stream" extends the long cold weather season.

SECONDLY, Michigan, along with a few neighboring states is located within the crossroads of major storm patterns. This mid-western zone is impacted by storm activitiy from the west, north and south in addition to occasional impact from the fringe of disturbances moving from south to north on the east coast.

FINALLY, the most unique and greatest impact is the "Great Lakes Influence." The expansive water areas almost surrounding the 3,100 mile long Michigan shoreline heat and cool far more slowly than the adjacent land masses. Thus a wide band of land mass (as far inland as Grand Rapids, 25 miles from Lake Michigan) responds slowly to changes in air temperature. These stabilizing temperatures cause a lag in the seasons resulting in long autumns and springs.

A summer lag within this perimeter shoreline zones can be cooled appreciably by the influence of the lake temperature which can be 20 to 30 degrees cooler in combination with the air movement over the water. Conversely, during the winter months the water temperature can be 20 to 30 degrees warmer, adding heat to the overlying air giving the land mass a comparatively warm breeze as late as December.

Moisture rising from the expansive waters of the Great Lakes forms clouds which diffuse solar heat during the daytime while the same cloud blanket holds in the earth's stored heat at night.

These lake generated clouds cause Michigan to be enjoying far less sunshine than comparable areas in the same latitude in other states beyond the influence of the Great Lakes. A community under this influence, Grand Rapids, receives only 50% of its possible sunshine assuming the skies were clear from sunrise to sunset every dya. Southeastern Michigan has on the average, 107 days per year which are partly cloudy 40 to 70% of the day, and 180 days which are 80 to 100% cloudy leaving only an average of 78 days which are clear. Cold air passing over the lakes during the winter is both heated and moistened causing much more winter cloudiness over the lakeshore areas. This will have an enormous influence on available solar energy in this region. Because of this, an accurate detailed measurement of the cloud cover and the diffused solar energy should be made and documented, by region, within the lakeshore zone.

The Great Lakes also have an effect on the amount of snow fall. For example, Grand Rapids, and even Battle Creek which is further inland, receives an average of 1½ to 3 times as much snow as Minneapolis, Minnesota. While the snow cover will vary in quantity considerably from the lower part of Michigan to the Upper Peninsula, the amounts are larger than land masses in comparable latitudes beyond the Great Lakes influence. A prolonged snow during the winter months will have a warming influence protecting the soil from deep frost which in lean snow cover years extends the cool temperature into mid-spring. Snow cover, its depth and duration has a considerable influence on ambient temperatures which in turn effect the amount of solar heat required for winter heating.

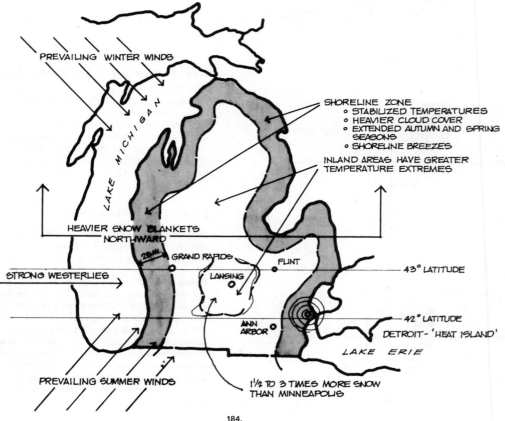

PREVAILING WINTER WINDS

LAKE MICHIGAN

SHORELINE ZONE
- STABILIZED TEMPERATURES
- HEAVIER CLOUD COVER
- EXTENDED AUTUMN AND SPRING SEASONS
- SHORELINE BREEZES

INLAND AREAS HAVE GREATER TEMPERATURE EXTREMES

HEAVIER SNOW BLANKETS NORTHWARD

25 MI. GRAND RAPIDS   FLINT

LANSING

STRONG WESTERLIES

43° LATITUDE

ANN ARBOR

42° LATITUDE

DETROIT- 'HEAT ISLAND'

LAKE ERIE

PREVAILING SUMMER WINDS

1½ TO 3 TIMES MORE SNOW THAN MINNEAPOLIS

SAGINAW FEDERAL BUILDING
Saginaw, Michigan

The Program and Objectives

Mr. Arthur F. Sampson, administrator of the General Services Administration, designated the Saginaw, Michigan Federal Building as an "Environmental Demonstration Project."

The objectives set by GSA were threefold:

To dramatize the commitment of both GSA and the Public Buildings Service to an improved environment via the design, construction and operation of Federal buildings.

To provide a testing ground for both recognized and innovative features and equipment.

To inspire all those engaged in the design and delivery processes of the construction industry toward the improvement of the environment.

The selection of Saginaw as the site of the demonstration project was dictated by the fact that the city was analogous and relevant to hundreds of other U.S. cities.

The building was to provide office space for eleven Federal agencies, a post office with loading dock area, plus parking for 100 official cars.

The Research/ Design Process

SHG/JJR conducted a state of the art survey of existing and emerging technologies that applied to the environmental objectives. It asked 700 universities to hold and report on, brainstorming sessions, resulting in a list of 140 items in two broad areas: conservation of resources and improvement of environment.

Concept evaluations were made measuring performance, cost/benefit, availability, reliability, and visibility, as well as compatibility with other systems, building function, and building scale.

An interdisciplinary team then reviewed all these evaluations to identify those concepts with the highest potential. This team then developed four very different concepts, utilizing those items with the highest potential. The four schemes were all developed to a point where their feasibility as the demonstration project could be examined.

Finally, a joint evaluation of the four proposals was made by GSA and SHG/JJR and the final solution was selected.

The Solution

The building is one-story, and partially recessed into the ground to provide an extension of the landscaped site onto the building roof. There are three levels of office space, as well as a loading dock level.

A number of energy and resource conservation features were included, most prominent of which is an 8,000 square foot flat-place solar energy collector slanting upward from the building at the optimum angle to the sun at this latitude.

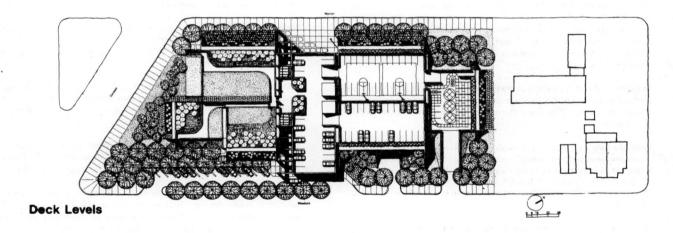

Deck Levels

Building Site Cross Section

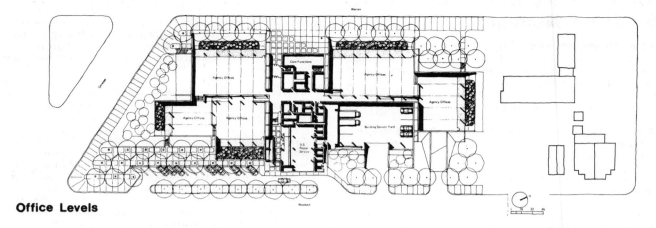

Office Levels

In addition to the overall aesthetic contribution, the landscape development, which covers approximately one half of the roof structure, also contributes to energy conservation by minimizing heat gain and loss. The plantings and lawn areas absorb energy, which is necessary for the process of photosynthesis, reflecting less heat or light than paved surfaces and, therefore, acts as a modifier to temperature fluctuations. The trees, shrubs and lawn areas will also provide an absorbing surface for airborne sounds while acting as a filter to remove dust and particles from the air.

The Building Plan

The building spans the entire site from east to west, and its stepped down configuration makes transition between the various deck and office levels easy. The deck levels contain the landscaped roof, seating, official parking and play areas, including two half-courts for basketball. Parking is shielded from street view by a 3½ foot parapet, and the surface of the parking area will have painted graphics suggesting the various recreaional activities it accommodates.

Entrance to the office levels is at grade from both east and west, into a core containing the post office, joint-use conference rooms, mechanical rooms and service facilities. This area is the only part of the interior that has full-height walls.

From the core area, three office levels step down into the building, each succeeding level two feet lower. These changes in floor and ceiling height help to break up into more human scale what might have been an overpoweringly large single contiguous space. Although 11 Federal agencies will have some space in the building, there will be no private offices nor dividing walls.

The window walls are sheltered by extremely deep (17 foot) overhangs which eliminate sun load on the glass. Offices will have exposed lighting and HVAC duct fixtures, with 75% of the light provided indirectly, bounced off the underside of the concrete roof.

The building will be of reinforced concrete construction, with post-tensioned concrete girders. A typical span would be 63 feet, as the bay size for the offices is 18 feet by 63 feet, which allows a maximum of column-free space.

Resource Conservation

One-floor building to eliminate need for vertical transportation.

Flat-plate solar energy collector in direct contact with water-filled metal pipes, and faced with a dead air space formed of two sheets of 1/8 inch tempered glass that makes the collector a heat trap. Water in the pipes will be heated as high as 180 degrees and piped to an insulated storage tank.

Closed loop heat pumps use this water to heat or cool the circulated air. Each bay is a separate heat pump unit, permitting maximum flexibility in shutting down unneeded units. When pumps are of different cycles, the heat given off by those on the cooling cycle will be added to that produced by the units on the heating cycle.

Indirect lighting provides about 75% of the general office space lighting, providing adequate light levels for all tasks, but using only two watts psf., instead of the usual four watts psf.

Open planning allows more efficient circulation of heating and cooling, as well as minimizing the amount of lighting that must be provided.

Insulation is provided in a number of ways. Earth berms are built-up against all solid walls, and there is an earth cover on the landscaped area of the roof. Windows will be double-glazed, tinted insulating glass, and will be shielded by overhangs as deep as 17 feet.

Water distribution to lavatories will be one-pipe, one-temperature water, reducing the energy needed to heat water that must then be tempered with cold water as the lavatory bowl. All faucets are spring loaded for automatic shut-off.

Irrigation water for the landscaping sprinkler system will be collected from the parking area runoff, stored in a tank, then filtered and reused. This system is estimated to reduce the use of city water for irrigation by 75%.

Toilet waste system will be a self-contained mineral oil treatment system, purified of wastes through filtration, returned as clear liquid to the toilet system. All solid residues will be incinerated via an initial and a secondary burning which will eliminate smoke and particles from dispersion into the environment. No city water nor connection to the city sewer system will be required by these toilets.

Recycled materials are demonstrated by crushed brick products, made from the brick in existing buildings to be demolished on the site. One large aggregate material will be bound with clear epoxy and used as a porous paving material on the roof and around planting areas, while a smaller size of crushed brick will be fused in a porcelain-like process into 12 inch by 18 inch panels to be used as both impervious paving and for lobby walls.

Environmental Enhancement

Landscaping will return the site as nearly as possible to its natural state, blending the man-made structure and its functions into the earth contours and the greenery.

Joint use of the facility by the various government agencies and the people of the community, by having the roof convert to a park and playground. The project is not a barrier between people and the CBD but an entrance point.

Shielded parking keeps the 100-plus parking spaces out of eyesight from adjoining sidewalks and streets, since the parking is higher than eye-level, and is further hidden behind a 42 inch parapet.

Courtyards, and their landscaping, are made a part of the office environment by locating them directly outside the full-lenght window walls.

Coordination with the development plans of the city make the Federal building a part of the long-range plans for the CBD, as well as an ad hoc object lesson for owners planning additional new buildings in the near future. When the street pattern is changed, and the CBD becomes an all-pedestrian area, the Federal building will fit into the proposed mall scheme without any needed alterations.

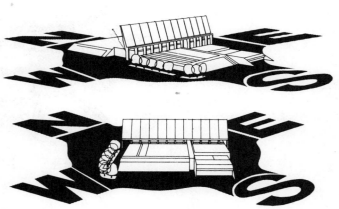

Solar Collector Orientation

The most efficient collection of solar energy would result if the collector surface can be continuously oriented to remain perpendicular to the direct incident solar radiation.

The final decision on fixed or movable collector for each project must be based on studies involving the following factors:

1. Geographical location
2. Intended use of collected energy (heating, cooling or both)
3. Increased cost for providing mobility for the collector
4. Variation in collection efficiencies with different angles of incident solar radiation

Pipe system conveys (1) rainfall from the parking deck area by drain inlets. All rainfalls of a lower intensity pass through this pipe system (2)-which in turn controls the flow of the rainwater to a point where oil and silt are removed and clear water is passed on to storm water storage tank. Higher intensity rainfalls where oil and silt concentrations are highly diluted (3) are conveyed directly to the storm water storage tank. On certain occasions when heavy rainfalls occur, a limited amount of rainfall may be wasted through the overflow pipe system. (4) The planting sprinkler system (5) is provided with water from the storm water storage tank. In the event that rainfalls are inadequate to meet sprinkler system demand, City water (6) may be used to supplement the system. At the end of the sprinkling season, tank is drained through this line. (7) Separated soil (8) is collected in a storage tank and pumped out as required. Sprinkler system water passes through the plantation area soil and any excess water (9) is wasted to the City sewer.

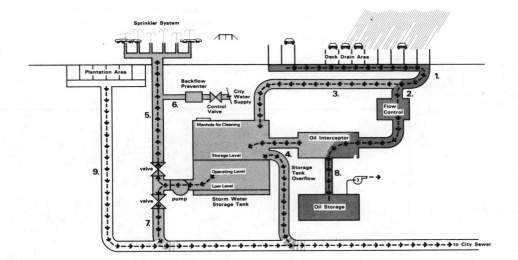

Solar Energy

Solar energy is absorbed by the flat plate collector and transferred to the fluid (water) flowing through the collector. This energy is directed to the low energy storage tank until the temperature level reaches about 80 F. If energy is still available for collection, it is then directed to the high energy storage tank up to temperature levels around 200 F.

Water from the low energy tank provides the heat sink-heat source fluid transfer media for the closed loop heat pump system in order to provide the heating or cooling of the space.

Water from the high energy tank provides the heat-energy for the low energy tank during those periods when solar energy is not available for collection and the low energy tank storage has been depleted. A boiler is cascaded with the high energy tank as a system back-up safeguard against the total energy depletion of the high energy tank. It also provides the heating energy for a small hot water radiation/unit heater system.

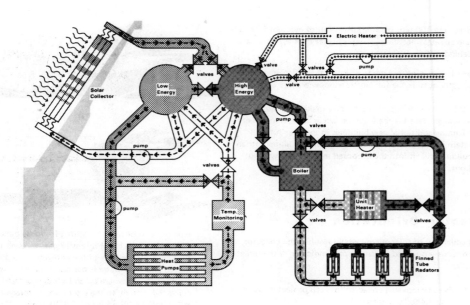

Newport West

Newport West, a low density cluster townhouse development, is located on the semi-rural northeast edge of Ann Arbor, Michigan. The sixty acre site is contiguous to the west edge of Ann Arbor's first and only park set aside as a forested preserve. One of the primary goals of the site plan was to set aside a substantial portion of buffer strip along the east property line which would protect the numerous and varied habitats edging the mature hardwood forest. In doing so the plan not only embodied protective measures for the park but also set up the natural framework where the 75% of housing units could take advantage of solar oriented slopes.

Glacial action shaped the topography of the site and the slopes are the expression of the terminal glacial moraines. The slopes are generally oriented to the east and south, facing into the morning and afternoon sun angles throughout the year.

The predominant soil on the site is a moist, heavy clay. Potential site development difficulties which might be expected to occur as a result of this soil condition (such as cold, wet basements in housing units) are essentially eliminated through several site factors which counter the effects of the heavy soils. The excellent natural drainage characteristics of the site assure efficient removal of surface water, so soils—at least on the slopes—are not waterlogged. The slope orientation to the warmth of east and southeast sun exposure somewhat alleviates the tendency to cold soils. Also, the 15 to 25% slopes extending the length of the entire site are ideal for two-level slope construction allowing many of the housing units to be constructed with exposed basements on the downslope side, facing the warm morning sun.

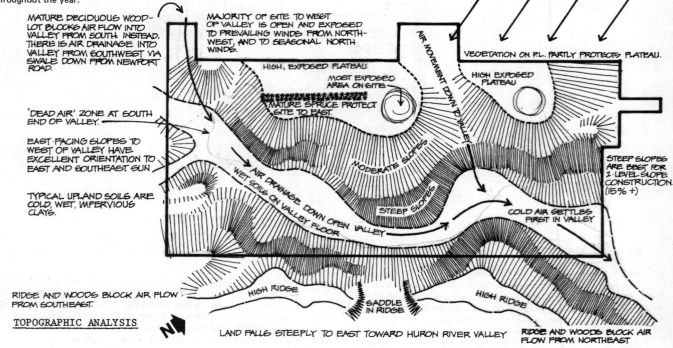

MATURE DECIDUOUS WOODLOT BLOCKS AIR FLOW INTO VALLEY FROM SOUTH. INSTEAD, THERE IS AIR DRAINAGE INTO VALLEY FROM SOUTHWEST VIA SWALE DOWN FROM NEWPORT ROAD.

'DEAD AIR' ZONE AT SOUTH END OF VALLEY.

EAST-FACING SLOPES TO WEST OF VALLEY HAVE EXCELLENT ORIENTATION TO EAST AND SOUTHEAST SUN.

TYPICAL UPLAND SOILS ARE COLD, WET, IMPERVIOUS CLAYS.

MAJORITY OF SITE TO WEST OF VALLEY IS OPEN AND EXPOSED TO PREVAILING WINDS FROM NORTHWEST, AND TO SEASONAL NORTH WINDS.

PREVAILING NORTHWEST WINDS

HIGH, EXPOSED PLATEAU.

MOST EXPOSED AREA ON SITE

MATURE SPRUCE PROTECT SITE TO EAST.

AIR MOVEMENT DOWN VALLEY

VEGETATION ON P.L. PARTLY PROTECTS PLATEAU.

HIGH EXPOSED PLATEAU

MODERATE SLOPES

STEEP SLOPES ARE BEST FOR 2-LEVEL SLOPE CONSTRUCTION. (15% +)

AIR DRAINAGE DOWN OPEN VALLEY

WET SOILS ON VALLEY FLOOR

STEEP SLOPES

COLD AIR SETTLES FIRST IN VALLEY

RIDGE AND WOODS BLOCK AIR FLOW FROM SOUTHEAST.

TOPOGRAPHIC ANALYSIS

HIGH RIDGE

SADDLE IN RIDGE

HIGH RIDGE

LAND FALLS STEEPLY TO EAST TOWARD HURON RIVER VALLEY

RIDGE AND WOODS BLOCK AIR FLOW FROM NORTHEAST

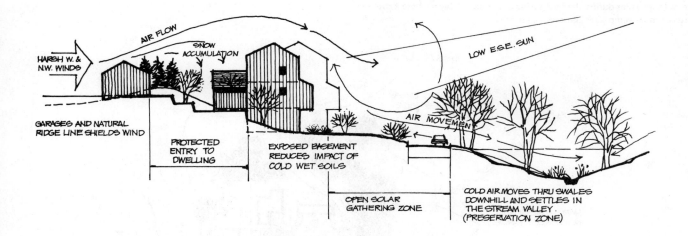

The valley floor, which parallels the toe of the primary slope, is an ideal dividing line and transitional zone between the housing community and the nature preserve. The low wetland valley, with an abundant variety of both flora and fauna, provides a natural air and surface water flow from higher elevations to tributary drainage swales. These swales are linked to the preserved wetland valley at intervals through the entire length of the site; the valley then connects to the expansive Huron River valley, ¾ of a mile to the northeast. The existing natural drainage system was preserved as a permanent open space in order to maximize these natural functions. The small exposed hilltop on the west edge of the site was planned to modify the force of harsh prevailing westerly and northwesterly winds by several means: arrangement of buffer-type plant materials, orientation of windows on buildings within the clusters, interior building use arrangement, and adequate insulation in those facades oriented to the harshest exposures.

The impressive woodland image of the site has been preserved by locating almost all of the housing units on the open, sun exposed slopes. One woodland pocket next to the valley was opened up to improve air flow and sun exposure. Other than this one small zone nearly all of the vegetation was preserved, including the meadows of native herbaceous plants with scattered native Hawthorn. A long double row of mature 80' high Norway Spruce (Picea abies) was preserved on the ridge line on the plateau, providing an excellent barrier from prevailing westerlies (an extremely harsh wind that extends the cool season and amplifies the cold during the winter months).

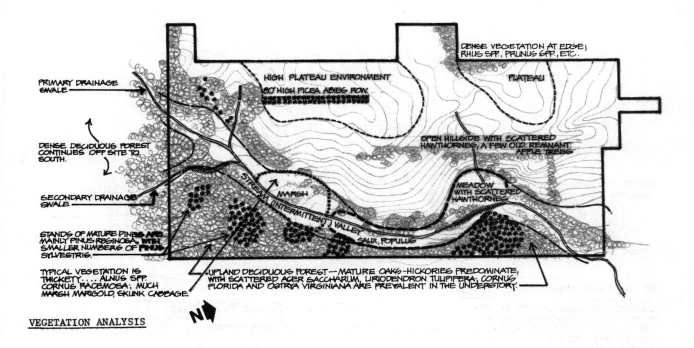

VEGETATION ANALYSIS

The housing units are arranged in a series of clusters adapting to the micro-physiography of the site, forming pockets of micro-climates which optimize available solar energy and provide protective buffers from the cold, windy, northern exposures. The resulting sunpockets function as comfortable outdoor areas during the cool spring and autumn days, where exotic plants of more temperate climates can survive.

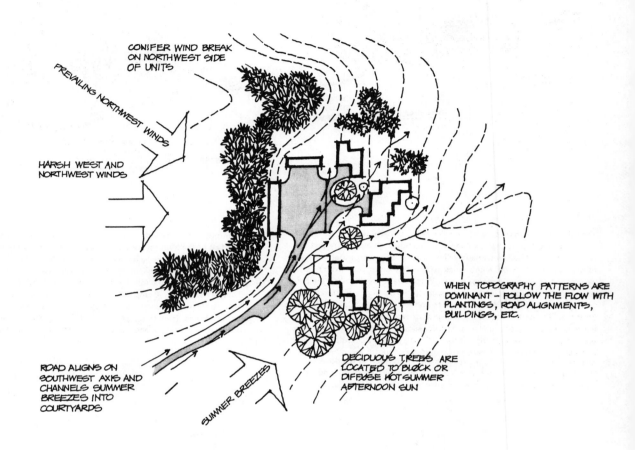

CONIFER WIND BREAK ON NORTHWEST SIDE OF UNITS

PREVAILING NORTHWEST WINDS

HARSH WEST AND NORTHWEST WINDS

WHEN TOPOGRAPHY PATTERNS ARE DOMINANT - FOLLOW THE FLOW WITH PLANTINGS, ROAD ALIGNMENTS, BUILDINGS, ETC.

ROAD ALIGNS ON SOUTHWEST AXIS AND CHANNELS SUMMER BREEZES INTO COURTYARDS

SUMMER BREEZES

DECIDUOUS TREES ARE LOCATED TO BLOCK OR DIFFUSE HOT SUMMER AFTERNOON SUN

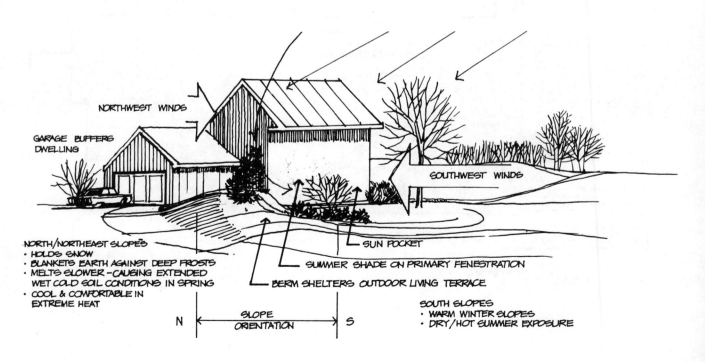

NORTHWEST WINDS

GARAGE BUFFERS DWELLING

SOUTHWEST WINDS

NORTH/NORTHEAST SLOPES
• HOLDS SNOW
• BLANKETS EARTH AGAINST DEEP FROSTS
• MELTS SLOWER -CAUSING EXTENDED WET COLD SOIL CONDITIONS IN SPRING
• COOL & COMFORTABLE IN EXTREME HEAT

SUN POCKET

SUMMER SHADE ON PRIMARY FENESTRATION

BERM SHELTERS OUTDOOR LIVING TERRACE

SOUTH SLOPES
• WARM WINTER SLOPES
• DRY/HOT SUMMER EXPOSURE

N |← SLOPE ORIENTATION →| S

190.

Large window areas, oriented to the optimum southeast sun, serve as auxiliary heat sources during the autumn, winter and spring. They enable more rapid heating of the units in the morning by receiving the low early southeast sun. A tendency to an uncomfortable degree of solar heat buildup in summer is reduced by the location of large deciduous trees in positions to diffuse the hot southwest summer sun from the south and west building facades; the east facades are kept open, receiving the comfortable morning sun and shaded by the buildings from heat of southwest afternoon exposure.

LARGE WINDOWS FACE SLIGHTLY SOUTH OF EAST, ABSORBING EARLY MORNING SUN AND RADIANT HEAT PRODUCED IN SOLAR HEATING OF THE SOUTH WALL.

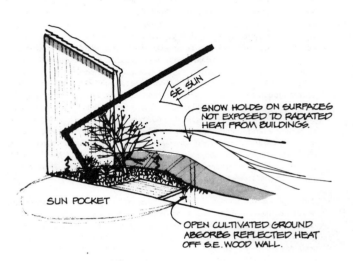

SNOW HOLDS ON SURFACES NOT EXPOSED TO RADIATED HEAT FROM BUILDINGS.

SUN POCKET

OPEN CULTIVATED GROUND ABSORBS REFLECTED HEAT OFF S.E. WOOD WALL.

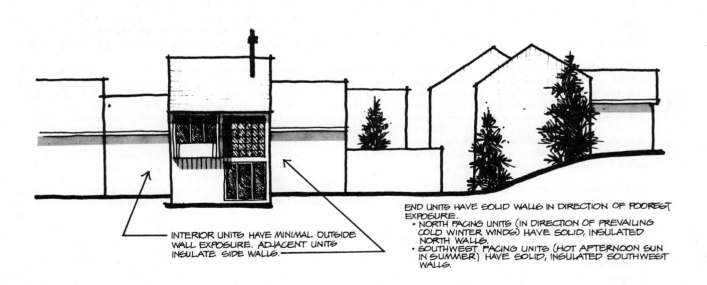

INTERIOR UNITS HAVE MINIMAL OUTSIDE WALL EXPOSURE. ADJACENT UNITS INSULATE SIDE WALLS.

END UNITS HAVE SOLID WALLS IN DIRECTION OF POOREST EXPOSURE.
• NORTH FACING UNITS (IN DIRECTION OF PREVAILING COLD WINTER WINDS) HAVE SOLID, INSULATED NORTH WALLS.
• SOUTHWEST FACING UNITS (HOT AFTERNOON SUN IN SUMMER) HAVE SOLID, INSULATED SOUTHWEST WALLS.

Case Studies
Cool and Temperate Regions
Sasaki Associates, Inc.
The Nova Scotia Experience

For the past two years SA has been actively involved in preparing a master tourism plan for the Department of Development and Department of Tourism in Nova Scotia. Although the growth of tourism has increased considerably over recent years, the Provincial accommodation base has not increased sufficiently to satisfy existing tourist demands and plans for future accommodation and recreation complexes were not available. Faced with a short tourist season the private sector was reluctant to develop facilities requiring large capital investments which would remain dormant for the winter and most of the spring and fall. Therefore, in an effort to stimulate the tourist industry the Provincial Government retained SA to identify a system of prime sites in the Province best suited for tourist complexes which were to include beside a diversity of accommodation types a wide range of both natural and man-made amenities. The Canadian and Provincial Governments deserve much credit for assembling a valuable and comprehensive data base which proved to be an invaluable planning tool and which quickly provided a means of understanding in broad terms the natural, social and economic characteristics of the Province.

Although a prime concern during the course of the study was not specifically the conservation of energy, the research conducted by SA could easily be applied to this objective especially if a large development such as a New Town were the project focus. This is particularly true since many large-scale developments receive aid in some form from the public sector which in turn is critically aware of energy shortages and is becoming increasingly responsive to meeting the new demands instilled by the problem.

Analysis on the macro-scale, i.e., on the Provincial, state-wide, or even multi-state scale can identify broad characteristics which can provide important indicators of energy requirements reflecting natural conditions. Obviously the decisions of locating a New Town or other large development involve the analysis of enormous numbers of complex considerations which ultimately include trade off evaluations and compromises, the aspect of energy conservation through the analysis of climatic data will be come increasingly more important in the future and may become more of a prime locational criterion than in the past.

The Province of Nova Scotia has dramatic natural extremes from flat inland lowlands to coastal mountain topography; from 70 degree saltwater summer temperatures to 55 degree saltwater summer temperatures; and from lush woodlands to barrens. The influence of the factors, therefore, plays a significant role in the tourist industry and has important weather breeding effects. In full recognition of these conditions, the Nova Scotia Department of Development and the Department of Regional Economic Expansion, through the Atmospheric Environment Service has prepared material extremely useful to planners including data related to:

1. High Summer—beginning, end and duration.

2. Snow Cover—February 28, maximum depth and duration.

3. Temperature—January, July, and water temperature.

4. Annual Conditions—heating degree days, precipitation, snowfall

5. Agro-Meterology: Growing Season—beginning, growing degree day, precipitation.

6. Agro-Meterology: Frost—last frost, first frost, frostfree period.

Perhaps the most important source of climatic data beyond that which is generally available in the Province is a report prepared by Mr. A. D. Gates entitled "The Climate of the Maritime Provinces As It Effects Tourism and Outdoor Recreation". This report takes a comprehensive view of climatic considerations and could serve as a prototype in a form tailored specifically for energy conservation in regions or states.

Obviously, the climatic requirements of tourism may not parallel the specific needs of energy conservation, but the analysis process may be similar or applicable. In this regard, the following process is offered as a "first cut" in a planning procedure which would provide raw data on a macro-scale and establish the foundation for further study on regional sub-regional, and project scales.

In keeping with normal planning practices, a first step would naturally relate to the formulation of project goals and objectives. For the purposes of this report, the identification of sites which through climatic analysis maximize energy conservation, is a prime objective. During the inventory and analysis phases of planning, a broad range of climatic data would be assembled and gaps in information identified. On the statewide or Provincial scale an important indicator of regional temperature ranges and one that has been used extensively for projecting energy requirements (oil) in the past is the use of information indicating Heating Degree Days such as illustrated on the map below for the Province.

A Heating Degree Day is a measure of the departure of the mean daily temperature from a given standard in this case 65°F. One degree day is counted for each degree of difference between the mean temperature for the day and the standard, where the mean temperature is less than 65°. It is assumed that for heating purposes, no artificial heat is required when the mean temperature is found by averaging the high and low temperature for the day. The calculated number of degree days may be accumulated for any period such as a month, a season or a year and the total has been found to be proportional to the relative fuel consumption for a given building.

MEAN MAY-SEPTEMBER HEATING DEGREE DAYS BASE 65°F

From an energy conservation point of view, it could be assumed that a housing complex or New Town located in the lightest tone illustrated on Map A would require the least amount of energy for heating requirement. Conversely the areas located within the darkest tones are historically colder and therefore require higher energy consumption for heating. Another factor which must also be taken into consideration is the reality that many areas in the Province require heat during summer months, and it is therefore advantageous to determine the temperature characteristics from May through September as well as for the winter season. The map below illustrates the summer heating degree day curves and provides an interesting comparison between annual and summer. Once again from the standpoint of energy efficiency the region illustrated in the light tone located in the southern portion of the Province is most desirable since moderate climatic conditions have been monitored over time.

Obviously the above determinations are only one input to the entire package of substudies which would be included in a full energy study—for example, accessibility and aesthetic desirability of an area would naturally play an important role in the analysis. As an example, travel to and from this zone to areas of employment or to recreational opportunities could become important energy consumption factors.

During the tourism study, other locational criteria based on climatic factors included the identification of fog zones, mean monthly temperatures, beginning, end, and length of high summer, precipitation, median snow cover for winter recreational opportunities, and wind characteristics. Certain other data could also be incorporated into the study to round out and support the heating degree day information. Although many factors are reflected in calculating Heating Degree Day information, wind direction/velocity and chill factor considerations are extremely important as in the proportion of days with full or partial sun.

Once the development sub-region has been identified through the use of broadly based climatic data, a regional or sub-regional study should be undertaken which would analyze factors influencing climate at a more finite level including topography, influences of water bodies, elevation, wind direction, vegetation, solar orientation, etc.

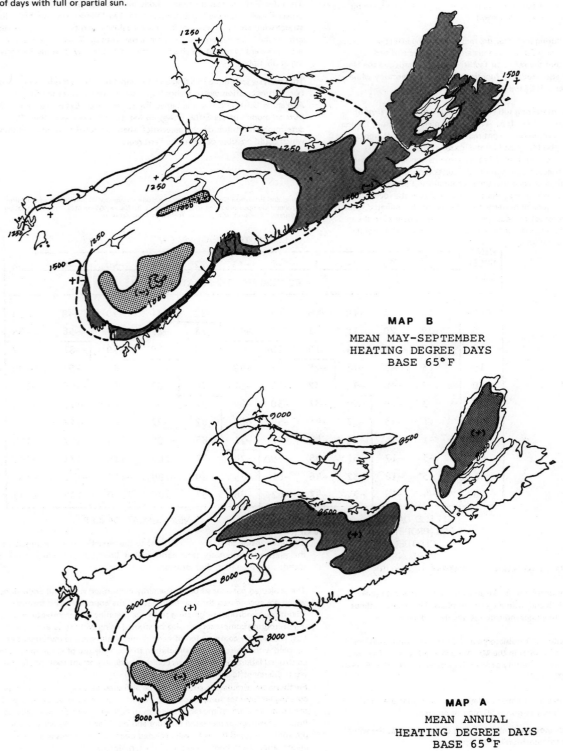

MAP B

MEAN MAY–SEPTEMBER
HEATING DEGREE DAYS
BASE 65°F

MAP A

MEAN ANNUAL
HEATING DEGREE DAYS
BASE 65°F

193.

S.U.N.Y. at Buffalo - Amherst Campus

In 1968 Sasaki Associates, Inc., (SA) was commissioned by the New York State University Construction Fund (SUCF) to prepare a comprehensive Campus Plan for a new 1,200-acre Amherst Campus of the State University of New York at Buffalo. The program was tailored to accommodate an ultimate daytime population of over 50,000 students, faculty and staff. Construction and occupancy of the 10.6 million gross square feet of building space is to be staged over a ten year period.

The major role of SA during and after the master planning effort was to coordinate the work of nearly 20 sub-campus architect teams and technical consultants to ensure that the resultant facilities not only satisfied the University's program, but also produced functional and visual harmony while fully realizing the potentials of the site.

Towards this end, one interesting sub-study* coordinated by the firm of Mt. Auburn Research Associates, Inc.,—Weather Dynamics Division, focused on the affects of wind motion on the entire campus and the air currents generated by individual buildings and by the juxtaposition of building groups. Through the use of wind tunnel measurements conducted on campus models simulated true external conditions, zones were identified which, due to building characteristics and locations, channel wind at velocities annoying or dangerous to pedestrians. The tests also included the identification of noxious fume dispersion patterns from flows which emanated from known sources. Utilization of this data provided broad guidelines to the individual architects and provided a general framework for architectural design and the grouping of buildings.

Recent annoying and sometimes dangerous wind problems created by multistoried contemporary architecture has re-vitalized the use of storm or prevailing wind simulation through the use of wind tunnels. When wind strikes the surface of a building, its velocity is transformed into pressure and the wind pressure flows down the building to the bottom of the structure. When an escape route is found such as around the corner of a tall building, the pressure is again transformed back to wind velocity.

The effects of wind on the human body has also been studied very carefully especially when coupled with temperature. The following chart which illustrates wind chill factors combining wind velocity and temperature reveals astounding information—for example, a day with 32°F temperatures has the equivalent temperature characteristics of a -1°F day if winds are blowing at 35 mph.

This analysis technique utilizing a wind tunnel on a scale model is not a tool borne from aviation design technology, but one which was used first by Alexander Gustave Eiffel, the famed French engineer. Extensive wind tunnel tests on models of the Eiffel Tower during its design phases in the 1880's determined its structural characteristics when subjected to wind conditions which could possibly occur in the Paris region.

*Wind and Pollution Control Study, The State University of New York at Buffalo—Amherst Campus prepared in cooperation with Sasaki Associates, Inc., by Weather Dynamics Division, Mt. Auburn Research Associates, Inc.

WIND SPEED (MPH)	LOCAL TEMPERATURE (°F)										
	32	23	14	5	-4	-13	-22	-31	-40	-49	-58
	EQUIVALENT TEMPERATURE (°F)										
CALM	32	23	14	5	-4	-13	-22	-31	-40	-49	-58
5	29	20	10	1	-9	-18	-28	-37	-47	-56	-65
10	18	7	-4	-15	-26	-37	-48	-59	-70	-81	-92
15	13	-1	-13	-25	-37	-49	-61	-73	-85	-97	-109
20	7	-6	-19	-32	-44	-57	-70	-83	-90	-109	-121
25	3	-10	-24	-37	-50	-64	-77	-90	-104	-117	-130
30	1	-13	-27	-41	-54	-68	-82	-97	-109	-123	-137
35	-1	-15	-20	-43	-57	-71	-85	- 99	-113	-127	-142
40	-3	-17	-31	-45	-59	-74	-87	-102	-116	-131	-145
45	-3	-18	-32	-46	-61	-75	-89	-104	-118	-132	-147
50	-4	-18	-33	-47	-62	-76	-91	-105	-120	-134	-148

LITTLE DANGER FOR PROPERLY CLOTHED PERSONS	CONSIDERABLE DANGER	VERY GREAT DANGER

WIND STUDY OF CAMPUS AT S.U.N.Y., AMHERST, NEW YORK

The buildings, and wind problems, in the area covered by this data sheet are unchanged in the 1975 and ultimate campus plans. The heavy outlines show the buildings, the fine lines indicate the ancillary features.

The stars show the location of troublesome and dangerous wind areas predicted using the results of studies in the Mt. Auburn wind tunnel. The data shown represent the average monthly probabilities based on Weather Bureau records for the years 1951-1960.

The first number indicates the number of days per month that the wind speed at head height at that position is 20 mph or greater.

The second number indicates the number of days per month that the wind speed at head height at that position is 10-19 mph.

The letter inside the star symbol refers to the table below which indicates the relative contributions from various wind directions to the high wind speeds encountered at that location.

The projected patterns of possible energy consuming wind chill conditions which are engendered through both natural site conditions and through man-made conditions originating from site planning and architectural massing of housing complexes can also be studied using wind simulation methods. Indeed, the combination of careful micro climatic site investigations coupled with environmental simulation through the use of study models in controlled laboratory conditions could provide important tools for formulating energy efficient plans and designs.

Furthermore, simulation analysis can be also used to identify means of increasing air flow for summer cooling. Architectural studies have illustrated the positive effects of movable openings in buildings which would take optimum advantage of increased pressure and resulting wind velocity as the pressure is released through well designed openings. Air flows generated by understanding and harnessing the characteristics and effects of wind velocities and pressure against masses, could substantially improve "natural" air conditioning within buildings.

194.

Once the planning framework and parameters have been established through model analysis, architectural and landscape design solutions could further the efficiency factor.

Illustration A is indicative of the method used to graphically interpret the results of the wind tunnel studies. The legend which accompanied the building diagrams reads as follows:

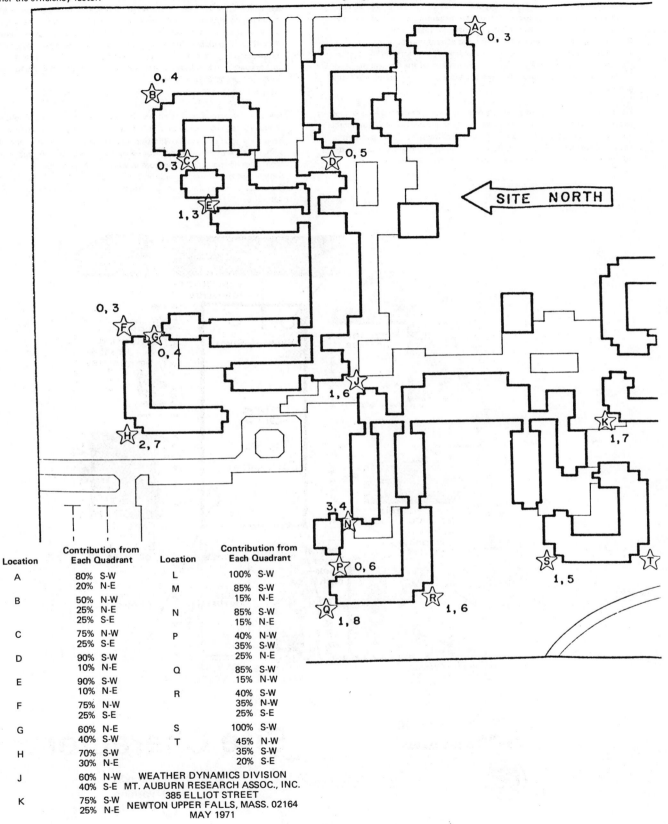

Location	Contribution from Each Quadrant		Location	Contribution from Each Quadrant	
A	80%	S-W	L	100%	S-W
	20%	N-E	M	85%	S-W
B	50%	N-W		15%	N-E
	25%	N-E	N	85%	S-W
	25%	S-E		15%	N-E
C	75%	N-W	P	40%	N-W
	25%	S-E		35%	S-W
D	90%	S-W		25%	N-E
	10%	N-E	Q	85%	S-W
E	90%	S-W		15%	N-W
	10%	N-E	R	40%	S-W
F	75%	N-W		35%	N-W
	25%	S-E		25%	S-E
G	60%	N-E	S	100%	S-W
	40%	S-W	T	45%	N-W
H	70%	S-W		35%	S-W
	30%	N-E		20%	S-E
J	60%	N-W			
	40%	S-E			
K	75%	S-W			
	25%	N-E			

WEATHER DYNAMICS DIVISION
MT. AUBURN RESEARCH ASSOC., INC.
385 ELLIOT STREET
NEWTON UPPER FALLS, MASS. 02164
MAY 1971

The Pioneer Gateway Experience

Sasaki Associates, Inc., has provided urban design services to the Pioneer Gateway project located in Minneapolis, Minnesota which is one segment of a total urban renewal program started approximately 14 years ago and known as the Gateway Center Urban Renewal Project. The intent of the 30-acre project was to eliminate blight and attract new business, residential and commercial life in the downtown area. The effort has had a considerable influence in the successful revitalization of the Central Business District and has helped provide a healthy atmosphere for major additional growth.

Pioneer Gateway was conceived as a self-contained residential environment within the urban core combining high residential densities with commercial, open space and recreational uses. The program included 1,165 condominium housing units on 5.63 acres, 56,00 square feet of commercial space, and 103,000 square feet of open space and indoor recreational facilities.

As shown in Illustration A—*Site Character*, the site is surrounded by existing buildings of varying heights which cast shadow patterns on the study area. An important consideration during the Pioneer Gateway planning process which is relevant to this study, therefore, was the investigations related to sun angles to determine the configuration of shadows cast by adjacent buildings as well as sun and wind characteristics. This data provided important tools in identifying prime locations for outdoor swimming pools and pedestrian plazas. The following description is extracted from the report prepared by Sasaki Associates and other consultants entitled *Pioneer Gateway—Urban Design Study*.

Climate

Climatic data shows that Minneapolis has a hardy climate with cold winter temperatures, augmented by a chill factor created by gusty northwest winds. There is a good deal of precipitation and a yearly average of only 58% sunshine. Temperature variations from season to season are extreme, ranging from lows of about -20°F or more in winter to high of about 90°F in the summer. In the winter, humidities are high and sunshine is at a minimum. Summers are generally warm, but usually comfortable due to low daytime humidity.

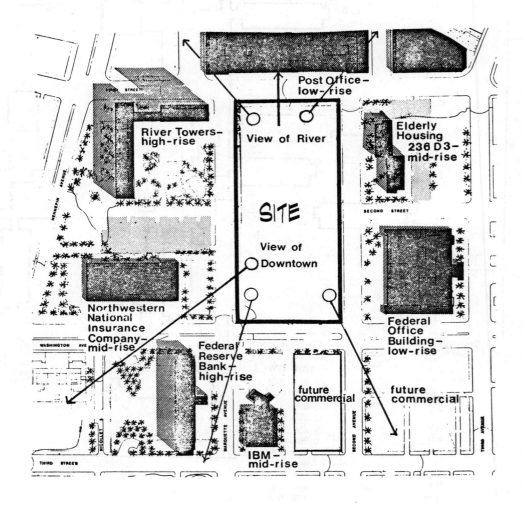

ILLUSTRATION A

Site Character

✳✳ Open Space
▒ Parking
○ Best Locations for Views

A. Sun and Wind

The sun conditions in the Minneapolis area (see Illustration B) indicate that primarily southern exposures of buildings receive solar radiational benefit during the winter. In summer, however, east, west, and south exposures receive solar radiation. Apartments with a southern exposure will receive direct sun during some portion of the day, year-round. Major open spaces should also be oriented toward the south and east to receive maximum sunlight and to offer protection from winter winds which flow from the northwest and are harsh and cold. To help buffer these winds, taller buildings should be located at the north and west edges of the site. In addition, walks along Marquette Avenue should also be protected by arcades.

Summer winds, from the southwest are cooling and pleasant, and therefore the massing should attempt to "capture" and utilize this flow by locating buildings with a low profile along this edge of the site.

B. Shadows

In winter the longest shadows are cast to the north and northeast as illustrated on Illustration C. Massing must take this fact into consideration to minimize shading the site, or adjacent property, in particular, The River Towers development. If towers were placed in the most ideal location, shadows would be imposed only on First Street and parts of Marquette Street, and Second Avenue, leaving the site free of the imposing shadows of the towers and exerting the lease possible adverse impact on surrounding properties. Low-rise buildings would best be located on the eastern and southern perimeter of the site to minimize shadows within the site.

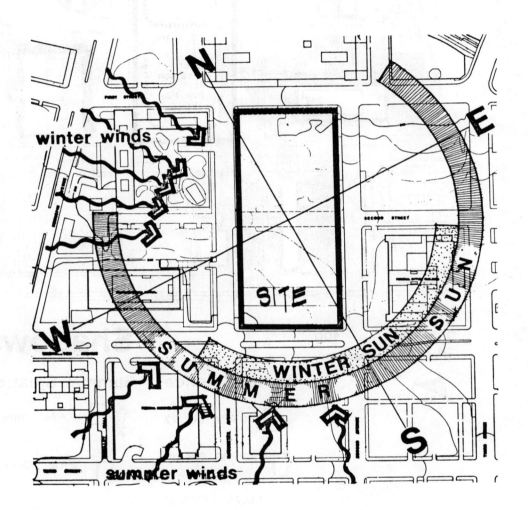

0   80   160   400

Sun and Wind

ILLUSTRATION B

SOURCE:
National Weather Records Center

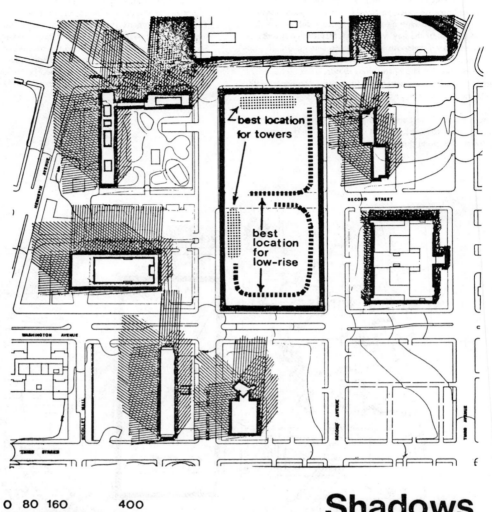

best location
for towers

best
location
for
low-rise

Shadows

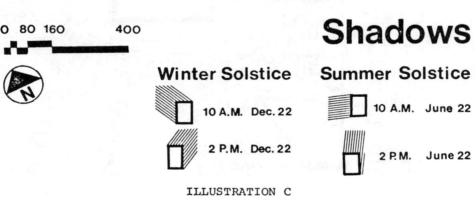

O 80 160   400

N

Winter Solstice

10 A.M. Dec. 22

2 P.M. Dec. 22

Summer Solstice

10 A.M. June 22

2 P.M. June 22

ILLUSTRATION C

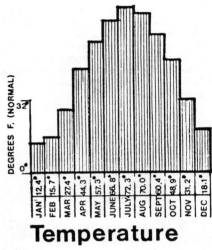

Temperature

JAN	FEB	MAR	APR	MAY	JUNE	JULY	AUG	SEPT	OCT	NOV	DEC
12.4°	15.7°	27.4°	44.3°	57.3°	66.8°	72.3°	70.0°	60.4°	48.9°	31.2°	18.1°

DEGREES F. (NORMAL)

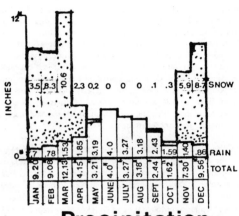

Precipitation

INCHES

SNOW: 3.5 8.3 10.6   2.3 0.2 0 0 0 .1 .3 5.9 8.7

RAIN: .7 .78 1.53 1.85 3.19 4.0 3.27 3.18 2.44 1.62 1.40 .86

JAN	FEB	MAR	APR	MAY	JUNE	JULY	AUG	SEPT	OCT	NOV	DEC
9.20	9.08	12.13	4.15	3.21	4.0	3.27	3.18	2.44	1.59	7.30	9.56

TOTAL

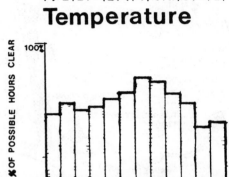

Sunshine

% OF POSSIBLE HOURS CLEAR

JAN	FEB	MAR	APR	MAY	JUNE	JULY	AUG	SEPT	OCT	NOV	DEC
50 %	57 %	54 %	56 %	58 %	62 %	70 %	67 %	61 %	57 %	39 %	40 %

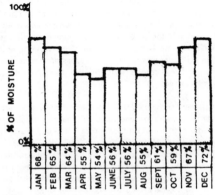

Relative Humidity

% OF MOISTURE

JAN	FEB	MAR	APR	MAY	JUNE	JULY	AUG	SEPT	OCT	NOV	DEC
68 %	65 %	64 %	55 %	54 %	56 %	56 %	55 %	61 %	59 %	67 %	72 %

Climate

SOURCE:
National Weather Records Center

ILLUSTRATION D

The Carrabassett Experience

For the past five years SA has provided plannint services to the Scott Paper Company which owns among its other vast holdings approximately one million acres of woodlands in the State of Maine. In realizing the potential value of these lands for recreational and tourism purposes Scott Paper Company retained SA to critically investigate the recreation potential of a large region located in north Central Maine and to identify those subregions worthy of more detailed study. One area identified was a subregion located near the town of Kingfield encompassing the Carrabassett Valley and its surrounding mountains. Existing land uses in the sparsely settled region were primarily forest-related industries and recreation which focused on an existing ski development—Sugarloaf Mountain.

Planning studies conducted by SA included identification of the recreational potential of the entire 52,000 acre region with a special emphasis on the interrelationships of the various potential recreational activities and facilities. One particular 2,000 acre parcel which proved to have exciting expansion possibilities was the existing ski comples at Sugarloaf Mountain and the linkage between Sugarloaf Mountain and Crocker Mountain located to the west of the existing slopes and trails. In an effort to study the physical characteristics of the mountains and their appropriateness for ski development, Mr. Shelden Hannah of Franconia, New Hampshire was retained. Mr. Hannah is an internationally known designer of ski trails and slopes for winter resorts and has exceptional knowledge of winter environments and a keen ability of "reading", the landscape to determine conditions which could affect ski development. Although our experience working with Mr. Hannah focused on ski resort development, the types of analyses could indeed by applied to locational studies performed for developments with energy conservation objectives.

Other studies which were conducted by SA as part of the Carrabassett Valley work program included shadow and sun angle analyses which indicated zones which would be in shadow cast by the mountains at various times of the year and at various times of day. The primary purpose of these studies was to aid in locating specific sites for the basic lodge facility and for seasonal housing which would receive winter sun. Land uses located in sunny areas would not only benefit from warmer temperatures but also benefit from the positive psychological effect of sunshine over shade.

The site visits by Mr. Hannah in the region slated for ski expansion were directed at analyzing the impact of prevailing northwest winds on the slopes and to verify the suitability of terrain by cross checking existing 200 scale topography for the area. The optimum time period for these on-site observations was immediately after a storm system had passed through the area and during the subsequent periods of high pressure and clear northwest winds which normally follows a storm in the region. One simple method used to identify areas protected from severe northwest winter winds which cause wind scouring on slopes and general discomfort to skiers was to search for zones where spruce and fir held snow on their branches due to wind protection. Conversely, coniferous trees in zones subjected to frigid storm following winds had lost the newly fallen snow from their branches. Adverse wind conditions were also recognized by the presence of snow scouring, caused by high winds following a storm which blows the newly fallen snow from the older base thereby exposing underlying old snow.

The identification of protected areas in this general region was very important since the two facing mountain ranges were bisected by a large stream with substantial grade changes from the upper stretches of the stream to the lower. A natural venturi effect was created by air currents passing along the stream valley between the two mountains as illustrated in Map A resulting in occasional detrimental high wind conditions.

Once these zones were identified, further investigations of snow characteristics on the slopes, gave further indications of adverse freeze-thaw problems which could cause icing conditions. Where freeze-thaw problems did exist, the snow had definite strata from the cycle created by snow thaw from heat generated by sunlight followed by freezing conditions. This is especially noticeable during the spring when the intensity of the sun is greater than in deep winter. The freeze-thaw zones which may be appropriate for solar energy generation could therefore be identified in this manner.

The relevance of these simple on-site investigations to energy conservation seems apparent since these same on-site analyses techniques coupled with other micro climatic studies and sunlight/shade studies could prevent excessive exposure of mountain located housing complexes to adverse climatic conditions. Therefore, in much the same way as vegetation types help to indicate both soil and moisture characteristics of an area, the analysis of winter environments can also provide important insights and guidelines for siting housing complexes in mountain environments to maximize energy efficiency.

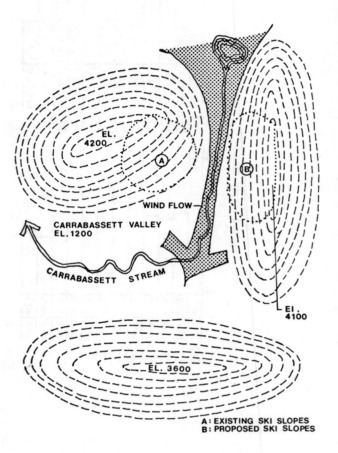

EL. 4200

WIND FLOW

CARRABASSETT VALLEY
EL. 1200

CARRABASSETT STREAM

EL. 4100

EL. 3600

A: EXISTING SKI SLOPES
B: PROPOSED SKI SLOPES

MAP A

Case Studies
Temperate Regions
Rahenkamp, Sachs, Wells and Associates

General Approach

Public concern for ecology in recent years has taught us valuable lessons that ought to be applied to energy conservation. Respect for energy conservation, like ecology, is integral to the land planning professions. What has now become apparent, however, is that energy, environmental and economic factors are closely interrelated. All are, in fact, part of a broader concern with responsible resource management.

Design and construction which are compatible with natural systems can result in large energy savings. We learned a long time ago that by working in conjunction with the natural environment, rather than in opposition to it, we would reduce the demands on our natural resource base, product a more livable product, and save our clients money. It becomes clear that the same design advantages apply in the area of energy conservation. Good siting, the efficient use of materials, and a sensitivity to microclimate can have a significant effect on energy conservation.

Pragmatically, sound resource management must also create positive benefits in the marketplace. This can occur either through reduced costs or improved market capture. As fewer and fewer people can afford new housing today, both the developer and the ultimate user are vitally concerned with the cost consequences of energy use. With increasingly high energy costs, a saving of even $10 or $20 a month becomes significant to a family's ability to pay for housing. Since modification of mean radiant temperature by 3 degrees will not only reduce energy demands but will also reduce annual heating and cooling costs by 6 to 8%, real savings are possible through sound energy conservation planning. Given the choice, they will prefer southern exposures with their associated lower heating costs, and through rather than back-to-back apartments units due to the greater air flow and lower air-conditioning costs. Cost efficiencies are also possible for the developer. By using natural vegetation to buffer buildings from temperature extremes, he can save several hundred dollars per acre in clearing costs. Responsible energy conservation should be clearly reflected in lower costs and greater demand for the project.

Specific Approach

Designing for energy conservation must consider three distinct aspects of development: 1) the pre-existing natural conditions of the site; 2) the actual construction; and 3) the subsequent user demands for energy. All are critical to an optimum energy conservation program. The significant natural systems of the site related to energy include solar radiation, precipitation, wind, physiographic features, drainage patterns and vegetation. Through appropriate site design and building location as well as architectural detailing, we can take advantage of the natural elements and spend less energy altering natural conditions. For instance we can cluster buildings to reduce the number of exterior, heat-losing surfaces and we can save existing plant material to buffer buildings from summer heat or winter winds.

The construction aspect involves the use of the least energy intensive building materials and the most energy saving maintenance design. Minimizing road networks, and substituting natural drainage swales for curbs and gutters, for instance, can reduce energy needs.

Subsequent user demands include the type and amount of transportation required to serve the site and the potential for full capacity use of structures and facilities. Mixing various housing types, commercial, recreational and even industrial uses in a single project minimizes transportation de-

mands and may permit the design of multi-use structures and sites. Large areas of natural vegetation can provide recreational and scenic areas with little maintenance or renewal efforts.

It should be apparent that a realistic design for energy conservation cannot simplistically concentrate on one or two isolated elements. In each of the following projects designed by our firm, attention has been given to the interrelationship of energy, environment and economics to ensure sound resource management. General background on the project is presented first, followed by a brief mention of the various energy conservation measures incorporated into the broad design process, after which a more detailed description of specific measures is provided.

Projects

PROJECT NAME: PINE HILLS
PROJECT LOCATION: TOWN OF BROOKHAVEN, SUFFOLK COUNTY, N.Y.

The general design problem was to develop a land use plan for a 1100 acre site on Long Island. Ground water recharge is a critical issue on Long Island due to the problem of salt water intrusion. Prevailing winds combined with the relatively flat topography created a major climatic factor.

The final design recommended a Planned Unit Development at a density of 2.35 units per acre, with 48% of the site in open space. (Figure 1) An industrial park was located along the adjacent expressway, avoiding the use of local roadways. An interior parkway collector system also eliminated the need for extensive local road connections and the consequent widening of local roads. This reduced the need for energy intensive concrete and asphalt materials.

Ground water recharge is usually achieved through costly sumps or similar devices. At Pine Hills this was replaced with a natural ponding system. The site was generally flat with the exception of a terminal glacial moraine running down the western edge of the site, and a definite valley system running diagonally across the site in a northeast to southwest direction. (Figure 2).

Soils in the valley and on the moraine were the most likely to hold and retain water. Consequently it was recommended that the valleys be retained as a natural drainage system, eliminating the need for costly storm sewer pipes. Roadways were designed to run diagonally across the valleys, creating dams to effect natural holding ponds. In addition by minimizing impervious cover, a maximum amount of open space was available to absorb storm water runoff generated by the new land uses. Thus the natural systems of the site were used in place of energy intensive, man-made facilities.

In addition, it was found that prevailing summer winds flow up the valleys. In the winter, the major wind pattern shifts 90 degrees. It was recommended that by not building in the valley or on the moraine, and by using the valley as a natural watering system, the natural growth of trees could be encouraged in these areas. Since the trees would be parallel to the summer breezes, air flow was permitted providing cooling in the summer. In the winter, the trees would provide shelter belts serving as wind buffers. The result is a general reduction in heating and cooling costs.

A comparison was made between the recommended PUD green space community, and standard development with the football field size, 20 foot deep sumps and a greater road area (Figure 3, 4).

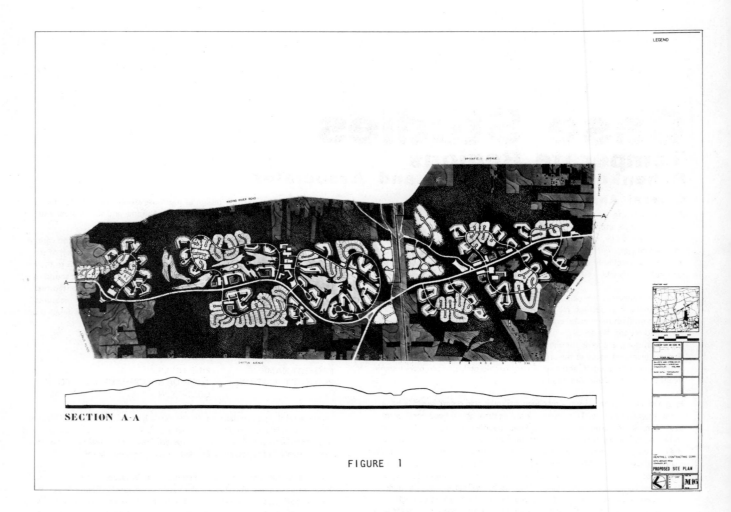

SECTION A-A

FIGURE 1

PINE HILLS   FIGURE 3

SUMMARY COMPARISON BETWEEN SINGLE FAMILY SUBDIVISION
UNDER EXISTING ZONING AND GREEN SPACE COMMUNITY

	Density DU/AC	Population (Total)	School Children	Tax Surplus	Land Coverage	Open Space
Single Family Subdivision	2.06	7,038	2,222	$1,944,000	25%*	3%
Green Space Community	2.35	4,243	867	$3,345,000	7%*	48%

*Primarily reflected in paved areas.

The above comparison points out that not only energy saving and environmental sensitivity but also substantial fiscal advantages are possible with a
design oriented toward sound resource management

PROJECT NAME: FLYING HILLS
PROJECT LOCATION: CUMRU TOWNSHIP, BERKS COUNTY, PA.

At Flying Hills, the design problem was to develop a land use plan for an
open space community on a 305 acre site. The site is very hilly (Figure 5)
with some low wet spots and a considerable variety of vegetative cover (Figure 6).

Approximately 30% of Flying Hills had been in actively used farmland and
meadows. In addition to wanting to keep this visual quality to the site, we
also wanted to provide recreational opportunities on-site. By providing a
golf course on the previous meadows, stocking the pond with fish, and en-
couraging the use of the pedestrian path systems, a reduction in gasoline

consumption by project users is achieved. When recreation amenities are
provided on-site, there is a reduction of vehicular movement necessary, and
at Flying Hills a family can literally spend a weekend at home without going elsewhere for recreation (Figure 7).

At Flying Hills, no curb or gutters are used. By allowing water to run into
grassed swales, a natural watering system is utilized and natural replenish-
ment of the water supply is obtained with considerable cost and energy savings. At a cost of approximately $5.50 per linear foot for 6" x 18" concrete
curb, the savings to the developer was nearly $174,800, about 50% of which
is materials. Not only is energy saved through reduced need for materials,
but also the water velocity is controlled. When water is channelized and
forced into a point output such as out of a pipe into a stream, the water
force can be considerable, contributing to flash flooding, and increasing
erosion. In addition, natural ponds were used to hold the excess water
generated by the project with sufficient capacity to deal with peak runoff
conditions resulting from overland flows.

Housing units were run along the contours of the land, molding construction into the site. By bending a linear cluster of buildings parallel to the
topographic contours, cut and fill as well as end wall construction is minimized. The result is considerable energy savings through both reduced grading activity and material requirements. The concept of contour clustering
also permits a major reduction in road lengths (Figure 8), with a sngnificant
saving in the amount of energy intensive paving materials. A linear road
length comparison study between the site designed as a single family layout
and as an open space PUD resulted in a 300% reduction in roads required
and a cost saving of $720,720.

202.

SECTION A-A

FIGURE 2

Category	Cost/ Linear Foot*	Open Space Community 3 mi of rd.	Single Family Plans 9 mi of rd.
Clearing	$ 3.00	47,520	142,560
Grading @ 12"	$ 2.00	31,680	95,040
Gravel base @ 6"	$ 4.50	71,280	213,840
Bituminous surface @ 2"	$10.50	116,320	498,960
Shoulder	$ 2.75	43,560	130,680
Total	$22.75	360,360	1,081,080

*Under each scheme 24 foot wide bituminous cartways were assumed.

PROJECT NAME: PINE RUN
PROJECT LOCATION: GLOUCESTER TOWNSHIP, CAMDEN COUNTY, N.J.

Pine Run provides another example of Planned Built Development design, on a relatively small 125 acre site (Figure 9). As with Flying Hills, there is a highly efficient road system with separation of pedestrian and vehicular transportation systems, and very limited use of curb. The result is considerable saving in paving required, and a safer and more easily used movement pattern is achieved. In addition to single family, townhouse and garden apartment residential uses, the PUD includes a lake an extensive open space system and a neighborhood shopping center. This provides needed services to the Pine Run community and adjoining residents, thereby reducing the necessity for automotive travel to remote commercial centers. The result is fuel savings. A 1972 user study of Pine Run* showed that indeed residents were likely to walk rather than drive to the stores for convenience items.

*People and Planning—Facts and Figures, Rahenkamp Sachs Wells and Associates, Inc. copyright 1974

203.

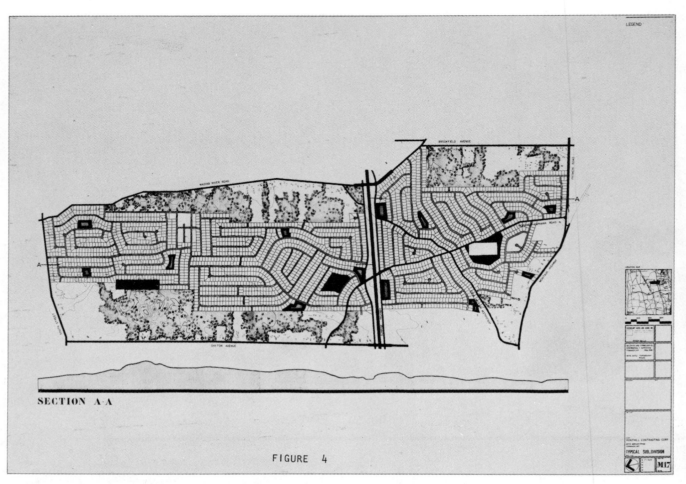

SECTION A-A

FIGURE 4

Due to a concern for the strong valley wall system (Figure 10), and a sensitivity to the natural air flow patterns (Figure 11), as they affect multifamily structures, the concept of contour clustering was used. A through-unit building module was designed, and a standard angle unit was utilized to bend the building with the natural contour, (Figure 12). This resulted in a loop parking lot on one side of the building and, on the other side, an open space system with pedestrian paths connecting each of the entranceways of the buildings to the open space and eventually to the community center. The advantage of this type of unit design, integrated into the valley wall system, was that considerable cooling effects could be experienced during the summers, while the existing vegetation in the southeast section of the site could provide protection from winter winds (Figure 12).

In addition, an important concern in designing Pine Run was the relationship of land use and energies expended in construction. It is reasonable to expect that when a developer achieves a higher return per acre because of increased density, he can afford to take greater care in how he works in high density areas. Consequently, it seemed logical to place the highest density residential uses in the steeper and more wooded areas and to place the lower density, single family detached homes on the flat farmland. This means that the typical amount of broad grading required for single family homes can be done quickly and efficiently without losing trees; and that a careful grading system can be worked out for the contour clustering of townhouses and garden apartments on the heavily wooded slopes.

In the multi-family areas, this meant that through the construction of cribbing retention walls, and a grading plan that was controlled and directed in a linear pattern along the length of the garden apartments, about 90% of trees beyond 15 feet from the building were saved. Here, energy producing natural fuel resources were conserved since the need for grading was minimized while the developer's energies were directed toward a more sensitive treatment of the landscape.

Some experimentation into energy conservation through design of single family units was also done at Pine Run to reduce the effect of winter winds. Essentially, a townhouse type unit with an attached garage was allowed to free stand on a small lot. A highly efficient, internal living space structure was achieved, and by utilizing fences around it, the exposed ends of the building were buffered from winds and outdoor rooms were, in effect, created. The buffering of the exterior end-walls resulted in greater efficiency in winter heating requirements, while at the same time providing an additional exterior living space directly associated with the unit.

Thus, by working with the natural environmental systems, various types of energy conservation were allowed to take place. Contour clustering of the mult-family units produced greater protection from winter winds, while allowing greater summer cooling effects by capitalizing on the air flow patterns. The energies expended by the developer during construction were more efficiently directed toward saving existing vegetation while minimizing the need for excessive grading. Paving material needs were reduced since a road hierarchy system was introduced into the design. The neighborhood shopping facility alleviated the necessity for project residents to use vehicular transportation for small shopping needs; and through experimentation in the design of single family homes, end walls were protected from winter winds through the design of a fencing system.

At Pine Run the use of contour clustering resulted in financial savings for the developer through reduced construction, and restricted clearing. The identified savings in clearing were found to vary between $250-$270 per acre; while building within 15 feet of existing vegetation cut necessary landscaping costs by $300-$500 per unit. Further savings also resulted with the reduced need to cover the steeply sloped areas with sophisticated ground covers, usually costing between $0.10 and $0.25 per square foot. The reduced need for vegetative cover also lead to maintenance savings in these sloped areas, roughly to the extent of $1,900 per year.

204.

LEGEND

- 0 - 5 %
- 5 - 10 %
- 10 - 15 %
- 15 - 20 %
- 20 + %

H.T.W.

RT. 10

INTERSTATE 176

FLYING HILLS

CLIENT BERKSHIRE GREENS

SLOPES

5

FIGURE   5

FIGURE 6

206.

FIGURE   7

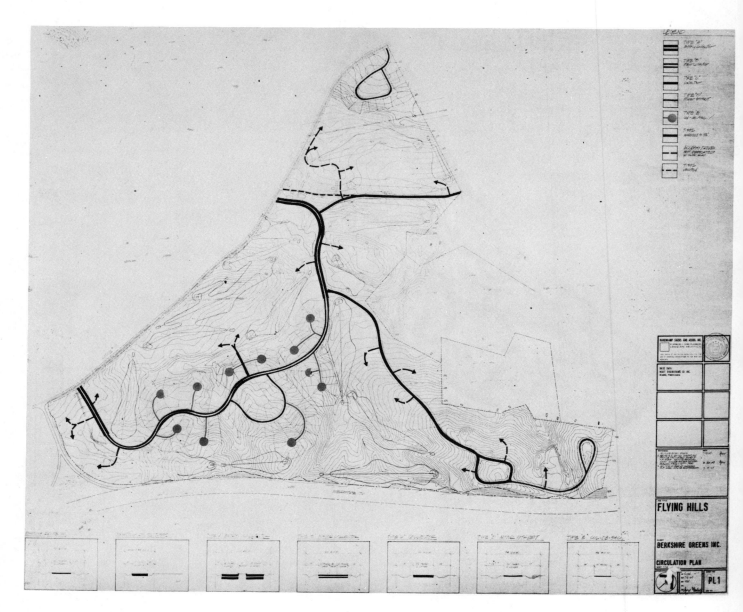

FIGURE 8

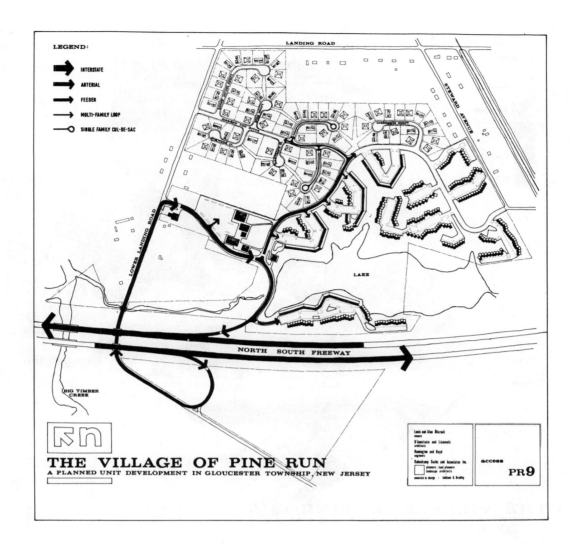

FIGURE 9

LEGEND

water

5% or less

5 to 15 %

15 to 25 %

25 % or more

LANDING ROAD

STEWARD AVENUE

LOWER LANDING ROAD

NORTH SOUTH FREEWAY

BIG TIMBER CREEK

THE VILLAGE OF PINE RUN
A PLANNED UNIT DEVELOPMENT IN GLOUCESTER TOWNSHIP, NEW JERSEY

Louis and Alan Bleznak
owners

D'Anastasio and Lisiewski
architects

Remington and Boyd
engineers

Rahenkamp Sachs and Associates Inc.
planners · land planners · landscape architects
associate in charge : Addison G. Bradley

slope analysis
PR2

FIGURE 10

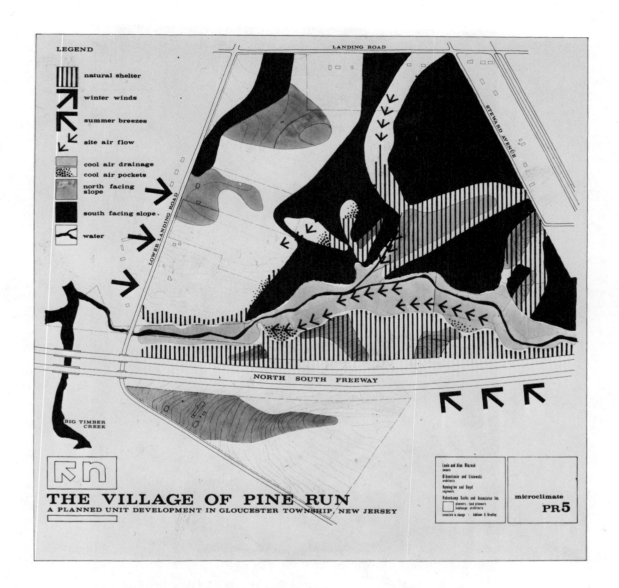

LEGEND

natural shelter
winter winds
summer breezes
site air flow
cool air drainage
cool air pockets
north facing slope
south facing slope
water

LANDING ROAD

STEWARD AVENUE

LOWER LANDING ROAD

NORTH SOUTH FREEWAY

BIG TIMBER CREEK

THE VILLAGE OF PINE RUN
A PLANNED UNIT DEVELOPMENT IN GLOUCESTER TOWNSHIP, NEW JERSEY

Louis and Alan Blaznek
owners
D'Anastasio and Lisiewski
architects
Hymington and Boyd
engineers
Rahenkamp Sachs and Associates Inc.
planners : land planners
landscape architects
associate in charge : Addison E. Bradley

microclimate
PR5

FIGURE 11

211.

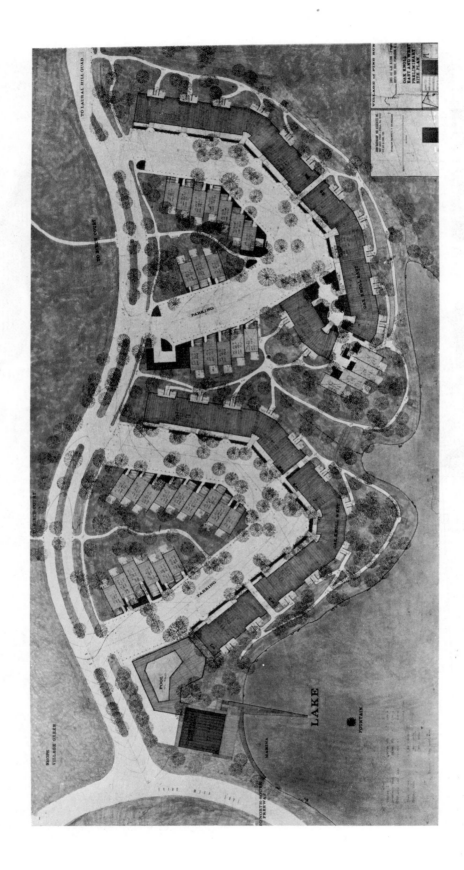

FIGURE 12

212.

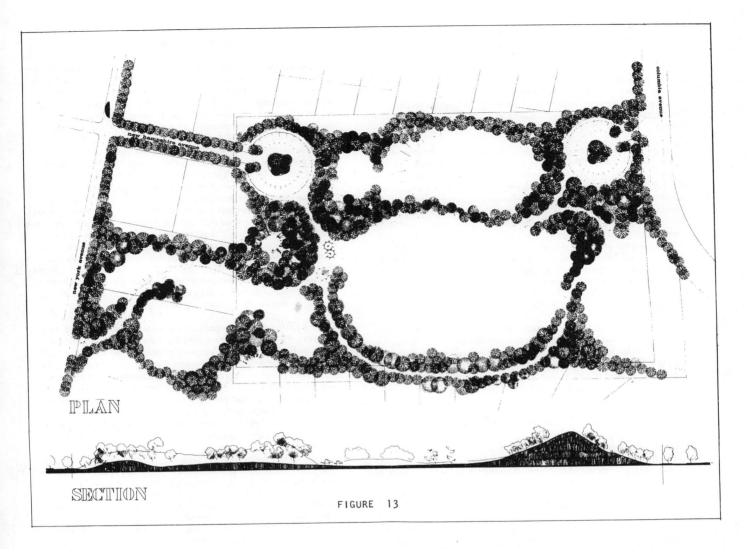

PLAN

SECTION

FIGURE 13

PROJECT NAME: PHOENIX APARTMENTS
PROJECT LOCATION: EDGEWATER PARK TWP., BURLINGTON
                     CO., N.J.

In the Phoenix Apartment project, the general design problem was developing a site plan for a virtually flat area with minimal vegetation. The energy component of the design constraint centered on the issue of how to give enclosure and provide a wind buffer for apartment structures. In contrast Pine Run and Flying Hills had variations in topography which could be utilized in design and could be worked with to save energy.

The solution was to stockpile the excess dirt excavated during construction and make hills on the site. The mounds were placed around each cluster of apartments, and plantings recommended for the crests. Together the mou mounds and matured plant material would provide adequate sheltering of the units to promote heating conservation. This, in addition, had the advant the units to promote heating conservation. This, in addition, had the advantage of creating outdoor spaces and some rhythm to the open space instead of being completely open and flat.

Thus, in the design, although extensive earth movement was required, the purpose was conservation of energy producing fuel resources in the long run. This was accomplished through creating a series of mounds around the apartment clusters, which, then the plant material mature, will provide buffering from the winter winds.

PROJECT NAME: HOLIDAY ON THE GREEN
PROJECT LOCATION: MT. LAUREL TOWNSHIP, BURLINGTON
                     CO., N.J.

In the design of Holiday on the Green, located in the farmlands of New Jersey, the premise was why tear down, destroy and, in effect, waste what energy has already been expended in constructing a barn, as opposed to making use of that energy. The design recommendation was not to burn the barn, but rather to use it initially as a sales center after rehabilitation. Once the project was sold out, it was to be turned over to the community as a recreation center.

The initial saving, and later multiple use of the barn, resulted in several types of energy conserved through use of an existing on-site facility. These savings involved using the already expended labor energies that originally built the barn, the materials required in original construction and the savings of energies required to tear down the structure. From an economic standpoint the barn now served two purposes—as a sales center and a community recreation center, and from a materials standpoint it has served three purposes (barn, sales and recreation).

In addition, increased efficiency in street lighting was achieved through an analysis of traffic patterns and critical visibility points. Based on this analysis, the conventional street lighting system of fixtures every few hundred feet was replaced by a system of lights located at intersections. Long expanses of roadways without pedestrian or vehicular access can be adequately illuminated by car headlights, without any increase in accident or crime rates. Thus by careful study of energy needs by project users, significant energy savings can be achieved.

213.

PROJECT NAME: ERLTON PARK
PROJECT LOCATION: CHERRY HILL TOWNSHIP, CAMDEN
                    CO., N.J.

The design problem was how to most efficiently extend the life of a dump-
ing site, closer to the origin of the trash, while creating a community park
(Figure 13). The energy component of the design centered on the multiple
use of land by turning the negative aspects of waste disposal into a product-
ive use.

Once filled the sanitary landfill became a flat sandy plot: eight acres of unwanted land, fenced off and standing vacant. Rather than purchasing a new
site at the expensive price of real estate in Cherry Hill and moving the landfill farther from the residents it served, the plan recommended continuing
landfill. The new refuse was used to create a new topography with hills,
some over 100 feet high, and valleys. Through design recommendations, the
sanitary landfill capacity of the site was extended seven years by having the
community create "mountains of trash" on the site. When the site reached
fill capacity, the entire area was to be seeded. For the next three to four
years, gases would be eliminated, and when the trash had sufficiently stabilized, a planting program was to be started. The money saved by the extended use of the landfill more than paid for the planting and construction required. In addition the Township did not have to fund the acquisition and
development of a park facility in this area.

Here a community facility was created, eliminating the need for people to
drive to other parks. In addition, it has economically extended the life of a
piece of land for sanitary purpose; and it has achieved double or triple purpose when only a single purpose was seen previously. The amount of energy
conserved by use of this multi-purpose area would be difficult to quantify,
however, the savings are quite evident.

PROJECT NAME: GLENSIDE NURSING HOME
PROJECT LOCATION: NEW PROVIDENCE, NEW JERSEY

The FHA financed Glenside Nursing Home entailed the design of a land use
plan suitable for a geriatric facility. The central design issue was the location
of the site on a floodplain of the Passaic watershed, along with the desire to
minimize extensive clearing and land fill of the heavily wooded site. The obvious but unsatisfactory solution would have been to clear the trees, fill the
ground above water level, and build on top.

Instead a special compacted fill was recommended under the building, parking lot and driveway. The existing drainage ways were cleaned and slightly
improved and a natural drainage swale system was added around the building. All pedestrial movement was achieved through a boardwalk system
(Figure 14) that went over the flood area instead of filling. Although the
boardwalk system was expensive to construct, the amount of money and
energy saved by not having to remove all the trees and fill the entire site
was greater than the costs of constructing beneficial facilities. Here a system had been developed which worked with a piece of land that could still
flood without incurring any damage.

The energy saving of the design was achieved through working with the
natural drainage system. A minimal amount of clearing and land fill was
required and therefore, less maintenance, most particularly grass mowing.
The result was a natural self-maintaining watering system for the lowlying
vegetative areas with these areas gradually developing into a type of wild-
life sanctuary.

PROMOTING GREATER ENERGY CONSERVATION

The previous projects indicate some of the ways in which concern for energy conservation influences design at Rahemkamp Sachs Wells and Associates. An understanding of the close interrelationship between energy, ecology and economics is central to our approach as well as an awareness of
the three phases of development critical to energy conservation: pre-
existing natural systems, construction, and subsequent user demands.

In our design process, we have been able to work with clients to determine
how these various concerns work together successfully. This was accomplished through convincing or educating a client to the fact that through
sensitivity to energy conservation techniques, the ultimate product will be
more marketable to the general public, and potentially yield a greater financial reward. An approach that we have taken toward making the client
more aware of energy needs and costs was the creation of the Development
Impact Energy Use Model. It has been designed to measure the consumption of gas, electricity and oil by the six major energy users in the residential dwelling unit: clothes drying, cooking, hot water, heating, base load,
and space cooling (Figures 15, 16 and 17). On the basis of survey information we have developed formulae which determine energy consumption by
fuel for various unit types. These formulae represent in each case the relationship of living space in square feet to energy use.

Of the six different energy needs, only space cooling and heating are significantly affected by climatic conditions. To account for climate differences, we use a set of geographical indices which reflect conditions in 36
major metropolitan areas in the country. Calculations for a specific project can be adjusted in accordance with climatic conditions in its specific
area. Aside from its value as an index of proposed energy use in a time of
impending and actual fuel shortages, the Energy Use Model also calculates
utility costs and expenses for the prospective homeowner and the rental
developer (Figure 18). This information has been found to be an extremely
effective marketing tool making the home buyer more aware of his total
housing cost.

Much more information is required, however, to quantitatively assess the
energy implications of design elements. There is an immense lack of data
on the actual magnitude of energy savings resulting from design sensitivities.
Monitoring a large residential development would provide much needed information but requires a great deal of time and funding. We have proposed
an energy study of the Pine Run project in Gloucester, New Jersey similar
to the socio-economic study completed in 1972. The procedure would involve monthly monitoring of utility consumption by type, size, location
and occupancy characteristics of each unit. In addition, periodic monitoring of wind intensity at strategic locations would be done. These pieces of
information could then be correlated with available planning and construction information to develop a quantitative data base indicating the relationship between fuel consumption, unit type, size, location and occupancy
characteristics at this project in southern New Jersey.

In the past, the decision on home purchase was very much a function only
of the cost of the structure and its related land. Due to shortages in natural
fuel resources, this decision is becoming highly influenced by the home's
fuel consumption or efficiency. Large savings can result in utility costs,
both in the winter through a decreased need for heating fuels, and in the
summer, by reduced need for air conditioning. With sources of heating and
cooling fuels becoming more scarce and expensive, it is our belief that additional information is increasingly demanded by residents, and will possibly
change certain emphases in site design.

ENERGY CONSUMPTION: SUMMARY

Project Years		1	2	3	4	5
Consumption						
Gas(MCF)		0.00	5100.82	10244.14	10244.14	5143.32
Cumulative		0.00	5100.82	15344.96	15344.96	30732.43
Electricity(KW-Hr)		0.00	892027.50	1791488.00	1791488.00	899461.00
Cumulative		0.00	892027.50	2683515.00	4475003.00	5374464.00
Oil (Gallons)		0.00	72752.69	146111.60	146111.60	73358.94
Cumulative		0.00	72752.69	218864.30	364975.90	438334.80
Cost						
Gas	($)	0.00	12088.94	24278.62	24278.62	12189.68
Cumulative	($)	0.00	12088.94	36367.56	60646.18	72835.81
Electricity	($)	0.00	50845.57	102114.80	102114.80	51269.28
Cumulative	($)	0.00	50845.57	152960.30	255075.10	306344.40
Oil	($)	0.00	26918.49	54061.30	54061.30	27142.80
Cumulative	($)	0.00	26918.49	80979.75	135041.00	162183.70
Total	($)	0.00	89853.00	180454.60	180454.60	90601.75
Cumulative	($)	0.00	89853.00	270307.60	450762.20	541364.00

	Operational Description	Input Data	Input Source	Output Data		
Figure 15						
	Energy Consumption					
	A. Gas					
	Gas used for	Clothes Drying Hot water	RSWA(EM,II)			
	Gas Cost/MCF	$ 2.37	Utility/ RSWA(EM,II)			
	Yearly Gas consumption in MCF per unit			Compute	SF	46.91
					TH	45.79
					GA	37.98
	Yearly Gas cost per unit			Compute	SF	$111.18
Figure 16	Operational Description	Input Data	Input Source	Output Data	TH	$108.51
					GA	$ 90.00
	Energy Consumption					
	B. Electricity					
	Electricity used for	Cooking Base Load Space Cooling	RSWA(EM,TI)			
	Electricity Cost/KW-hr	$ 0.06	Utility/ RSWA(EM,TI)			
	Yearly electricity consumption in KW-hr per unit			Compute	SF	8490.35
					TH	7511.85
					GA	7022.60
	Yearly electricity cost per unit			Compute	SF	$ 483.95
					TH	$ 428.18
					GA	$ 400.29
Figure 17	Operational Description	Input Data	Input Source	Output Data		
	Energy Consumption					
	C. Oil					
	Oil used for	Heating	RSWA(EM,TI)			
	Oil Cost/Gallon	$ 0.37	Supplier/ RSWA(EM,TI)			
	Yearly oil consumption in gallons per unit			Compute	SF	728.20
					TH	615.30
					GA	558.85
	Yearly oil cost per unit			Compute	SF	$269.44
					TH	$227.66
					GA	$206.78

Footnotes

Introduction

1. *Weatherwise Gardening,* Ortho Book Division, Chevron Chemical Company, San Francisco, 1974, p. 80.

2. *Ibid.,* p. 77

3. *Ibid.,* p. 76

4. *Ibid.,* p. 45

5. *Ibid.,* p. 74

6. Adapted from: Rudolf Geiger, *The Climate Near the Ground* (Cambridge, Mass., Harvard University Press, 1965), p. 409.

7. *Weatherwise Gardening,* p. 43.

8. C.A. Federer, "Effect of Trees in Modifying Urban Microclimate" in *Trees and Forests in an Urbanized Environment,* Cooperative Extension Service, University of Massachusetts, U.S. Dept. of Agriculture, Amherst, Mass., March 1971, pp. 23-24.

9. *Ibid.*

10. William E. Reifsnyder and Howard W. Lull, *Radiant Energy in Relation to Forests,* Technical Bulletin No. 1344, U.S. Department of Agriculture, Forest Service, December 1965, p. 93.

11. *Ibid.,* pp. 63-64

12. *Ibid.,* p. 65

13. *Ibid.,* p. 66

14. *Ibid.,* p. 67

15. *Ibid.,* p. 66

16. *Ibid.,* p. 68

17. *Ibid.,* pp. 68-69

18. *Ibid.,* pp. 69-70

19. *Ibid.,* pp. 70-71

20. *Ibid.*

21. *Ibid.,* p. 76

22. *Ibid.,* pp. 75 & 78

23. Joseph Kittredge, *Forest Influences* (New York, McGraw-Hill, 1948) pp. 48-49.

24. Federer, p. 25

25. *Ibid.*

26. *Ibid.,* p. 26

27. *Ibid.,* pp. 25-27

28. *Ibid.,* p. 25

29. N. Tani, "On the Wind Tunnel Test of the Model Shelter-hedge," *Bulletin of the National Institute of Agricultural Sciences* (of Japan), Series A, No. 6, March 1968, p. 32.

30. Figure courtesy of Gary O. Robinette.

31. Arthur E. Ferber, *Windbreaks for Conservation,* U.S. Department of Agriculture, Soil Conservation Service, Agriculture Information Bulletin 339, October 1969, p. 3.

32. *Ibid.*

33. Johan van der Linde, "Trees Outside the Forest," *Forest Influences,* Food and Agriculture Organization of the United Nations, Rome, 1962, p. 162.

34. *Ibid.,* p. 167

35. *Ibid.,* p. 172

36. *Ibid.*

37. Kittredge, *Forest Influences,* p. 63

38. *Ibid.,* p. 65

39. *Ibid.,* pp. 66-67

40. *Ibid.,* p. 67

41. *Ibid.,* p. 68

42. *Ibid.,* p. 69

43. *Ibid.,* p. 70

44. *Ibid.,* pp. 95-96

45. *Ibid.,* p. 97

46. *Ibid.,* pp. 97-98

47. *Ibid.,* p. 98

48. *Ibid.*

49. N. Tani, "On the Wind Tunnel Test of the Model Shelter-hedge," *op. cit.,* p. 28

50. *Ibid.,* p. 29

51. *Ibid.,* p. 31

52. *Ibid.,* p. 44

53. Robert F. White, "Landscape Development and Natural Vegetation: Effect of Moving Air on Buildings and Adjacent Areas," *Landscape Architecture,* June 1953, pp. 75-76.

54. *Ibid.,* p. 76

55. *Ibid.*

56. *Ibid.*

57. Gary O. Robinette, *Plants/People and Environmental Quality,* U.S. Department of the Interior, National Park Service, Washington, D.C., 1972, p. 77.

58. *Ibid.*

59. Victor and Aladar Olgyay, *Design with Climate* (Princeton, New Jersey, Princeton Univ. Press, 1963), p. 99.

60. J. M. Caborn, *Shelterbelts and Windbreaks* (London, Faber and Faber, 1965), p. 27.

61. Robinette, *op. cit.,* pp. 78, 79.

62. *Ibid.,* p. 79

63. N. M. Gorshenin, *Principles of Shelterbelt Layout on Arable Slopes,* U.S.S.R. All-Union Institute of Forest Culture and Forest Amelioration Excerpts (Moscow, 1934), U.S. Forest Service Translation No. 64, 1946, p. 49.

64. B. I. Logginov, *Principles of Field Protective Forestation,* All-Union Academy of Agricultural Science Translations, v. 3 (Moscow, 1941), p. 28.

65. H. Iizuka, "On the Width of Shelterbelts," Forestry Experimental Station, Bulletin 56 (Meguro, 1952), p. 23

66. Geiger, p. 312.

67. Y.A. Smal'Ko, "Range of Wind Sheltering in Forest Stips of Different Structures," Izvestiya AN SSSR, Geographic Series 5 (1954), p. 5.

68. N.P. Woodruff and A. W. Zingg, "Wind Tunnel Studies of Shelterbelt Models," *Journal of Forestry,* v. 51 (1953), pp. 173-178.

69. Robinette, *op. cit.,* pp. 80, 81.

70. *Ibid.,* p. 81

71. *Ibid.,* p. 82

72. *Ibid.*

73. *Ibid.,* p. 83

74. *Ibid.*

75. Y. Panfilov, "A Contribution to the Problem of the Effect of the Shelterbelts on Wind Velocity on the Steep Slopes," *Sovetska Agronomiska,* v. 1 (1940), pp. 11-17.

76. Werner von Nageli, *Uber die Wind verhältnisse im Bereich gestaffelter Windschutzreiten,* v. 23, p. 14.

77. Robinette, *op. cit.,* p. 84.

78. *Ibid.*

79. *Ibid.,* p. 85

80. *Ibid.*

81. *Ibid.*

82. Werner von Nageli, *Untersuchungen uber die Windverhältnisse im Bereich von Schilfrohkaunden,* v. 29 (Ebenda, 1953), p. 82.

83. W. Kreutz, *Der Windschutz* (Dortmund, 1952), p. 83.

84. J. van Eimern, "Kleineklimatische Geländeautnahma in Quickbom/ Holstein," *Annalen der Meteorologie,* v. 4 (1951), pp. 259-269.

85. Carlos G. Bates, "The Windbreak as a Farm Asset," *Farmers Bulletin,* No. 1405 (1944), p. 4.

86. Geiger, *op. cit.,* p. 497 ff.

87. M. Jensen, *Shelter Effect: Investigations into the Aerodynamics of Shelter and its Effect on Climate and Crops* (Copenhagen, 1954), p. 38

88. J. M. Caborn, *Shelterbelts and Windbreaks* (London, Faber and Faber, Ltd., 1965), p. 60.

89. Wilfred Bach and Edward Mathews, "The Importance of Green Areas in Urban Planning," Paper prepared for Bioclimatology and Environmental Health Workshop, Public Health Service, U.S. Dept. of Health, Education and Welfare, Cincinnati, July 14-16, 1969.

90.

90. Kittredge, *Forest Influences, op. cit.,* p. 71

91. Robinette, *op. cit.,* p. 86

92. *Ibid.,* p. 87

93. *Ibid.,* pp. 86, 87

94. *Ibid.,* p. 87

95. Geiger, *op. cit.,* p. 335.

96. *Ibid.,* p. 336

97. *Ibid.,* p. 334

98. *Ibid.,* p. 336

99. Robinette, *op. cit.,* p. 99

100. *Ibid.*

101. Kittredge, "The Influence of the Forest on the Weather," *op. cit.,* pp. 99, 100

102. *Ibid.*

103. Robinette, *op. cit.,* p. 89.

104. *Ibid.,* p. 90

105. *Ibid.,* p. 87

106. *Ibid.,* pp. 89, 90

107. *Ibid.,* p. 90

108. Marvin D. Hoover, "Water Action and Water Movement in the Forest," *Forest Influences,* Food and Agriculture Organization of the United Nations, Rome, 1962, p. 72.

109. Kittredge, "The Influence of the Forest on Weather," *op. cit.,* pp. 1 122-123.

110. Robinette, *op. cit.,* p. 91.

111. *Ibid.*

112. *Ibid.,* p. 92

113. *Ibid.,* pp. 91-94

114. *Ibid.,* pp. 93, 94

115. Kittredge, *Forest Influences, op. cit.*

116. Robinette, *op. cit.,* p. 95

117. *Ibid.*

118. *Ibid.*

119. Kittredge, *Forest Influences, op. cit.,* p. 62

120. Robinette, *op. cit.,* p. 95

121. *Ibid.,* pp. 95, 96

122. *Ibid.,* p. 96

123. Kittredge, *Forest Influences, op. cit.,* p. 53

124. *Ibid.,* p. 55

125. *Ibid.,* pp. 57-58

126. *Ibid.,* p. 61.

127. Kittredge, "The Influence of the Forest on Weather," *op. cit.,* pp. 105, 108.

128. *Ibid.,* p. 89

129. *Ibid.,* p. 93

130. Aldo Pavari, "Introductory Remarks," *Forest Influences,* Food and Agriculture Organization of the U.N., p. 10.

131. *Ibid.,* pp. 8, 9.

132. *Ibid.,* pp. 12, 13

133. *Ibid.,* pp. 15, 21

134. *Ibid.,* p. 24

135. Robinette, *op. cit.,* p. 96

136. *Ibid.,* p. 97

137. *Ibid.*

138. *Ibid.,* pp. 96, 97

139. S. Elwynn Taylor and Gerald Pingel, "Green Grass That's Not So Green," *New York Times,* April 11, 1976.

140. "Nature's Own Cooling System," *Nursery Business* (September-October 1970), p. 59

141. Roy A Mecklenberg, *et. al.,* "The Effect of Plants on Microclimate, Dust Concentration and Noise Reduction in the Urban Environment: A Preliminary Report," Michigan Agricultural Experiment Station Journal, Article No. 5552, Michigan State University, East Lansing, Michigan.

142. *Ibid.*

143. *Ibid.*

144. *Ibid.*

145. Robinette, *op. cit.,* p. 98

146. *Ibid.*

147. *Ibid.*

148. *Ibid.*

149. *Ibid.*

150. *Ibid.,* p. 99

151, *Ibid.,* p. 100

152. *Ibid.*

153. Douglas S. Way, Unpublished Student Paper, University of Wisconsin, 1967, Madison, Wisc.

154. *Ibid.*

155. *Ibid.*

156. *Ibid.*

157. *Ibid.*

158. *Ibid.*

159. *Weatherwise Gardening,* p. 78.

160. *Ibid.,* p. 82

161. *Ibid.,* p. 74

162. *Ibid.,* p. 78

163. *Ibid.*

164. *Ibid.*

165. H. M. Vernon and T. Bedford, "Environmental Warmth and Human Comfort," *British Journal of Applied Physics* (February 1950), pp. 33-38.

166. *Ibid.*

167. S. F. Markham, *Climate and the Energy of Nations* (London, 1947) p. 143.

168. C.E.P. Brooks, *Climate in Everyday Life* (London, 1950), p. 53.

169. C. A. Federer, "Effects of Trees in Modifying Urban Microclimate," *op. cit.,* pp. 23-24.

170. Robinette, *op. cit.,* pp. 67-68

171. *Ibid.,* p. 68

172. *Ibid.,* p. 67

173. Geiger, *op. cit.,* p. 5

174. Cliff Tandy (ed.), *Handbook of Urban Landscape* (London, The Architectural Press, 1972), p. 59.

175. Olgyay, *Design With Climate, op. cit.*

176. Garrett Eckbo, *The Art of Home Landscaping* (New York, F.W. Dodge Corp., 1956), p. 23.

177. Don Koberg and Jim Bagnall, *The Universal Traveler* (Los Altos, California, William Kaufmann, Inc., 1972), p. 17.

178. *Ibid.,* pp. 20, 21

179. *Ibid.,* p. 26

180. *Ibid.,* p. 27

181. *Weatherwise Gardening, op. cit.,* p. 74.

182. *Ibid.*

183. *Ibid.,* pp. 74, 75

184. Wallace, McHarg, Roberts and Todd, *Woodlands New Community: An Ecological Plan* (Houston, Texas, Woodlands Development Corp. 1973), p. 25.

185. *Ibid.*

186. Regional Planning Council, Baltimore, Maryland, *Environmental Characteristics Planning: Preparation of a Plan,* Baltimore, 1972, pp. 8, 9.
187

187. AIA Research Corporation, *Energy Conservation in Building Design* (Washington, D.C., American Institute of Architects, 1974), pp. 24-29.

188. Figure courtesy of Richard E. Browne Associates, Engineers, Planners, Architects, Columbia, Md.

189. Figure courtesy of Richard E. Browne Associates.

190. Jack Kramer, *Your Homemade Greenhouse and How to Build It* (New York: Cornerstone Library, 1975), p. 12.

191. Olgyay, *Design With Climate, op. cit.,* p.

192. *Ibid.*

193. *Ibid.*

194. *Ibid.*

195. *Ibid.*
196. *Ibid.*

197. *Ibid.*

198. *Ibid.*

199. *Ibid.*

200. *Ibid.*

201. *Ibid.*

202. *Ibid.*

203. Paul Jacques Grillo, *What is Design?* (Chicago, Paul Theobald and Co., 1960), p. 25.

204. *Ibid.,* p. 95

205. *Ibid.*

206. *Ibid.,* p. 94

207. *Ideas for Landscaping,* Sunset Book (Menlo Park, California, Lane Books, Books, 1972), p. 10.

208. *Ibid.* p. 11

209. *Weatherwise Gardening, op. cit.,* p. 43.

210. Craig Johnson, Unpublished Paper on Planting Design, Utah State Univ., Logan, Utah, 1974.

211. *Ibid.*

212. *House and Home,* Vol. 46, No. 2 (New York, McGraw-Hill), Feburary 1974, pp. 202-204.
213. *Ibid.*

214. Kevin Lynch, *Site Planning* Cambridge, Mass. M.I.T. Press, 1962), pp. 90-98.

215. Eckbo, *op. cit.,* p. 47.

216. *Weatherwise Gardening, op. cit.,* pp. 31-41

217. *Ibid.,* p. 31

218. *Ibid.*

219. *Ibid.,* p. 32

220. *Ibid.,* p. 33

221 Robinette, *op. cit.,* p. 71

222. *Weatherwise Gardening, op. cit.,* p. 35

223. *Ibid.*

224. *Ibid.,* p. 36

225. *Ibid.* p. 38

226. *Ibid.* p. 37

227. *Ibid.* p. 38

228.

229.

230. *Weatherwise Gardening, op. cit.,* p. 58.

231. *Ibid.,* p. 38

232. *Ibid.,* p. 58

233. *Ibid.,* p. 38

234. *Ibid.,* p. 58

235. Adapted from *Weatherwise Gardening, op. cit.,* p. 36

236. *Ideas for Landscaping,* p. 12

237. *How to Build Fences and Gates,* Sunset Book (Menlo Park, California, Lane Books, 1971), pp. 8, 9.

238. *Ibid.*

239. *Ibid.,* p. 16

240. *Ibid.,* p. 17

241. *How to Build Patio Roofs,* Sunset Book (Menlo Park, California, Lane Books, 1956), p. 50.

242. *Ibid.,* p. 49

243. *Outdoor Building Book,* (Sunset Book (Mel

243. *Outdoor Building Book,* Sunset Book (Menlo Park, California, Lane Books, 1953), p. 13.

244. *How to Build Patio Roofs, op. cit.,* pp. 3-5.

245. *Weatherwise Gardening, op. cit.,* p. 37.

246. *Ibid.,* p. 44

247. *Ibid.,* p. 45

248. *How to Build Patio Roofs, op. cit.,* p. 6.

249. *Ibid.*

Bibliography

The following is a partial listing of publications and references dealing with energy conservation, solar energy utilization and site development for these purposes. It is not definitive but does give some sources for gaining additional information on these subjects.

THE SUN

The Sun, Sunlight, and Solar Radiation. Technical studies and measurements.

BOOKS:

Living with the Sun. Phoenix, Arizona, 1958. Association for Applied Solar Energy.

Control of Sun Penetration. CEBS Researches and Facilities Commonwealth Experimental Bldg. Station, November 1971.

Neubauer, L.W. *The Solaranger.* Davis, University of California, 1949.

Walter Schwagenscheidt, *Vergleichende Besonnungsuntersuchungen.* (Comparable Examination for the Orientation of Dwellings.) Beilage zu Heft der Zeitschrift "Das Neue Frankfurt." April, May, 1930. A very exact interpretation of sun angle requirements for dwellings under specific circumstances. In German, French, and English, it is very complicated.

Solarometer. Toledo, Libby-Owens-Ford Glass Co., 1948.

Sun Angle Calculator. Toledo, Libby-Owens-Ford Glass Co., 1950.

PERIODICALS:

C. C. D. Brammall, "Solarscope and Artificial Sky." Toronto: *Journal of the Royal Architectural Institute of Canada.* Ser. No. 297, Vol. 27, No. 5, May 1950, pp. 177-178. Description of the work at the Australian Commonwealth Experimental Building Station in Sydney, N.S.W., as well as a brief history of previous research elsewhere on the relation of sun to buildings.

Fritz, S. "The Albedo of Ground and Atmosphere," *Bulletin of the American Meteorological Society,* v. 29, no. 6, 1948.

Gates, D. M. and W. Tantrapron. "The Reflectivity of Deciduous Trees and Herbaceous Plants in the Infrared to 25 Microns," *Science,* v. 94, no. 11, 1948.

Hand, Irving F. "Charts to Obtain Solar Altitudes and Azimuths," *Heating and Ventilating,* v. 38, October 1948.

Penndorf, R. "Luminous Reflectance (Visual Albedo) of National Objects." *Bulletin of the American Meteorological Society,* v. 37, no. 5, 1956.

THE HISTORY OF ARCHITECTURAL ADAPTATION TO CLIMATE

Description of adaptations made in various times and places to deal with the climate.

BOOKS:

Henry R. Aldridge, *The Case for Town Planning,* London: National Housing and Town Planning Council, 1915. Shows in part how ancient civilizations, and more recent ones, too, have accounted for the climate in their building practices.

Canada, National Research Council, *Swedish Test Hut Research Program.* (Translated from the original of Nils Holmquist.) (Translation TT-96.) Ottawa: January, 1950. Mostly regarding the use of different materials as far as the weather is concerned.

Banister Fletcher (Sir), *A History of Architecture.* London: B. T. Batsford, Ltd., 1946. Excellent analysis of the history of architecture on the comparative method. Each period has a small discussion on the manner in which climate influenced the architecture.

Jiro Harada, *The Lesson of Japanese Architecture.* London: The Studio, Ltd., 1936. Contains a number of comments on the Japanese consideration for climate in their building practices.

Hideto Kishida, *Japanese Architecture.* Tokyo: Board of Tourist Industry, Japanese Government Railways, May, 1935. Contains a few interesting comments on the influence of climate on Japanese architecture. A small, paper bound, pocket-book-size volume.

Moholy-Nagy, Sibyl. *Native Genuis in Anonymous Architecture.* New York: Horizon Press, Inc., 1957.

Vitruvius (trans. by Frank Granger), *On Architecture.* (From Harleian MS. 2767.) London: William Heinemann, Ltd., 1931. A wealth of historical thought on planning with the climate is contained herein.

Arnold Whittick, *European Architecture in the Twentieth Century.* London: Crosby Lockwood and Son, Ltd., 1950. Discusses early-modern attempts to conform to the climate when designing buildings.

R. R. Wycherley, *How the Greeks Built Cities.* London: MacMillan and Co., Ltd., 1949. An occasional mention of climate and how the ancient Greeks provided for it in their town planning.

PERIODICALS:

Atkinson, R. "The Genesis of Modern Tropical Architecture." *Royal Society of Arts Journal.* July, 1969, pp. 546-61.

Jens Mollerup, "Denmark and Danish Architecture." Toronto: *Journal of the Royal Architectural Institute of Canada,* vol. 26, no. 8, August 1949, pp. 227-30. One or two evidences of how the Danish architects account for climate.

WIND—AERODYNAMICS

Wind, aerodynamics, and air circulation.

BOOKS:

Aynsley, R.A. The Growing Awareness of Environmental Aerodynamics. Dept. of Architectural Science, University of Sydney, Moael Report MR5 1971.

Edinger, James G. *Watching for the Wind.* Anchor 549.

PERIODICALS:

Harris, R. I. The response of structures to gusts. *Wind Effects on Buildings and Structures.* 1963.

Hoydysh, Walter G. "Aerodynamics: New Design Criteria." *Microclimate,* pp. 58-61.

Hoydysh, Walter G. "Aerodynamics: Town Design Criteria, Pollution, Climate Control." *Architectural Forum,* p. 58-61, Sept. 1972.

Ramadas, L. A., and S. L. Malurkar, "Surface Convection and Variation of Temperature Near a Hot Surface," *Indian Journal of Physics*, v. 7, no. 7, 1927.

Sealey, A. "Climate Studies: Local Airflow & Building." *Architects Journal*, vol. 144, pp. 983-90, October 27, 1965.

Wiener, S. Imanuel, "Solar Orientation: Application of Local Wind Factors," *Progressive Architecture*, February 1955, pp. 114-118.

ORIENTATION

Orientation. Site layout, building shape, and building orientation.

BOOKS:

The American Public Health Association, Committee on the Hygiene of Housing, "Planning the Neighborhood," Public Administration Clearing House, Chicago, 1948 (*Standards for Healthful Housing*, vol. 1).

Norman J. Jenkins, A.A.S.A., A.R.A.I.A., A.N.Z.I.A., *The Australian House*. Sydney: W. J. Nesbit (no date). The discussion on orientation on pages 10 to 16 is quite good.

Illinois, University of (Agricultural Experiment Station and Extension Service in Agriculture and Home Economics, College of Agriculture), *Planning the Illinois Farmstead for Efficiency, Health and Enjoyment*. Urbana: 1947. A fine booklet on orientation of farm buildings with respect to the sun, winds, and odors.

Malone, Thomas and Friedman, D., "Solar Radiation and Heat Transmission in Dwellings," Research Report, Massachusetts Institute of Technology, Cambridge, 1952.

Horace, Miner, *St. Denis, A French-Canadian Parish*. Chicago: The University of Chicago Press, 1939. Describes, in part, the architectural side of rural Quebec and how the inhabitants shelter themselves from the climate.

Stamo Papadaki, *The Work of Oscar Niemeyer*, New York: Reinhold Publishing Corp., 1950. Among the many points brought out are the Niemeyer techniques for protecting buildings from the sun.

Royal Institute of British Architects, London: Joint Committee on the *Orientation of Buildings*, 1933.

Harold Spence-Sales, *How to Subdivide*. Ottawa: Community Planning Association of Canada. 1950. Planning for sunlight and daylight in the layout of housing developments; framed in a colorful format.

United States of America, Department of the Interior (Housing Authority of), *Planning the Site*. (Bull. no. 11.) Washington: U.S. Government Printing Office, 1939. Page 26 offers a few remarks on problems of orientation in the design of low-rent housing projects.

Arnold Whittick (in collab. with Johannes Schreiner): *The Small House: Today and Tomorrow*. London, Crosby Lockwood and Sons, Ltd., 1947. Contains some interesting points on orientation (pp. 59-62, 142, 155-156, 160, 170, 174, 202) on trees for shade purposes (pp. 62, 123, 158, 160, 169), and on smoke pollution (pp. 60, 90, 91, 208).

PERIODICALS:

Beckett, H. E., "Orientation of Buildings," *Journal of the R.I.B.A.*, vol. 40, 1933, p. 62.

"Fitting a House to the Land," *Saturday Review of the Arts*, 55, no. 37, 41 (Sept. 1972).

Givoni, B.; Newman, E.; Hoffman, M., "Effect of Orientation and Shading of Classrooms on Thermal and Illumination Conditions," research report to Ford Foundation Building Research Station/Haifa, August, 1968.

Madill, H. H., "The Architect and the Hospital Board." Toronto: *Journal of the Royal Architectural Institute of Canada*, vol. 28, no. 4, April 1951, p. 98-99. Professor of Architecture at the University of Toronto gives his views on orientation.

Olgyay, Victor and Olgyay, Aladar, "Environment and Building Shape," *Architectural Forum*, August 1954, pp. 104-108.

Olgyay, Aladar, "Solar Control and Orientation to Meet Bioclimatic Requirements," *ibid.*, pp. 38-46.

Olgyay, Victor and Olgyay, Aladar, "The Theory of Sol-Air Orientation," *Architectural Forum*, March 1954, pp. 133-137.

Beckett, H. E., "Orientation of Buildings," *Journal of the R.I.B.A.*, vol. 40, 1933, p. 62.

"Orientation of Buildings," *Journal of the R.I.B.A.*, vol. 40, 1933.

Waldraum Percy, "The Orientation of Buildings," London: *Journal of the Royal Institute of British Architects*, vol. 40, 3rd series, Nov, 11, 1932, pp. 23-24. Waldrum letter to *Journal*.

West, H. G., "The House if a Compass." *Landscape*, 7, no. 2, 24-27 (Autumn 1951).

MATERIALS

Materials. The physical performance of building materials and their uses.

BOOKS:

Anson, M.; Kennedy, W. B.; & Spencer, J. W., "Effect of Envelope Design on Cost Performance of Office Buildings," Australia, CSIRO: Division of Bldg. Res., Reprint no. 572 (from U.S. National Bureau of Standards Specifications Publication no. 361, vol. 2, 395-406).

Close, P. D., *Sound Control and Thermal Insulation of Buildings*, 1966.

Hasan, A., *Boxed Walls for Buildings in Tropics*, West Pakistan, Building Research Station, 1969, unpriced.

Koenigs Berger, Otto & R. L., *Roofs in the Warm Humid Tropics*, London, Published for the Architectural Association by London Humphries, 1965.

Malone, Thomas and Friedman, D., *Solar Radiation and Heat Transmission in Dwellings*, Research Report, Massachusetts Institute of Technology, Cambridge, 1952.

Tyler Stewart Rogers, *Design of Insulated Buildings for Various Climates*. F. W. Dodge Corporation, 1951. A thin, attractive volume concerning the influence of climate on the selection of building materials.

Schaupp, Wilhelm, "External Walls: Cladding Thermal Insulation, Damp-Proofing," New York: Transatlantic Arts, 1967.

Simpson, J. W., and P. J. Horrobin, *The Weathering and Performance of Building Materials*, Aylesbury, Medical and Technical Publishing Co., 1970, pp. 286. Abstract: This book is intended for the architects who design and detail buildings, and for the surveyors, engineers, and contractors who are responsible for their erection and maintenance. Some of the sections in the book were originally presented at a course on "The Performance of the External Surfaces of Buildings" (Manchester, 1970).

Hubert R. Snoke and Leo J. Waldron, *Survey of Roofing Materials in the Northeastern States*. (National Bureau of Standards of), Building Materials and Structures Report BMS 29. Washington: U.S. Government Printing Office, October 11, 1939.

Hubert R. Snoke and Leo J. Waldron, *Survey of Roofing Materials in the South Central States*. U.S. of America, Department of Commerce (National Bureau of Standards of), Building Materials and Structures Report BMS 84. Washington: U.S. Government Printing Office, May 5, 1942.

J. F. vanStraaten: *Thermal Performance of Buildings*, Elsevier Publishing Co., Ltd., 1968. Bldg. Res. Institute of South Africa.

PERIODICALS:

Button, D. A., "Control of Daylight, Noise and Heat by the Building Structure," *Bldg., Res. Technol.*, 1970.

Baker, M. C., "Flashings for Membrane Roofing," *Canadian Building Digest*, Sept. 1965.

Buchberg, H., Naruishi, J.: *Build*. A rational evaluation of thermal protection alternatives for shelter. Scientific, 1967, 21 (1) 37-57. A procedure is described for the quantitative evlauation of thermal protection alternatives that can account for environmental, structure, and occupancy factors. Over fifty design alternatives which represent iterations on three basic constructions are examined in connection with the achievement of low cost shelter in a climate characterized by severe summers and less than mild winters. The three basic constructions were solid concrete block, hollow concrete block, and wood frame. Digital computer programmes were used to calculate the solar inputs, shape moduli, and diurnal temperature distributions.

Givoni, B., The effect of roof construction upon indoor temperatures, *Proceed. 1st Inst. Congress of Bio-Climatology,* Pergamon Press, 1962, pp. 237-245.

Sutton, G. E., Roof spray for reduction in transmitted solar radiation, *ASHVE Journal Section, Heat, Piping and Air Conditioning,* September 1950, pp. 131-137.

Wells-Thorpe, J. A., "Exterior Colour," "Colour in Architecture," *Building Materials,* 1966, July, pp. 43-93.

"Wind, Sun, Rain and the Exterior Wall." *Architectural Record,* pp. 205-216 (Sept. 1967), vol. 142, no. 3.

Yellott, J. E., "How Materials React to Solar Energy." *Architectural Record,* pp. 196-198 (May 1966), vol. 139, no. 5.

SHADING

Shading and Sun Shielding.

BOOKS:

American Society of Heating, Refrigerating and Air Conditioning Engineers. *Shade Factors.* New York, 1960.

Danz, Ernst, *Sun Protection: an International Architectural Survey.* New York: F. A. Praegern, 1967.

Olgyay and Olgyay, *Solar Control and Shading Devices,* Princeton University Press, Princeton, N. J. 1957.

Olgyay, Aladar. *Shading and Insolation Measurement of Models.* Austin, University of Texas Press, 1953.

PERIODICALS:

"Free Standing Sun Roof Shields Entire Building." *Architectural Forum,* p. 9 (Jan./Feb. 1970), vol. 132, no. 1.

"Exterior Screening Wall, Low Rise Office Building." *Architectural Forum,* p. 41 (Dec. 1966) vol. 125, no. 5.

"Sun Hoods Reduce Cooling Loads Without Cluttering Facade." *Progressive Architecture,* p. 178-180 (Aug. 1966), vol. 47, no. 8.

Northeast, J., "Office Staff in the Shade," *Building Maintenance,* 1971, 5 CD (Jan.), 16-FS.

"Precast, Prestressed Concrete Sunscreen." *Progressive Architecture,* p. 106 (Oct. 1972), vol. 53, no. 10.

Sato, S., "Calculations of the Received Solar Radiation in the Shade of Windbreak at Miyazaki City," *Journal of Agricultural Meterology* (Tokyo), vol. 11, no. 6, 1955.

"Sculptured Concrete Screens." *Architectural and Engineering News,* p. 32-35 (Feb. 1968), vol. 10, no. 2.

"Shade Factors," American Society of Heating, Refrigerating, and Air-Conditioning Engineers, *Heating Ventilating Air Conditioning Guide,* 1960.

Smith, Neill, "Dental Clinic Glare Screen." *Progressive Architecture,* p. 144 (Nov. 1968), vol. 49, no. 11.

Van der Linde, R. J., and J. P. M. Woudenberg. "A Method of Determining the Daily Variations in Width of a Shadow in Connection with the Time of the Year and the Orientation of the Overshadowing Object," *Journal of the Netherlands Meteorological Institute,* vol. 102, no. 2, 1946.

TROPICAL ARCHITECTURE

Tropical Architecture. Emphasis on the Hot-Humid Climate.

BOOKS:

Y. M. and J. Drew. *Tropical Architecture in the Dry and Humid Zones.* New York: Reinhold Pub. Record., 1964.

Fullerton, R. L., *Building Construction in Warm Climates,* vol. 1E (Oxford Tropical Handbooks), London and Oxford University Press, 1967.

House Design for Hot Climates, Commonwealth Building Station, Notes on the Science of Building, October 1971.

Lee, Douglas H. K., *Physical Objectives in Hot Weather Housing.* Washington: Housing and Home Finance Agency, Office of International Housing, 1963.

National Research Council BRAB, *Housing and Building in Hot-Humid and Hot-Dry Climates,* Conference No. 18 and 19, 1953, conducted by BRAB, Washington, D.C., 1953.

Lee, D. H., *Physiological Principles in Tropical Housing with Especial Reference to Queensland.* 1944 (published by University of Queensland).

Oakley, David. *Tropical Houses: A guide to their Design.* London, Batsford, 1961.

PERIODICALS:

Holshausen, C. G., "A Selective Bibliography of Physiological Studies and Problems of Building Design and Indoor Climate Control in Hot Humid and Hot Dry Environments," *Architectural Scientific Review* (Sydney), Dec. 1966, pp. 130-138, vol. 9, no. 4.

"Building for Warm Climates: A Bibliography." *Overseas Building Notes,* June 1967, pp. 2-12.

"Building Overseas in Warm Climates." B.R.S. Digest, 1968, no. 92 (second series).

Progressive Architecture, Vol. XXX, no. 3, pp. 46-50. "Dormitory, Pondichery, India." New York: Reinhold Publishing Corporation, March, 1949. A description of Antonin Raymond's design for this Indian dormitory, which shows so well the influence of the climate.

B. C. Raychaudhury and N. K. D. Chaudhury: Thermal performance of dwellings in the tropics, *Indian Construction News,* Dec. 1961, pp. 38-42.

S. J. Richards, J. F. van Straaten and E. N. van Deventer. Some ventilations and thermal considerations in public design to suit climate. *S. A. Architectural Record,* vol. 45, no. 1, Jan. 1960.

Sperling, R., "Non-Traditional Building for Warm Climates: A guide to Good Design and Practice," *Overseas Building Notes,* April 1967.

TROPICAL ARCHITECTURE

Tropical Architecture. Emphasis on the Hot-Dry Climate.

PERIODICALS

"Designing for a Dry Climate," *Progressive Architecture,* Aug. 1971, no. 8, pp. 50-52.

Raychandhur, B & C and Others, "Indoor Climate of Residential Buildings in Hot Arid Regions: Effect of Orientation." *Building Science,* pp. 79-88 (Jan. 1965).

Sehgal, J. L., "Effect of Climate on Building Design in Hot-Arid Regions." *National Buildings Organization Journal* (India), vol. 8, no. 2, pp. 47-49, (April 1963).

COLD CLIMATE

Cold Climate. Polar and High Altitude Architecture.

BOOKS:

Givoni, B., "The Effect of Roof Construction Upon Indoor Temperatures," *Proceedings 1st Intereducational Congress of Bio-Climatology,* London: Sept. 1960.

Proksch, Viktor, *Houses in the Alps.* Innsbruck/Tyrol: Pinguin Verlage, 1964.

Styles, D.F. and Melbourne, W. H., *Outline Design to Minimize Drift Accumulation,* Antarctic Directory of Supply, Australian National Antarctic Expeditions, Melbourne, Australia (1968).

Army, Department of the: *Arctic Construction,* Technical Manual TM 5-349, Headquarters Army, Washington, D.C.

PERIODICALS:

Erskine, R., "Architecture and Town Planning," *Polar Record,* vol. 14, no. 89, University of Cambridge (May, 1968) pp. 165-171.

GENERAL

BOOKS:

Carson, Arthur, *How to Keep Cool.*

Garrett Eckbo, *Landscape for Living.* New York: Architectural Record with Duell, Sloan and Pearce, 1950. An ecological study of climate landscaping, and architecture. Suggested reading.

James Marston Fitch, "Buildings Designed for Climate Control in the United States of America," National Research Council (Building Research Advisory Board of), *Weather and the Building Industry.* Research Conference Report No. 1). Washington: 1950. A popular report about climate control.

Herrington, L. P., *Human Factors in Planning for Climate Control,* Building Research Advisory Board, Washington, D.C., 1950, pp. 85-91 (Research Conference Report No. 1).

L. Hilberseimer, *The New City, Principles of Planning.* Chicago: Paul Theobald, 1944. A first-rate book about good city planning. Describes the remarks of Xenophon and Socrates, practices of Pueblo dwellers, designs of Miljutin, Le Corbusier, and F. L. Wright among others. Excellent descriptions of orientation problems, sun angles, smoke pollution, winds, etc.

Knsen, B., & Sharp, H., *Environmental Technologies in Architecture.*

Rudofsky, *Architecture without Architects.* Garden City, New York: Doubleday Co., Inc., 1964.

Jose Luis Sert, *Can Our Cities Survive?* Cambridge: The Harvard University Press, 1942. Considerable space devoted to sun, smoke, winds, orientation, etc., in planning.

Simonds, John. *Landscape Architecture,* New York: F. W. Dodge Corporation, 1961.

Maron J. Simon, ed., *Your Solar House,* New York: Simon and Schuster, 1947. A popular analysis of the solar house problem, with a house plan for each of the 48 states and the District of Columbia.

C. Strock. *Handbook of Air Conditioning, Heating and Ventilating,* 1959.

Design of Residence for Climatic Comfort, Cambridge, Mass., MIT, Albert Farwell Bemis Foundation, 1954.

PERIODICALS:

Ghaswala, S. K., "Elements of Heliotherma Planning," *Indian Builder,* 11, 143-150 (December 1963).

Sanoff, Henry "Seven Acres of Underground Shelter: an Italian Tunnelee in Fresno . . . Invents an Ingenious Response to Climate." *AIA Journal,* 47, 2, 66-68 (Feb. 1967).

Kenneth C. Welch, "Indoor-Outdoor Living." *Illuminating Engineering,* vol. XLVII, no. 6, pp. 297-299. New York: Illuminating Engineering Society, June 1952. A description by the architect-owner of a very nice climatically-designed house in Grand Rapids, Michican. Emphasis placed here on lighting problems.

Julian Whittlesey, "New Dimensions in Housing Design." *Progressive Architecture,* vol. XXXII, no. 4, pp. 57-68. New York: Reinhold Publishing Corporation, April 1951

Wright, Henry, "Environmental Technology as a Design Determinant." *Canadian Architect,* vol. 12, pp. 41-46 (January 1967).

ENERGY CONSERVATION

BOOKS:

Heery & Heery, AIA/Research Corp., Dublin-Mindell-Bloome Associates. Energy Conservation Design. Guidellines for Federal Buildings Workshop, pp. 18-20, November 1973.

Technical Options for Energy Conservation in Buildings, prepared for joint emergency conservation in buildings. National Conference of States on Building Codes and Standards, June 19, 1973.

PERIODICALS:

Darayam, Sital, "Energy Conservation in Building Design." *Progressive Architecture,* vol. 52, no. 10, pp. 130, October 1971.

Dubin, Fred S. "Saving Energy in Architectural Design." *AIA Journal,* pp. 18-21, December 1972, vol. 58, no. 6.

Dubin-Mindell-Bloome Associates, Consulting Engineers and Planning. "Life Cycle Costs for Various Building Sub-Systems," April 2, 1973.

"Energy Conservation Needs New Architecture and Engineering." *Public Power,* vol. 30, no. 2, pp. 20 (Mar./Apr. 1972).

Yaglou, C. P., "Radiant Cooling, Investigated at Harvard," *Heating and Ventilating,* May 1947, pp. 102-104.

INTERFACE—WEATHER

Weather, general discussion of the effect of weather upon man and vice versa, emphasizing the climate of cities.

BOOKS:

Drysdale, J. W., *Climate and House Design; Physiological Considerations,* Sydney, Australia Commonwealth Experimental Building Station, 1948, 15 pp. (Duplicated Document No. 25.)

Drysdale, J. W., *Climate and Design of Buildings; Physiological Study No. 2,* Sydney, Australia, Commonwealth Experimental Building Station, 1950, 22 pp. (Duplicated Document No. 32.)

Fitch, James M., *American Building, The Forces That Shape It,* Boston: Houghton Mifflin Company, 1948.

Lee, Douglas, H. K., *Physiological Objectives in Hot Weather Housing,* U.S. Housing and Home Finance Agency, Washington, D.C. 1953, p. 2.

Jaros, *Solar Effects on Building Design.* National Research Council Building Research Institute, Conference, Spring 1962. Pub. No. 1007.

Aronin, Jeffrey Ellis, *Climate and Architecture,* New York: Reinhold Publishing Corp.

Givoni, B., *Man, Climate and Architecture,* London: Elsevier Publishing Co., Ltd., 1969.

Koenigsberger, Otto; Carl Mahoney & Martin Evans, *Climate and House Design,* Vol. 1, Design and Low Cost Housing and Community Facilities. United Nations, New York, 1971.

Landsberg, H. E., The climate of towns, in "Man's Role in Changing the Face of the Earth," p. 584, University of Chicago Press, Chicago, Illinois, 1956.

Maunder, M. J., *The Value of the Weather.* Published University paperback.

Olgyay, Victor & Aladar Olgyay. *Design with Climate.* Princeton, New Jersey: Princeton University Press, 1963.

Pleijel, G. V., "Investigation Regarding Daylight and Solar Heat Radiation for the Town of Karachi, Pakistan," Stockholm: reprinted from the Report on Central Directorate and Local Office Karachi, for the State Bank, February, 1950. A discussion concerning the influence of the sun on the design of buildings in Karachi. Well-presented.

Wright, Henry N., *Solar Radiation as Related to Summer Cooling and Winter Heating in Residences.* John B. Pierce Foundation Report. New York, John B. Pierce Foundation, 1936.

PERIODICALS:

Allen, W. A., "The Control of Micro-Climate, *Official Architecture,* May 1969, pp; 609-610.

Architectural Forum, The Magazine of Building, vol. 89, no. 5, pp. 97-1160. "Measure," New York Times, Inc., November, 1948. A discussion of the influences of heat, atmosphere, light, sound, etc., on architecture and planning. An interesting general account.

Lowry, W. P., The climate of cities, *Scientific America,* 217, no. 2. 15-23 (1967).

Oldyay, Victor, "The Temperate House," *Architectural Forum,* vol. 94, March 1951, pp. 179-94. Mostly about the sun and temperatures, there is considerable data here. However, it is unfortunate that the graphs are very difficult to interpret, unless one spends a lot of time on them.

O'Sullivan, P. E., "Climate and the Building Environment." *Planning Outlook,* vol. 5, pp. 14-23 (Spring 1969).

O'Sullivan, P. E., "Integrated Environment Design og Building." Proceedings EDRA #3, vol. 2, pp. 25-2-1.

National Research Council, Building Research Institute. *Solar Effects on Building Design.* Report of a program held as part of the Building Research Institute 1962 Spring Conferences Washington, Building Research Institute, 1963, AIA.

Richards, S. J., "Sunlight and Buildings," *South African Architectural Record,* 1967, Dec., pp. 19-23.

Architectural Record, vol. 108, no. 2, p. 161. "Weather's Effect on Design Studied by Texas Colleges." New York: F. W. Dodge Corp., August, 1961. Summarizes the investigations being done in Texas.

CLIMATOLOGY

The gathering of climatoligical data, and the application of technical data.

BOOKS:

Abbruzzese, Marieg. *Climate and Architecture: Selected References.* Housing and Home Finance Agency, Washington, D.C., March 1951

Crenshaw, Richard. *Climate.,* unpublished student paper, University of Pennsylvania, Philadelphia, 1970.

Geiger, Rudolf. *The Climate Near the Ground.* Cambridge: Harvard University Press, 1950. This is about the only text on microclimatology. Extremely detailed, some of it applies to architecture, some does not. However, a "must" for all interested in pursuing this field further. An exhaustive bibliography, with special attention given to German works.

Geiger., R., "The Climate Near the Ground." (4th ed.), Harvard University Press, Cambridge, Massachusetts, 1965.

Hardy, Alex and T. S. Wiltshire, *Environmental Simulation Laboratory,* School of Architecture, Newcastle University.

Housing and Home Finance Agency, Div. of Housing Research, *Application of Climatic Data to House Design,* January 1954.

Kusuda, T. &. F. J. Powell, "Use of Modern Computer Programs to Evaluate Dynamic Heat Transfer and Energy Use Processes in Buildings," *Performance Concept in Buildings,* Proceedings of the Joint BLLEUASTM-CIB Symposium, held May 2-5, 1972, National Bureau of Standards (U.S.). Spec.

Olgyay., Victor and Olgyay, Aladar and Associates, "Application of Climatic Data to House Design," U.S. Housing and Home Finance Agency, Washington, D.C. 1953.

The Climate of Cities: A Survey of Recent Architecture, U.S. Department of HEW, Consumer Protection and Environmental Health Service.

Teaching the Teachers on Building Climatology, vol of reprints, CIB, Stockholm, September 1972.

Trewartha, G. T., *An Introduction to Climatology,* New York: McGraw Hill Company, 1968.

PERIODICALS:

Bernatzky, Aloys, "Climatic Influences of the Greens and City Planning," *Anthos,* vol. 5, no. 1, 1966.

Burberry, P. (ed.), *A. J. Handbook: Building Environment* (Sec. 4: climate and topography). *Architects Journal,* 1968, October, pp. 747-63.

Burberry, P. (ed.), *A. J. Handbook: Building Environment. Architects Journal:*
October 9, 1968, pp. 815-30
October 16, 1968, pp. 879-98
October 23, 1968, pp. 949-68
October 30, 1968, pp. 1017-36
November 13, 1968, pp. 1147-68
November 20, 1968, pp. 1215-34
November 27, 2968, pp. 1283-98
December 11, 1968, pp. 1343-64
December 11, 1968, pp. 1411-30
December 18, 1968, pp. 1483-1600
January 8, 1969, pp. 107-35

Chandler, T. J., "Urban Climates and their Relevance to Building and Urban Design," Proceedings, American Concrete Institute Symposium: Our Physical Envornment in 1980. The Hague, November 1971.

Christi, H. R., "Vertical Temperature Gradients, in a Beech Forest in Central Ohio," *Ohio Journal of Sciences,* vol. 52, no. 6, 1952.

Everetts, John Jr., "Analysis and Influence of Climatology Upon Air Conditioning Design," BRAB Conference Report No. 1, Washington, D.C., 1950, pp. 123-131.

Gates, David M. "Energy Exchange in the Biosphere." Harper and Row Bio Monographs.

Landsberg, H., "Bioclimatology of Housing." Meterol. Monograph 8, 81. American Meterol. Society, 1954, Boston, Massachusetts.

Landsberg, Helmut, "Climate and Planning of Settlements," Convention Symposium I, *Urban and Regional Planning,* The American Institute of Architects, Washington, D.C. May 1950.

Landsberg, Helmut, "Microclimatic Research in Relation to Building Construction." BRAB Conference Report No. 1, *Weather and the Building Industry,* Washington, D.C., January 1950.

Landsberg, Helmut, "Microclimatic Research in Relation to Building Construction," *Architectural Forum,* March 1947, pp. 114-119.

Landsberg, Helmut, E., "Microclimatology and House Building" (speech delivered at the First Annual Technical Forum of *House Beautiful's* Climate Control Project), September 20, 1949. Like most of Landsberg's papers, this one is not only interesting reading, but filled with a lot of worthwhile information.

Landsberg, Helmut, E., "Use of Climatological Data in Heating and Cooling Design," *Heating, Piping and Air Conditioning,* vol. 19. no. 9, pp. 121-125. September 1947. An excellent account of what climatic data to look for, how to get it, and what to do with it.

Manley, G., "Microclimatology—Local Variations of Climate Likely to Affect the Design of Siting of Buildings." London: *The Journal of the To Royal Institute of British Architects.* Third Series, vol. 56, no. 7, pp. 317-323. To be recommended. Illustrates some of the phenomena in the small-scale variations of the climate. Has small, but good, bibliography.

Myrup, Leonard O., "A Numerical Model of the Urban Heat Island." *Journal of Applied Meterology,* vol. 8, pp. 917-918, December 1968.

Regional Climate Analysis and Design Data, *Bulletin of the American Institute of Architects,* edited from November 1949 on.

Sealey, A., "Microclimatology: Climate Studies by Birmingham School of Architecture." *Architects Journal,* No. 24, 1965, pp. 1229-32, vol. 144.

Waterhouse, F. L., "Microclimatological Profiles in Grass Cover in Relation to Biological Problems," *The Quarterly Journal of the Royal Meterological Society,* vol. 81, no. 5, 1955.

LANDSCAPE DEVELOPMENT FOR ENERGY CONSERVATION AND SOLAR ENERGY UTILIZATION

American Institute of Architects. *Site Planning for Solar Energy Utilization,* a report prepared for the Solar Energy Program of the Office of Housing and Building Technology, National Bureau of Standards, U.S. Department of Commerce and the Office of Policy Development and Research, U.S. Department of Housing and Urban Development, ASLA Foundation,

McLean, Virginia, 1976, available through N.T.I.S. Springfield, Virginia. 326 pp. This is the publication on which this based. It contains much of the same information but is oriented more toward solar energy utilization.

AIA Research Corporation for the U.S. Department of Housing and Urban Development, Office Policy Development and Research, *Solar Dwelling Concepts,* 1976, 146 pp.

"Economic Evaluation of Large-Scale Residential Energy Systems." Decision Sciences Corporation, Summary of Current Activities. Report done in conjunction with Meridian Engineering, Inc. for the Office of Research and Technology, Department of Housing & Urban Development.

E. F. L. Report, "The Economy of Energy Conservation in Educational Facilities." (July 1973).

Oak Ridge National Laboratory Energy Team, "Electrical Energy and Its Environmental Impact—Progress Report of December 31, 1972". ORNL-NSF Environmental Program.

U.S. Department of Commerce. *Eleven Ways to Reduce Energy Conservation*

U.S. Department of Commerce. *Eleven Ways to Reduce Energy Consumption and Increase Conservation in Household Cooling,* NBS/IAT/BRO. Prepared in cooperation with the Office of Consumer Affairs.

Energy Conservation Design Guidelines for Office Buildings. AIA/Research Corporation Dubin-Mindell-Bloom Associates, P. C., Heery & Heery, Architects. General Services Administration/Public Buildings Services. (1974).

General Services Administration, *Energy Conservation Guidelines for Existing Office Buildings,* Public Building Service, G.S.A., 1975, 140 pp.

Griffin, C. W. Jr., *Energy Conservation in buildings: techniques for economical design.* Washington Construction Specifications Inst., 1974. 183 pp.

National Association of Home Builders, *The Builders Guide to Energy Conservation.* N.A.H.B., Washington, D.C., 1974, 63 pp. A concise guide to what the builder needs to know about energy conservation, this book does contain sections on building siting and design building orientation and land use planning.

Ortho Books *Weather Wise Gardening,* Ortho Division, Chevron Chemical Co., San Francisco, California, 1974. 96 pp.

Robinette, Gary O. *Plants, People and Environmental Quality: A Study of Plants and Their Environmental Functions,* Department of the Interior, National Park Services, (1972).

Rogers, Tyler Stewart. *Thermal Design of Buildings.* (New York, Wiley and Sons, Inc., 1964).

. *State Building Design Guidelines for the Conservation of Energy.* Department of Administration, State Bureau of Facilities Management, (Madison, Wisconsin). (September 1973).

Seidet, Margins R., Platkin, Steven M., and Beck, Robert O. "Energy Conservation Strategies, May 1973". Implementation Research Division, Office of Research and Monitoring, U.S. Environmental Protection Agency (EPA-R5-73-021).

Stein, Richard G., FAIA. "Architecture and Energy". New York Chapter, American Institute of Architects, Committee on the Natural Environment.

. "Energy and Architecture", *The Architectural Forum,* Vol. 139, No. 1. (July/August 1973), pp. 38-58.

Sunset, *How to Build Fences and Gates,* Lane Books, Menlo Park, California, 1974, 96 pp.

Sunset, *How to Build Patio Roofs,* Lane Publishing Co., Menlo Park, California, 1959, 96 pp.

Sunset *How to Build Patio Roofs,* Lane Books, Menlo Park, California, 1964, 96 pp.

Sunset, *Landscaping for Modern Living,* Lane Publishing Co., Menlo Park California, 1958, 190 pp.

Sunset, *Sunset Ideas for Landscaping,* Lane Books, Menlo Park, California, 1974, 96 pp.

"Task Force on Practices and Standards: Conservation of Energy (Draft Report)". National Power Survey Technical Advisory Committee. (May 7, 1973).

Technical Options for Energy Conservation in Buildings. National Bureau of Standards Technical Note 789, (U.S. Government Printing Office No. C13.46:789, July 1973).

The Potential for Energy Conservation. A staff Study, Executive Office of the President, Office of Emergency Preparedness, (U.S. Government Printing Office No. C13.46:789, July 1973).

University of Texas at Arlington, *Alternatives in Energy Conservation: The Use of Earth Covered Buildings,* Proceedings and Notes of A Conference held in Fort Worth, Texas, July 9-12, 1975. Funded by the National Science Foundation, Sponsored by the University of Texas at Arlington; including the Institute of Urban Studies (through the Center for Energy Policy Studies and the School of Architecture and Environmental Design. NSF-RA-76006.) This collection of papers is the most definitive and up-to-date exposition of the state-of-the-art in earth covered buildings and the impact of the earth and landforms in building design.

Villecco, Marguerite, Editor. *Energy Conservation Opportunies in Building Design.* A Report to the Ford Foundation Energy Policy Project, (Washington, D.C., AIA/Research Corporation, October 1973).